Evolution and Human Behaviour

Evolution and Human Behaviour

Darwinian Perspectives on Human Nature

JOHN CARTWRIGHT

palgrave

Library of Congress Cataloging-in-Publication Data

Cartwright, John, 1953–
Evolution and human behavior : Darwinian perspectives on human nature / John Cartwright.
p. cm.
A Bradford Book
Includes bibliographical references and index.
ISBN 0-262-03281-3 (alk. paper) — ISBN 0-262-53170-4 (pbk. : alk. paper)
1. Behavior evolution. 2. Genetic psychology. 3. Darwin, Charles Robert, 1809–1882.
I. Title.

BF701,.C37 2000
155.7—dc21   99-087600

Editing and origination by
Aardvark Editorial, Mendham, Suffolk

Printed in Great Britain

To those nearest:
Winnie, Margaret, Chloe, Laurence and Ffion

Contents

List of Figures

List of Tables

Preface

This book explores the application of Darwinian ideas to the way in which humans think, feel and behave. It is intended primarily to serve as a text for undergraduates studying courses of which the evolutionary approach to behaviour forms a significant part. Such courses go by a variety of names: human behavioural ecology, sociobiology, evolutionary psychology, biological anthropology and, occasionally, human ethology. Such diversity reflects partly the spreading reach of a powerful set of ideas, as well as the historical fact that animal behaviour, and certainly 'human nature', was never the sole intellectual property of a single discipline. Given that Darwinism is once again a powerful force shaping our intellectual agenda, it is hoped that the book will also appeal to a wider audience of anyone eager to study the bearing of scientific ideas on the human condition.

Over the past 30 years, the image of evolutionary theory has been transformed. From something once associated in the public mind with fossils, extinction, bones and dusty museums, it has emerged, retooled, as a dynamic set of ideas that promises for a second time round to transform our self-image. Darwinism is setting the pace for the questions we can ask and the answers we can give. Why are men, on average, taller than women? Why does the pornography industry cater largely for men? Why do people spend so much of their time gossiping? Why do women not know when they are ovulating? Why are divorces early in a marriage initiated usually by women and those later on by men? Why are roughly equal numbers of boys and girls born? Why is sexual jealousy such a common cause of male violence? These are just a few of the many specific questions that modern Darwinism addresses.

At a deeper level, Darwinism tackles more fundamental issues such as the origin of sexual reproduction, patterns of violence in families, the existence of altruistic behaviour, the mating behaviour of men and women, what the sexes find attractive in each other, the phenomenon of consciousness, the relationship between genes and culture and so on. Many people refuse to countenance a biological approach to some of these social and cultural issues on ideological grounds, preferring instead, it sometimes seems, to adopt the stance of that often-quoted Victorian lady, Lady Ashley, who hoped that Darwin's ideas were not true or, if they were true, that at least they would not become widely known. But the progress of ideas cannot be arrested and, as evolutionary thinking cuts deeper into the human psyche, we need once again to take stock of what it means to be human.

Textbooks are traditionally supposed to be concerned with conventional and stable knowledge within a field. This has proved to be a daunting task for the evolutionary approach to human behaviour. The subject matter spans a number of disciplines, and the whole field rapidly changes and shifts in direction. Pinning down a body of universally accepted wisdom would be an elusive if not impossible task. All this of course adds to the excitement of witnessing an enterprise that is unfolding and taking shape. In this book, therefore, I have sought to present core concepts, principles and theories, most of which should be robust enough to be of service for many years, together with recent research findings that appear to hold future promise.

Chapters 2–5 deal with the fundamentals of natural and sexual selection, and with the genetic basis of inheritance, subjects on which there is now a strong consensus. In these early chapters, although humans remain a central concern and source of reference, numerous allusions will be made to the behaviour of non-humans. The point of this is to demonstrate the power and veracity of the essential concepts as they are applied and tested on simpler animals. Humans are troublesome organisms: they are highly intelligent, respond perhaps more than any other animal to learning from the environment, and have an enormously complex culture that surrounds, shapes and infuses their lives. Once the reader is convinced, however, that the essential principles of Darwinism offer the best explanation we have for the behaviour of non-humans, it is assumed that he or she will then be in a more receptive frame of mind to explore their potential in human contexts.

There is also a more profound point at stake. It was always Darwin's intention to apply his theory to the minds of human and non-human animals alike. In this respect, he was right: to suppose that only humans can have an evolutionary psychology is to erect an untenable species barrier. In the same way as genetics began with studies on peas and *Drosophila* flies, so the insects, fish and mammals used here as examples are useful in demonstrating the foundations of evolutionary theory.

Chapters 6–11 are much more concerned with *Homo sapiens* as a unique species: our large brains, our capacity for language and culture, our highly developed sense of physical beauty, patterns of marriage and divorce, co-operation and conflict within groups and so on.

The scientific meat of this book is sandwiched, as it were, between historical and philosophical introductory and final chapters that examine the interaction between evolutionary thinking and the wider social context. The concerns raised here are regarded as important for a number of reasons. The reception of a set of ideas is, sadly, not a simple function of its merits: there are social and political forces that also shape and select what is allowed into the canon of accepted wisdom. Second, the very subject matter of this book defies a neat compartmentalisation into some technical branch of biology or psychology, affecting how we view ourselves as a species and our sense of ourselves as moral animals. To avoid these issues would be both a neglect of duty and contrary to the wider ambitions of the Darwinian project.

A Nobel prize-winning scientist once said that the aim of physics was to find a grand theory of everything that could be fitted on a T-shirt. This is more serious than might at first be thought. Certainly, one of the driving forces behind science, and an important criterion for judging ideas, is that fundamental theories should, as the bottom line, be simple and, preferably, elegant. Darwinism is not quite a theory of everything, but it comes pretty close. Darwin, I think, would be flattered that even physicists, who in his day were slow to grasp the significance of evolution, are now employing his ideas on natural selection to conjure up multiple universes that survive by natural selection. Mercifully, such speculations lie outside the scope of this book. What I do hope becomes clear, however, is that Darwinism is simple, profound and elegant, and that its core theories could (just) be fitted on a medium–large T-shirt. As Darwin himself said, however, 'all the labour consists in the application of the theory' (Darwin, 1858), and that shows no signs of ending just yet.

Acknowledgements

My acknowledgements must start with a confession: I was originally attracted to Darwinian accounts of human behaviour because of a set of motives completely different from those which now sustain my interest. As a rather naïve graduate student in the late 1970s, I succumbed to what was then a fashionable pursuit of trying to deconstruct the social and political roots of scientific ideas. In that heady atmosphere, it seemed obvious that 'culture' was the great causal factor in determining not only how humans behaved, but also how humans interpreted their behaviour and fashioned scientific accounts designed to explain it. Sociobiology, and the early shoots of what was to become evolutionary psychology, seemed an easy target for this type of analysis, so in I plunged. But instead of an ideologically tainted science, I found there a world of ideas united by a seriousness of purpose and integrity alongside which the ambitions of the sociology of science seemed facile and trivial. I was, in short, converted to the evolutionary perspective on human nature. My interest in the history of science remained, but at least I had avoided wasting too much time on what Maynard Smith was so accurately to describe later as 'a recent fashion in the history of science to throw away the baby and keep the bathwater – to ignore the science, but to describe in sordid detail the political tactics of the scientists' (Foreword to Cronin, 1991).

I now believe that the science in this area is too reliable and too important to ignore. My initial debt, therefore, must be expressed to all those contemporary scientists who have laboured to develop and expand the initial insights that Darwin laid down over 100 years ago, and whose work I have relied upon to produce this book, as well as to those writers and researchers who have helped to sift through this mass of new material and point to its significance. This list is a long one, but included among those most notable in helping me to formulate my own understanding are Laura Betzig, David Buss, Leda Cosmides, John Tooby, Helena Cronin, Martin Daly, Margo Wilson, Richard Dawkins, Robin Dunbar, William Hamilton, John Maynard Smith, Matt Ridley, Donald Symons, Robert Trivers and Edward Wilson. The list could be extended, but I hope the references in the book will suffice to express my debt to all the other workers.

The book grew partly out of a frustration that, whereas exciting new developments in human behavioural ecology, sociobiology and evolutionary psychology were being communicated to the research community, few authors had, until recently, responded to the need for an accessible student textbook. I can only hope that this book does justice to the ideas and caters for the needs of students who wish to explore this fascinating field.

Several people commented on draft sections of this book, and I express my thanks to them. They include Lottie Hosie, Juliet Leadbeater, Andrew Lilley and

Caroline Harcourt. My thanks also go to Roger Davies, who helped to conceive the original idea of the book and who offered advice and encouragement as well as supplying some useful references. The library at Chester College is a highly efficient organisation and assisted enormously in locating books and articles. Special thanks to Peter Williams, Carol Thomas and Angela Walsh in this respect and also, not least, for overlooking the fact that, at any one time, several shelves of books seemed to be issued on my ticket. I also owe special thanks to my editor, Frances Arnold, whose calm professionalism and wise counsel helped to steer the project to completion. Gratitude also goes to Jo Digby, forever a cheerful voice at the other end of a telephone call to the publishers.

Finally to Margaret, who bore the strains of a distracted husband with fortitude.

The author and publisher would like to thank the following organisations and individuals for their help in obtaining images and their permission to use these:

Dr Lindsay Murray for photograph of common chimp and Dr Allison Fletcher for photograph of gorilla, both in Box 6.1; Alan Bamford for Figure 7.4; David Perrett and colleagues at The Perception Laboratory, School of Psychology, University of St Andrews, Scotland for Figure 9.13; Francis Arnold for Figure 5.7.

The following images were researched by Image Select International:

Academic Press for Figure 1.2 from Sternglanz *et al.*, 1977, Adult preferences for infantile facial features, *Animal Behaviour*, **25**: 108-115; A.P. Watt Ltd on behalf of The National Trust for Places of Historic Interest or Natural Beauty for Figure 2.4; Elsevier Science for Figure 2.6 adapted from Crawford, C. B., The future of sociobiology, *Trends in Ecology and Evolution*, **8**:184–7, 1993; *Scientific American* and P. J. Wynne for Figure 3.13 adapted from Wilkinson, G. 1990, Food sharing in vampire bats, *Scientific American*, **262**: 76–82; Baillière Tindall for Figure 4.8 from Wiley, R. H., 1973, Territoriality and non-random mating in the sage grouse, *Centrocercus urophasianus*, *Animal Behaviour Monographs*, **6**: 87–169; Macmillan Magazines and Ulrich Mueller for Table 4.6, data taken from Mueller, U., 1993, Social status and sex, *Nature*, **363**: 490; University of Chicago Press for Figure 5.2 from Le Boef, B. J. and Reiter, J. 1988, Lifetime reproductive success in northern elephant seals, in Clutton-Brock, T. H. (ed.) *Reproductive Success*; Wadsworth Publishing, a division of Thompson Learning for Figure 5.3 from Alexander, R. D. *et al.*, 1979, Sexual dimorphisms and breeding systems in pinnipeds, ungulates, primates and humans, in Chagnon, N. I. A. and Irons, W. (eds) *Evolutionary Biology and Human Social Behaviour: An Anthropological Perspective*; Academic Press for Figure 5.8 from Petrie, M. *et al.*, 1991, Peahens prefer peacocks with elaborate trains, *Animal Behaviour*, **41**: 323–31; Macmillan Magazines for Figure 5.9 from Petrie, M. 1994, Improved growth and survival of offspring of peacocks with elaborate trains, *Nature*, **371**: 598–99; Macmillan Magazines for Figure 5.10 from Andersson, M., 1982, Female choice selects for extreme tail length in a widowbird, *Nature*, **299**: 818–20; Cambridge University Press for Figure 6.6 adapted from Friday, A. E., 1992, Human evolution: the evidence from DNA sequencing in Jones, S. *et al.*, 1992, *The Cambridge Encyclopaedia of Human Evolution*; Oxford University

Press for Figure 6.11 from Young, J. Z., 1981, *The Life of Vertebrates;* Cambridge University Press for Figure 6.12 from Deacon, T. W., 1992, The human brain, in Jones, S. *et al., The Cambridge Encyclopaedia of Human Evolution;* David Bygott for Figure 6.14 from Byrne, R., 1995, *The Thinking Ape,* Oxford University Press; Oxford University Press for Figures 6.16 and 6.18 adapted from Byrne, R., 1995, *The Thinking Ape;* Cambridge University Press for Figures 6.17 and 6.19 from Dunbar, R., 1993, Coevolution of neocortical size, group size and language in humans, *Behavioural and Brain Sciences,* **16**: 681–735; Elsevier Science for Figure 7.2 reprinted from *Cognition,* **31**: Cosmides, L., The logic of social exchange: Has natural selection shaped how humans reason? Studies with the Wason selection task, 187–276, 1989; Figure 7.3 redrawn from *The Human Primate* by Passingham, R. © W.H. Freeman; University of Chicago Press for Table 7.1 adapted from Dunbar, R. and Aiello, L. C., 1993, Neocortex size, group size and the evolution of language, *Current Anthropology,* **34**(2): 184–93; Academic Press for Figure 8.1 from Smith, R. L., 1984, Human sperm competition, in Smith, R. L. (ed.) *Sperm Competition and the Evolution of Mating Systems;* Blackwell Scientific for Figures 8.4 and 8.5 redrawn from Harvey, P. H. and Bradbury, J. W., 1991, Sexual selection, in Krebs, J. R. and Davies, N. B. (eds), 1991, *Behavioural Ecology;* Academic Press for Figure 8.6 from Birkhead, T. R. and Moller, A. P., 1992, *Sperm Competition in Birds: Evolutionary Causes and Consequences;* Cambridge University Press for Figure 8.7 from Short, R. V. and Balaban, E., 1994, *The Differences between the Sexes;* John Murray for the opening lines from *A Subaltern's Love Song;* Blackwell for Figure 9.2 from Kenrick, P. T. *et al.,* 1996, Evolution, traits and the stages of human courtship: Qualifying the parental investment model, *Journal of Personality,* **58**: 97–116; *New Scientist* for Figure 9.3 from Dunbar, R. 1995, Are you lonesome tonight, *New Scientist,* 11.2.95; Devendra Singh for Figures 9.4 and 9.5 from Singh, D., 1993, Adaptive significance of female attractiveness, *Journal of Personality and Social Psychology,* **65**: 293–307, © 1993 by the American Psychological Association; Devendra Singh for Figures 9.7 and 9.8 from Singh, D., 1995, Female judgement of male attractiveness and desirability for relationships: role of waist to hip ratios and financial status, *Journal of Personality and Social Psychology,* **69**(6): 1089–101, © 1995 by the American Psychological Association; Elsevier Science for Figure 9.10 from Manning, J. T. *et al.,* Fluctuating asymmetry, metabolic rate and sexual selection in human males, *Ethology and Sociobiology,* **18**: 15–21, 1997; Kluwer Academic Publishers for Figure 9.11 from Figure 11.2, p. 198 of Voland, E. and Engel, C. 1989, Women's reproduction and longevity in a premodern population, in Rasa, E. *et al.* (eds), *The Sociobiology of Sexual and Reproductive Strategies,* London, Chapman & Hall; Aldine de Gruyter for Tables 10.2 and 10.3 and Figures 10.5, 10.7 and 10.8 from Daly, M. and Wilson, M., 1988, *Homicide,* New York, © Aldine de Gruyter, 1988; Elsevier Science for Figures 10.10 and 10.11 from Buckle, L. *et al.,* Marriage as a reproductive contract: patterns of marriage, divorce and remarriage, *Ethology and Sociobiology,* **17**: 363–77, 1996; Elsevier Science for Table 11.2, data taken from Cowlinshaw, G. and Mace, R., Cross cultural patterns of marriage and inheritance: A phylogenetic approach, *Ethology and Sociobiology,* **17**: 87–97, 1996; Elsevier Science for Tables 11.3 and 11.4 adapted from Smith, M. *et al.,* Inheritance of wealth as human kin investment, *Ethology and Sociobiology*, **8**: 171–82, 1987; Image Select for Figures 9.14, 11.6 and 12.1; Science Photo Library for Figures 1.1, 1.3 and 3.4 (© A. Barrington Brown); Still Pictures for Figures 4.7 (© Jorgen Schytte), 5.4 (© Mark Carwardine), 5.12 (© Nigel Dickinson), 8.2 (© Mark Edwards) and 10.9 (© Shehzad Noorani); Oxford Scientific Films for Figure 6.13 (© Clive Bromhall); Ann Ronan Picture Library for Figure 2.3; Bridgeman Art Library for Figure 8.3.

Every effort has been made to trace all the copyright holders but if any have been inadvertently overlooked the publishers will be pleased to make the necessary arrangements at the first opportunity.

1

Historical Introduction: Evolution and Theories of Mind and Behaviour, Darwin and After

The reason why psychologists have wandered down
so many garden paths is not that their subject is resistant
to the scientific method, but that it has been inadequately
informed by selectionist thought. Had Freud better understood
Darwin, for example, the world would have been spared such
fantastic dead-end notions as Oedipal desires and death instincts.

(Daly, 1997, p. 2)

Psychology and biology are neighbouring disciplines, and across their boundary there should be a productive two-way traffic of ideas: a revolution in one should impact on the other. It is all the more surprising and regrettable then that for much of the 20th century, while biology became more securely based on deepening evolutionary foundations, psychology failed lamentably to exploit the potential of Darwinian thought. There were some exceptions – William James being the most notable – but many psychologists either ignored Darwinism or, more damagingly, misunderstood the message that it held. Psychology was poorer as a result, underwent a number of false starts and turns and, in the eyes of many, suffered from a singular lack of direction.

The reasons for this missed opportunity, and the background to more recent developments in animal ecology, sociobiology and evolutionary psychology that now promise to reinstate lines of communication, are the subjects of this chapter. The first half deals with efforts to construct a science of animal behaviour and examines the early hopes, expressed by Darwin and his followers, that the study of animal minds would throw light on human psychology. It turned out that parallels were by no means easy to establish, and the promise of Darwinism was delayed for over a hundred years. The second half of the chapter reviews attempts to apply evolutionary ideas directly to human behaviour in the years following Darwin and up to the early 1970s. Early efforts in this direction were weakened by an incomplete understanding of heredity and genetics. When, by the 1930s, a better understanding came along, it was too late: powerful ideologies had corrupted the whole enterprise. The final theme of this chapter deals with the social context of scientific ideas. Although science has its own internal logic and momentum, there are always social forces at work shaping any scientific undertaking. The history of evolutionary theorising about man must necessarily, therefore, take into account the social and political factors that have clouded the study of human nature.

Darwinism began in 1859 when Darwin, in his fiftieth year, finally published his masterwork *On the Origin of Species by Means of Natural Selection*. The book, originally intended as an abstract of a much larger volume, contained concepts and insights that had occurred to Darwin at least 15 years earlier, yet he had wavered and delayed before publishing. The larger volume never appeared, and Darwin was forced to rush out his *Origin* following a remarkable series of events that began in June of the previous year. It is 1858 therefore that serves as a convenient starting point.

1.1 The origin of species

On 18 June 1858, Darwin received a letter from a young naturalist called Alfred Russel Wallace, then working on the island of Ternate in the Malay Archipelago. When Darwin read its contents, he felt his world fall apart. In the letter was a scientific paper in the form of a long essay entitled 'On the Tendency of Varieties to Depart Indefinitely from the Original Type'. Wallace, innocent of the irony, wondered whether Darwin thought the paper important and 'hoped the idea would be as new to him as it was to me, and that it would supply the missing factor to explain the origin of species' (Wallace, 1905, p. 361). The ideas were far from new to Darwin: they had been an obsession of his for half a lifetime. Wallace had independently arrived at the same conclusions that Darwin had reached at least 14 years earlier, and the demonstration of which Darwin saw as his life's work. Darwin knew that the essay must be published and, in a miserable state, exacerbated by his own illness and fever in the family, wrote for advice to his geologist friend and scientific colleague Sir Charles Lyell, commenting that he 'never saw a more striking coincidence' and lamented that 'all my originality, whatever it may amount to, will be smashed' (Darwin, 1858).

Fortunately for Darwin, powerful friends arranged a compromise that would recognise the importance of Wallace's ideas and simultaneously acknowledge Darwin's previous work on the same subject. A joint paper, by Wallace and Darwin, was to be read out before the next gathering of the Linnean Society on 1 July 1858. The reading was greeted by a muted response. The President walked out, later complaining that the whole year had not 'been marked by any of those striking discoveries which at once revolutionise, so to speak [our] department of science' (Desmond and Moore, 1991, p. 470). At Down House, Darwin remained in an abject state, coping with a mysterious physical illness that plagued him for the rest of his life and nursing a nagging fear that it might seem as if he had stolen credit from Wallace. He was also grieving: his young son Charles

Waring had died a few days earlier. As the Linnean meeting proceeded, Darwin stayed away and attended the funeral with his wife Emma. By the end of the day, the theory of evolution by natural selection had received its first public announcement, and Darwin had buried his child.

After the Linnean meeting, Darwin set to work on what he thought would be an abstract of the great volume he was working on. The abstract grew to a full-length book, and his publisher, Murray, eventually persuaded Darwin to drop the term 'abstract' from the title. After various corrections, the title was pruned to *On the Origin of Species by Means of Natural Selection*, and Murray planned a print run of 1250 copies.

Darwin, amid fits of vomiting, finished correcting the proofs on 1 October 1859. He then retired for treatment to the Ilkley Hydropathic Hotel in Yorkshire. In November, Darwin sent advance copies to his friends and colleagues, confessing to Wallace his fears that 'God knows what the public will think' (Darwin, 1859a). Many of Darwin's anxieties were unfounded. When the book went on sale to the trade on 22 November, it was already sold out. It was an instant sensation, and a second edition was planned for January 1860. Thereafter, man's place in nature was changed, and changed utterly.

1.1.1 New foundations

In the *Origin*, Darwin was decidedly coy about the application of his ideas to humans, but the implications were clear enough and, in the years following, both Darwin and Huxley began the process of dissecting and exposing the evolutionary ancestry and descent of man. It was towards the end of the *Origin*, however, that Darwin made a bold forecast for the future of psychology:

> In the distant future I see open fields for far more important researches. Psychology will be based on a new foundation, that of the necessary acquirement of each mental power and capacity by gradation. Light will be thrown on the origin of man and his history. (Darwin, 1859b, p. 458)

On the origin of man, Darwin was right, and light continues to be thrown with each new fossil discovery. On psychology, the new foundation that Darwin foresaw has been slow in coming. Over the past 20 years, however, there have been signs that a robust evolutionary foundation is being laid that promises to sustain a thoroughgoing Darwinian approach to understanding human nature. The bulk of this book is about those foundations.

1.2 The study of animal behaviour

A number of disciplines have lain claim to providing an understanding of animal behaviour, including ethology, comparative psychology, behavioural ecology and, emerging in the 1970s, sociobiology. The problem for the historian is that these terms were not always precisely defined and the disciplines frequently overlapped. It is appropriate, therefore, to consider the origins of comparative psychology and ethology together.

1.2.1 *Comparative psychology and ethology: the 19th-century origins*

Comparative psychology

For Darwin, it was clear and indisputable that all life on earth had evolved from lowly origins. Behaviour, morphology and physiology had all been shaped by the twin forces of natural and sexual **selection**. In this sense, Darwin subscribed to what can be called 'psychoneural monism' – the idea that mind and body are not separate entities. Scientific **naturalism**, in accounting for life and its origins, had seemed finally to square the circle. In respect of man, Wallace agreed with Darwin that in bodily structure, humans had descended from an ancestor common to man and the anthropoid apes, but he objected to the view that natural selection could also explain man's mental faculties. Darwin was, however, determined to push through his programme. He stressed the essential unity between animal and human minds, noting that 'there is no fundamental difference between man and the higher animals in their mental faculties' (Darwin, 1871, p. 446).

It was Darwin's *Expressions of the Emotions in Man and Animals,* published in 1872, that more than any other book acted as a spur to the study of both ethology and comparative psychology. In this work, Darwin, in the traditions of the day, described the behaviour of animals using terms that were extrapolated from the mental life of humans. For Darwin, no human mental function was unique: the way to understand human minds was by invoking processes found in the minds of other animals. Although he observed the behaviour of his pets, animals at the zoo and his children, Darwin made few experiments of his own on animal behaviour and for the most part relied upon anecdotal information given to him by naturalists, zoo-keepers and other correspondents. With hindsight, the anecdotal approach appears fatally flawed. Darwin's lasting contribution to the ethological approach was, however, to provide an evolutionary framework to the study of behaviour, and to highlight the value of observing animals in their natural settings.

There were many who sympathised with the misgivings of Wallace and, to defend Darwin's approach, there grew up what has been called the 'anecdotal movement' (Dewsbury, 1984). Anecdotalists such as George John Romanes were so called because they relied upon reports and unsystematic observations of animal behaviour to provide an empirical base for their writings. To demonstrate the continuity between animal and human minds, it was thought necessary to show that animals could reason, show complex forms of social behaviour and display human-like emotions. Romanes is important to the history of animal psychology in the sense that it was the reaction against these ideas that was to shape future developments. In his book *Mental Evolution in Man* (1888), Romanes advanced the doctrine of levels of development that could be expressed on a numerical scale. In this system, humans were born at level 16, reaching the level of insects at 10 weeks and that of dogs and the great apes at about 15 months. Romanes also suggested that human emotions were common to other animals according to the complexity involved. Fish, for example, he thought capable of experiencing jealousy and anger, birds pride and resentment, and apes shame and remorse (Romanes, 1887, 1888).

In 1894, the year that Romanes died, two books were published that were pivotal in defining the way ahead. The first was Wilhelm Wundt's *Lectures on Human and Animal Psychology*, in which Wundt (1832–1920) criticised the anecdotal method of Romanes. The second book, perhaps more significant, was *An Introduction to Comparative Psychology* (1894) by Conwy Lloyd Morgan (1852–1936). Morgan was a pioneer of comparative psychology and did much to establish the scientific credentials of the emerging discipline and place it on an evolutionary basis. He became the first scientist to be elected to the Royal Society of Great Britain for work in psychology. In the book mentioned above, he rejected the **Lamarckian inheritance** of acquired traits and outlined his famous canon:

> In no case may we interpret an action as the outcome of the exercise of a higher psychical faculty, if it can be interpreted as the outcome of the exercise of one which stands lower in the psychological scale. (Lloyd Morgan, 1894, p. 53)

Morgan's canon is of course a sort of law of **parsimony** in the style of what is called Occam's razor; that is, where two explanations are possible, we choose the simplest and resist entertaining unnecessary hypotheses. In this instance, scientists were urged to avoid interpreting animal behaviour in terms of the thoughts and feelings experienced by humans.

Many of the early comparative psychologists were concerned with learning. Important and pioneering work in trial and error learning was conducted by the American Edward Lee Thorndike (1874–1949). A typical experimental arrangement set up by Thorndike was that an animal would receive a reward for a type of behaviour that was initially discovered accidentally. The time taken to repeat the behaviour was taken as an indication of learning. Predictably, behaviours that met with favourable consequences for the animal were learnt quickly. This learning process was termed **'operant conditioning'**, and the acceleration of learning through positive reward became known as the law of effect. After World War I, Thorndike became almost exclusively concerned with human psychology. In rejecting the anecdotal approach of Romanes and moving towards laboratory studies of caged animals, Thorndike and others were beginning a trend that was to dominate comparative psychology for the next 50 years.

Ethology

The word 'ethology' is derived from the Greek word *ethos* meaning character or trait. In Britain, there was a long and, by the late 18th century, popular tradition of natural history involving the observation and documentation of the behaviour of animals. However, the science of ethology probably began in France, its early pioneers including Jean Baptiste Lamarck (1744–1829), Etienne Geoffroy-Saint-Hilaire (1772–1844) and Alfred Giard (1846–1908). In contemporary biology, the name of Lamarck is still well known as that of someone who provided the only serious rival to the mechanism of natural selection as a way of accounting for the preservation of favoured traits. In 1809, the year in which Darwin was born,

Lamarck published *Philosophie Zoologique*, in which he advanced the view that species could change over time to new species (transmutationism). One mechanism for such changes that Lamarck suggested was the idea that organisms could, through their own efforts, modify their form, and these modifications could be passed to the next generation. Although this was a small part of Lamarck's entire theory, the notion that acquired characters can be inherited has become indelibly associated with his name (see Chapter 2). Lamarck made few friends among French scientists. He acquired a reputation for publishing inaccurate weather forecasts, which did his career no good; more tellingly, his views on biology were denigrated by the French anatomist George Cuvier (1769–1837). Lamarck died blind and poor in 1829.

In the late 19th century, the concerns of comparative psychology and ethology overlapped, and the careers of some scientists straddled what were only later to become divergent disciplines. Lloyd Morgan, for example, is often cited as being one of the founding fathers of both ethology and comparative psychology. It was during the 20th century that differences in training, methodologies and even fundamental assumptions about the nature of animals led to a schism.

1.2.2 *Ethology and comparative psychology in the 20th century*

Ethology 1900–70

One of the giants of 20th-century ethology was the Austrian Konrad Lorenz (1903–1989) (Figure 1.1). Lorenz originally trained as a doctor but was influenced by the work of Oscar Heinroth at the Berlin Zoological Gardens on the behaviour of birds. Heinroth exploited the analogy between animals and humans in both directions. Animals could be understood using concepts drawn from the mental life of humans, and this understanding could then be reapplied to understand the human condition. Lorenz frequently expressed his debt to Heinroth's approach.

By conducting experiments at his home on the outskirts of Vienna, Lorenz observed numerous features of animal behaviour that have become associated with his name. In one classic study, he noted how a newly hatched goose chick will 'imprint' itself on the first moving object it sees. In some cases, this was Lorenz himself, and chicks would follow him about, presumably mistaking him for their mother. Lorenz stressed the importance of comparing the behaviour of one **species** with another related one and argued for the importance of understanding evolutionary relationships between species. In this respect, Lorenz unashamedly drew parallels between the behaviour of humans and other animals. In his most popular work, *King Solomon's Ring*, for example, he suggested that the 'war dance of the male fighting fish… has exactly the same meaning as the dual of words of the Homeric heroes, or of our Alpine farmers, which, even today, often precedes the traditional Sunday brawl in the village inn' (Lorenz, 1953, p. 46).

One of Lorenz's early concepts was that of the **fixed action pattern,** which referred to a pattern of behaviour that could be triggered by some external stimulus. Using a term that was later to prove so troublesome for ethology,

Figure 1.1   Konrad Lorenz (1903–1989)

The Austrian zoologist is here shown being followed by a group of ducklings as if he were their mother. This is an instinct Lorenz called imprinting. Lorenz and his Dutch colleague Niko Tinbergen helped to found the science of ethology

Lorenz regarded these action patterns as 'instincts' forged by natural selection and common to each member of the species (Box 1.1). Fixed action patterns have the following characteristics:

- Their form is constant, that is, the same sequence of actions and the same muscles are used
- They require no learning
- They are characteristic of a species
- They cannot be unlearnt
- They are released by a stimulus.

Evidence of a fixed action pattern that is often cited is the observation that a female greylag goose (*Anser anser*) will retrieve an egg that has rolled outside her nest by rolling it back using the underside of her bill. Lorenz noticed that this action continued even when the egg was experimentally removed once the behaviour had begun. Once it started, the behaviour had to finish whether it was effective or not. The stimuli that trigger fixed action patterns became known as sign stimuli or, if they were emitted by members of the same species, releasers. An interesting example is to be found in the behaviour of the European robin (*Erithacus rubecula*), documented by the British ornithologist David Lack in the 1940s. Lack showed that the releaser for male aggression in this species is the patch of red found on the breast of the bird. A male robin will attack another

Box 1.1

One of the few areas in which the concept of fixed action patterns has been successfully transferred from ethology to human psychology concerns the response of adults to babies and the early reactions of newborn babies. Newborn humans show a number of potential fixed action patterns, such as the grasping motion of hands and feet if something touches the palm or sole of the foot. There is also a programmed search for the nipple, which consists of a sideways movement of the head if the lips are touched. Up until about 2 months of age, even eye-sized spots painted on cardboard will elicit a smiling reaction from a child. Lorenz suggested that the facial features of human babies, such as a large forehead, large eyes, small chin and bulging cheeks, serve as social releasers that act upon **innate releasing mechanisms** to activate nurturant and affectionate behaviour. This intriguing idea could in part explain why humans find baby-like faces, be they on humans, young animals or even toys such as teddy bears, alluring. There does appear to be considerable experimental support for Lorenz's idea (Eibl-Eibesfeldt, 1989; Archer, 1992). Sternglanz et al. (1977) investigated the effect of varying the features of a baby's face on its perceived attractiveness as ajudged by American college students. By altering various parameters of line drawings, the overall conclusion was, as Lorenz suggested, a marked preference for faces with large eyes, large foreheads and small chins. A composite drawing combining all the features with high attractiveness ratings is shown below.

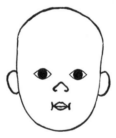

Figure 1.2   Composite drawing of the ideal
infantile face (redrawn after Sternglanz et al., 1977, Figure 7)

male that it finds in its territory, but it will also attack a stuffed dead robin and even a tuft of red feathers (Lack, 1943).

Once the essence of the stimulus has been identified, it becomes possible in some cases artificially to exaggerate its characteristics and create supernormal stimuli. If a female oyster catcher (*Haemotopus ostralegus*), for example, is presented with a choice of egg during incubation, it will choose the larger. Even if an artificial egg twice the size of its own is introduced, the oyster catcher still prefers the larger one, even though common sense (to an outsider) would indicate that it is unlikely to be an egg actually laid by the bird.

Lorenz had little interest in the individual variation of instincts displayed by different members of the same species, a subject now of great interest to behavioural ecologists. Burkhardt suggests that this neglect of intraspecific variation in behaviour was partly a reflection of the fact that Lorenz wished to distance himself from animal psychologists and their work on captive animals. Lorenz distrusted inferences from laboratory and domesticated animals on aesthetic grounds and on his concern that captive animals showed too much variability in the behaviours they had learnt. This variability was, to Lorenz, a hindrance (Burkhardt, 1983).

Lorenz believed that animal instincts could be used to reconstruct the all-important evolutionary phylogeny of individual species. This approach can be illustrated by the gift-giving behaviour of species of flies in the family Empididae. In one species, *Hilara sartor*, the male presents the female with a present of an empty silken bag, which the female attempts to unravel, while the male copulates with her. This seemingly pointless behaviour is best understood by comparing it with behaviours displayed by related species. One problem for males in this family is that the female is likely to eat an approaching male. In the species *Hilara quadrivittata*, the male avoids this by presenting a food parcel wrapped in a silken balloon. As the female unravels and eats her gift, the male is able to mate with a reduced risk of being eaten. It seems that *Hilara sartor* has evolved one step further and the male has dispensed with the gift.

It was one of Lorenz's students, Nikolaas Tinbergen (1907–1988), who finally completed the establishment of ethology as a serious scientific discipline. Tinbergen joined Lorenz in 1939 and helped to develop methods for studying behaviour in the wild. In 1949, he moved to Oxford and led a research group dedicated to the study of animal behaviour. Tinbergen studied how fixed action patterns interact to give a chain of behavioural reactions. In his classic study of the stickleback, Tinbergen showed how, during the courtship ritual, males and females progress through a series of actions in which each component of female behaviour is triggered by the preceding behaviour of the male, and vice versa, in a cascade of events (Tinbergen, 1952). The culmination of the sequence is the synchronisation of gamete release and fertilisation.

Both Tinbergen and Lorenz developed models to conceptualise the patterns of behaviour they observed. Lorenz interpreted his observations as consistent with a psycho-hydraulic model, sometimes called, somewhat disparagingly, the flush toilet model. If behaviour is interpreted as the outflow of water from a cistern, the force on the release valve can be interpreted as the trigger. The model was more sophisticated than suggested by its comparison with a domestic flush toilet, but its essential feature was the accumulation of 'action-specific energy' in a manner analogous to the accumulation of a fluid in a cistern. Freud employed similar hydraulic metaphors in his thinking about drives and repression. Despite their obvious shortcomings as accurate analogues of mental mechanisms, they are still commonplace in everyday speech. To 'explode with rage' or 'let off steam' are both echoes of the type of models utilised by Lorenz and Freud.

Tinbergen developed an alternative model that, while retaining the concept of accumulating energy that drives behaviour, suggested a hierarchical organisation of instincts that are activated in turn. The models of Lorenz and Tinbergen met with much subsequent criticism. It proved difficult, for example, to correlate the features of the model with the growing body of information from neurobiology about real structures in the brain.

One of Tinbergen's most lasting contributions was a clarification of the types of question that animal behaviourists should ask. In 1963, in a paper called 'On the Aims and Methods of Ethology', Tinbergen suggested that there were four 'whys' of animal behaviour:

1. What are the mechanisms that cause the behaviour? (Causation)
2. How does the behaviour come to develop in the individual? (Development or **ontogeny**)
3. How has the behaviour evolved? (Evolution)
4. What is the **function** or survival value of the behaviour? (Function)

It has been noted that a useful mnemonic for these is ' ABCDEF: **A**nimal **B**ehaviour, **C**ause, **D**evelopment, **E**volution, **F**unction' (Tinbergen, 1963).

To appreciate the application of these questions, it may be useful to consider an example. In many areas of the northern hemisphere, birds fly south as winter approaches. One such species is the wheatear (*Oenanthe oenanthe*). Wheatears migrate to Africa in the winter even though some groups have moved from their European breeding grounds and have established new populations in Asia and Canada. We could ask of this behaviour: What triggers flocks to take to the air? How do they 'know' when the time arrives to depart and which way to fly? These questions address the **proximate causes** of the behaviour and relate to Tinbergen's first question of causation. The answer lies in specifying the physiological mechanisms that are activated by environmental cues, possibly day length, temperature, angle of the sun and so forth.

Probing further, we could ask how the ability to fly over such vast distances in a species-typical manner is acquired by an individual. Do animals know instinctively which way to fly and how far to travel, or do they have to learn some components from a parent or an older bird? These questions belong to Tinbergen's second category dealing with the development and ontogeny of behaviour.

We could also ask about the evolution of the behaviour to its present form. Is the behaviour found in related species? If so, has it been acquired by descent from a common ancestor? A particular question concerns why even Canadian and Asian wheatears travel to Africa. If the aim is simply to move south, those in Canada and Asia could save themselves thousands of miles of travel. Is the move to Africa by the new populations outside Europe a 'hangover' from when the wheatear only lived in Europe? These questions refer to the evolutionary origin of the behaviour as raised in the third 'why' question above.

Finally, in relation to the movement to Africa, we could ask questions about **ultimate causation**. Why do birds make such arduous and perilous journeys? How does flying to Africa increase the survival chances of those making the journey? It must carry some advantage over not moving, otherwise a mutant that appeared and did not fly away would leave more survivors, and non-migration would gradually become the norm. The type of answer given to this last functional 'why' of Tinbergen would presumably show that the benefits of travelling in terms of food supply and then securing a mate outweigh the drawbacks in terms of risks and energy expenditure. Ultimately, we would have to show that migrating is a better option than staying in one place as a means of leaving offspring. We will then have demonstrated the function or adaptive significance of the behaviour.

Taking an overview of Tinbergen's four 'whys' as a very broad generalisation, psychologists (dealing with humans or animals) have tended to be interested in the 'whys' of proximate causation and ontogeny, and less interested in questions about evolution and adaptive significance. The growth of sociobiology and evolutionary psychology is a direct attempt to reverse this trend and supply a unifying paradigm for the behavioural sciences based on the understanding of ultimate causation and evolutionary function (Barkow *et al.*, 1992).

The work of Lorenz and Tinbergen is usually classified as being central to the tradition of classical ethology. Classical ethology was forced to adjust its ground as a result of telling criticisms that appeared in the 1950s and 60s from comparative psychologists and, to be fair, discoveries by the ethologists themselves. Lehrman (1953) in particular was a forceful critic of the use of the term 'innate' in ethology. The attempt to classify behaviour as either innate or learned was soon seen to be too simplistic (Archer, 1992). The deprivation experiments suggested by Lorenz, in which an individual is reared in isolation from other individuals and hence sources of learning, simply isolate an individual from its social environment rather than from temperature, light and nutrition. Isolation experiments beg the question: From what has the animal been isolated? Peking ducklings (*Anas platyrhynchos*), for example, are able to recognise the call of their species if it is heard during embryonic development while the chick is still in the shell (Gottlieb, 1971). Moreover, individuals create their own environment as a result of their actions. Aggressive or assertive people create a different environment from shy people, resulting in a different set of feedback. It came to be realised that all behaviour must be the result of both influences.

Important work by Thorpe (1961) on song development in chaffinches demonstrated the mutual interdependence between heredity and the environment. Thorpe demonstrated that whereas the ability of a chaffinch to sing was to some degree 'innate', the precise song pattern depended upon exposure to the song of adults at critical times during the development of the young bird. The form the song took also depended on the ability of the chaffinch to hear its own song. Chaffinches would not, however, learn the songs of other birds, even if exposed from birth. Furthermore, once song development occurred, a chaffinch would not learn other variants. Work such as this showed that the interaction between innate templates of behaviour and the environment is more complex than hitherto thought.

Eventually, the classical ethologists had to accept that behaviour that they had often labelled as innate could be modified by experience. It did not follow from this of course that all behaviour was learnt and unconstrained by genetic factors, as some of the behaviourists seemed to imply. The history of comparative psychology is also one in which fundamental assumptions had to be revised.

Comparative psychology 1900–70

An early exponent of the methods that came to be associated with comparative psychology was Ivan Petrovich Pavlov (1849–1936). Pavlov was the son of a priest who began his career studying medicine, and was awarded the Nobel prize in 1904 for his work on digestion. Pavlov showed that if the sound of a bell accompanied the presentation of food to a dog, the dog would learn to associate the sound with food. Eventually, the dog would salivate at the sound of the bell even in the absence of food. Pavlov thereby produced the first demonstration of what later became known as classical conditioning. By focusing on the observable reactions of animals without presupposing what went on in their minds, Pavlov stressed the objectivity and rigour of these methods in contrast to a psychology that dwelt upon putative inner experiences. Pavlov's work on conditioning became known to Western psychology around 1906 and, although his more ambitious claims for the establishment of a new brain science went unheeded, his methodology proved highly influential.

The focus on the observable reactions of animals and humans under controlled conditions came to be regarded as the hallmark of comparative psychology and what later became known as behaviourism. It is easy to overestimate the influence of behaviourism on 20th-century psychology. Smith (1997) notes that it served the polemical interests of cognitive psychologists in the 1960s to represent psychology between the years 1910 and 1960 as a behavioural monolith from which they were seeking liberation. As a broad generalisation, however, a pattern was emerging. By the middle of the 20th century, the European approach to animal behaviour and extrapolations to humans was dominated by ethology, while in the United States the experimental approach using laboratory animals was pre-eminent.

One figure who more than any other seemed to symbolise the behaviourist approach in its early days was John Broadus Watson (1878–1958). In his later years, Watson came to be reviled by ethologists as the architect and archetype of an alien approach to the study of behaviour. By adopting a positivist approach to knowledge, Watson claimed that psychology would be retarded in its development unless it ditched its concern with unobservable entities such as minds and feelings. Similarly, both animal and human psychology must abandon any reference to consciousness. A psychology that deals with inner mental events is clinging, he claimed, to a form of religion that has no place in a scientific age. For Watson, the brain was a sort of relay station that connected stimuli to responses.

Watson issued his manifesto in a series of lectures delivered in 1913 at Columbia University. Before World War I, the reaction was lukewarm: some welcomed his objective approach but warned against its excesses. Others feared that a sole concern with behavioural phenomena as opposed to human consciousness would reduce psychology to a subset of biology. It was after the war that behaviourism became more deeply embedded in American scientific culture. The war had demonstrated the value of objective tests applied in the classification of military personnel. By 1930, behaviourism had become the dominant viewpoint in experimental psychology. In stressing the importance of environmental conditioning, Watson's whole approach was profoundly anti-evolutionary and anti-hereditarian.

He denied that such qualities as talent, temperament and mental constitution were inherited. Perhaps his most famous remark on the effect of environmental conditioning, and one of the most trenchant and extreme statements of environmentalism in the literature, is his claim for the social conditioning of children:

> Give me a dozen healthy infants, well-formed, and in my own specified world to bring them up, and I will guarantee to take any one at random and train him to become any type of specialist I might select – doctor, lawyer, artist, merchant-chief and yes, even beggar-man and thief, regardless of his talents, penchants, tendencies, abilities, vocations and race of ancestors. (Watson, 1930. p. 104)

Behaviourism sought philosophical credibility by allying itself with a philosophy of science known as logical **positivism** and articulated by a group of philosophers known as the Vienna Circle. The logical positivists argued that statements are only meaningful, and hence part of the purview of science, if they can be operationally defined. A statement about the world then only becomes meaningful if it can be verified. The aims of this approach were to outlaw religious and metaphysical claims to knowledge. It was in this approach to epistemology that American behavioural psychology, with its emphasis on empirical, quantifiable and verifiable observations, found a natural ally.

The decline of behaviourism in the 1960s coincided with the downfall of logical positivism as a reliable philosophy of science. Philosophers such as Popper and historians such as Kuhn showed that the verifiability criterion of meaning espoused by the Vienna Circle was untenable both in theory and as a realistic description of the way in which science actually worked. The irony of the linkage between behaviourism and positivism is neatly summed up by Smith (1997, p. 669):

> It appeared as if the behaviourist enterprise had emptied psychology of its content in order to pursue an image of science that was itself a mirage.

One movement in animal psychology related to behaviourism but deserving a different title is Skinner's operant psychology. Skinner studied at Harvard and was particularly influenced by the work of Watson and Pavlov. Skinner's programme adopted a number of key principles. One was that he believed that science should be placed on a firm foundation of the linkages between empirical observations rather than speculative theory. For Skinner, and in this respect he resembles Watson, theoretical entities such as pleasure, pain, hunger and love were meaningless and should be expunged from laboratory science. Another essential feature of Skinner's work was that he thought that all behaviour could be resolved and reduced to a basic principle of reinforcement. One typical schedule of reinforcement devised by Skinner was to reward pigeons in a box with grain. By rewarding some forms of behaviour and not others, he was able to make the rewarded behaviour more probable, an approach that became known as operant conditioning.

While behaviourists in America were feverishly attempting to jettison from psychology excess metaphysical baggage, Freud in Europe was weaving a psychology replete with rich and colourful complexes, emotions and subconscious forces. Skinner was both an admirer and critic of Freud. For Skinner, Freud's great discovery was that human behaviour was subject to unconscious forces. This accorded well with the view of behaviourism that conscious reason was not in the driving seat of human behaviour. Freud's mistake was that he encumbered his theory with unnecessary mental machinery, such as the ego, the super ego, the id and so on. For Skinner, such entities were not observable and hence were not justified in scientific enquiry.

By 1960, a large number of psychologists in America had been trained in Skinner's methods. Their output was influential in some areas, such as the inculcation of desirable habits in the development of children, but few behaviourists were willing to go as far as Skinner in suggesting that organisms were empty boxes. Skinner faced his most difficult hurdle when, in his book *Verbal Behaviour* (1957), he attempted to interpret language development in terms of operant conditioning. There were already signs that many behaviourists were realising that language threatened to be the 'Waterloo' of behaviourism. Skinner marched on, however, and argued that there was nothing special about language, denying any fundamental difference in verbal behaviour between humans and the lower animals. Skinner had now pushed behaviourism too far, and its weaknesses were fatally exposed.

In 1950, a linguist called Noam Chomsky, then relatively obscure, reviewed Skinner's *Verbal Behaviour* and, in showing that behaviourism failed lamentably when tackling language, also undermined some of its basic pretexts. Chomsky argued that behaviourism could hardly begin to account for language acquisition. He showed that Skinner's attempts to apply the language of stimulus and response to verbal behaviour lapsed into vagueness and finally hopeless confusion. For Chomsky, behaviourism could not be improved or modified: it was fundamentally flawed and had to go. Chomsky's review, and his own positive programme for linguistics stressing the creativity of language and its foundation on inherent, deep-seated mental structures, sparked off a revolt against behaviourism that tumbled it into terminal decline. This roughly coincided with increasing reports from animals researchers that animals trained according to operant conditioning methods would occasionally revert to behaviours that seemed to be instinctive.

1.2.3 *Interactions between comparative psychology and ethology*

Between the years 1950 and 1970, there were a variety of approaches to the study of animal behaviour, although ethology and comparative psychology were the dominant players. The exact boundaries of these disciplines and the extent of their interactions were often more complex than is conveyed by a simple 'warfare model'. There was, however, a frank exchange of critical comments and an atmosphere of intellectual rivalry. The ethologists often identified the whole of comparative psychology with the extreme environmentalism of Watson's later years. They saw comparative psychology as a discipline obsessed with what rats do in mazes, bereft of a unifying theory and ignorant of the adaptive and evolutionary basis for behaviour. In return, psychologists saw ethology as lacking in scientific rigour, employing dubious concepts such as innateness and burdened by unsubstantiated models.

An outsider looking into this mêlée would also note that Lorenz and Tinbergen achieved a rapport with the reading public by writing fluent and popular works. In contrast, comparative psychology appeared to be more introspective and plagued with self-doubt. As late as 1984, the comparative psychologist Dewsbury lamented that 'we seem to suffer an identity crisis' (Dewsbury, 1984, p. 6). Correspondingly, some of the harshest portrayals of comparative psychology came from the psychologists themselves. It seemed to be a discipline set on a course for self-destruction. In a classic paper called 'The Snark Was a Boojum', Beach (1950) demonstrated how, between the years 1911 and 1945, the output of research on animal behaviour from the perspective of comparative psychology had increased while the number of species actually studied had dwindled. By 1948, the vast majority of articles in this field was concerned with one single species, the Norway rat, and was dominated by reports of learning and conditioning experiments. It was as if, as Lorenz complained, the word 'comparative' was a sham.

The fondness for the Norway rat stood in stark contrast to the much wider diversity of organisms studied by the ethologists. There were other characteristic differences too. In 1965, McGill published a table that captured in broad terms the contrast between ethology and comparative psychology (Table 1.1).

Table 1.1 Differences between comparative psychology and ethology (adapted and modified from McGill, 1965, p. 20)

	Comparative Psychology	**Ethology**
Geographical focus	North America	Europe
Background training of researchers	Psychology	Zoology
Animal subjects studied	Norway rat and pigeons	Variety of species
Emphasis	Learning: the development of behaviour as moulded by environmental stimuli	Instinct: behaviour as the expression of evolution-derived **characters**
Methods	Laboratory work Statistical analysis of the effect of different variables	Careful observation of natural behaviour. Experiments in the field
Attitude to animal subjects	Objective and dispassionate	Close familiarity, even emotional attachment

Further surveys involving analyses of the content of journals, in the style of Beach, were conducted at regular intervals after 1950. In the 1970s, several observers concluded that the earlier scenario outlined by Beach, of a subject blinkered by its own concern with a few species, was still very much in evidence (Scott, 1973; Lown, 1975). To be fair to comparative psychology, this narrowness of focus was a conscious decision and did not overly concern its practitioners. They were searching for universal laws of behaviour that, once established for any one species, could be applied to others. However, the day that such laws could be transferred to illuminate human behaviour seemed to recede as fast as it was chased.

By the late 1970s, comparative psychology was in a sorry state. The field seemed to suffer a crisis of confidence, faced the ignominy of being identified (however unjustly) with a discredited behaviourism, and the epithet 'compara-tive' sounded alarmingly hollow. The neglect of an adaptive approach to behaviour, as well as an underappreciation of phylogenic differences in learning and intelligence, was increasingly seen as a damaging shortcoming. In 1984, Dewsbury wrote a sustained defence of the discipline that challenged a number of misconceptions and tried to give it a respectable history (Dewsbury, 1984), but even then it was clear that the study of behaviour was moving in radically new directions. The attempt to patch up the image of psychology contrasted sharply with the hybrid vigour demonstrated by the merger of ethology and ecology to yield behavioural ecology and sociobiology.

A rapprochement was eventually established between ethology and compara-tive psychology, and each side learnt valuable lessons from the other. Two people in particular who sought to bridge the disciplines and effect a reconcilia-tion were the American psychologist Donald Dewsbury and the Cambridge ethologist Robert Hinde. In 1982, Hinde took a sanguine view and concluded that, despite a few remaining differences, the two approaches had broadened in their concerns and 'on the whole the distinction between the two groups barely exists' (Hinde, 1982, p. 187). Dewsbury was also, in 1990, calling for reconcil-iation between causative and functional (adaptive) approaches to behaviour. By this time, behavioural ecology, and sociobiology in particular, was well involved in exploiting new insights into the adaptive basis of behaviour generated by concepts such as parental investment and reciprocal altruism, kin selection theory and a revival of Darwin's own theory of sexual selection. Dewsbury in fact complained that the concentration on the functional approach threatened to imbalance the study of behaviour (Dewsbury, 1990).

Before examining the remarkable efflorescence of ideas associated with this movement, which is after all the main subject of this book, we will examine the background of applying evolutionary ideas to human behaviour, for it is precisely in this area that behavioural ecology and sociobiology succeeded in bridging the human–animal divide where comparative psychology failed.

1.3 Evolution and theories of human behaviour: Darwin and after

1.3.1 *Man the moral animal*

In the *Origin*, Darwin tactfully avoided a reference to humans. This was a skilful move: Darwin knew that, for his theories to be accepted by the scientific community, he must proceed slowly. Above all else, he wanted to avoid the damaging controversies that had engulfed previous work on the evolution of man.

One problem that particularly troubled evolutionists was how civilised conduct could arise in a world where the uncivilised were just as successful at breeding. The moralist and political writer William Rathbone Greg noted the paradox when, in 1868, he observed that:

> The careless, squalid, unaspiring Irishman, fed on potatoes, living in a pig sty, doting on superstition, multiplies like rabbits or ephemera: the frugal, foreseeing, self-respecting ambitious Scot, stern in his morality, spiritual in his faith, sagacious

and disciplined in his intelligence passes his best years in struggle and celibacy, marries late, and leaves few behind him. (Quoted in Richards, 1987, p. 173)

Greg's views, predictably enough those of a Scottish gentleman, were known to Darwin, but Darwin in his later works clung to the idea that natural selection would still do its work. Jailed criminals, for example, would bear fewer children, and the poor crowded into towns would suffer a higher mortality. Those like Greg and Galton, who were less sanguine about the reforming potential of natural selection, sought comfort in other solutions. Greg hoped that high standards of morality would spread from the worthy downwards. Galton hoped that once the truths of heredity and evolution were appreciated, appropriate social action would follow.

Equally troubling was the scientific problem of how to account for the emergence of sophisticated mental faculties from a world of brutal struggle. What possible survival value could Victorian virtues confer on primitive man? The whole problem of the origin of moral conduct threatened to derail the evolutionary project as soon as it started.

It was now that Wallace parted company with Darwin. By 1869, he was arguing that evolution could not by itself account for the origin of consciousness or the higher mental faculties of man. What seems to have prompted Wallace's apostasy from the cause of naturalism was his conversion around 1866 to spiritualism. Like many of his British contemporaries, including Francis Galton, Oliver Lodge and William Crookes, and some in America, such as William James, Wallace attended seances. Like others, he became convinced that the various apparitions observed could only be explained by the operation of supernatural forces. Darwin himself and his more level-headed colleagues, such as Huxley and Tyndall, remained sceptical. For Wallace, spiritualism provided the final proof of a conviction that had been maturing for years. What possible survival could follow from abstract mathematical abilities, or what Wallace called the 'metaphysical faculties'? It followed that there could be no Darwinism of the mind. Some higher power must have shaped and directed the mental evolution of man. No wonder Darwin wrote to Wallace in 1869 warning that 'I hope you have not murdered too completely your own and my child' (quoted in Desmond and Moore, 1991, p. 569).

Darwin was conscious of these reservations held by Wallace and many others when he worked on *The Descent of Man* (1871). In this work, Darwin examined four types of activity held to be distinctive characteristics of humans: tool use, language, an aesthetic sense and religious ardour. The question then to be asked was whether or not such attributes could be found in other species. Darwin thought that they could and demonstrated to his own satisfaction at least a primitive capability for all four in other species.

1.3.2 *Herbert Spencer (1820–1903)*

Herbert Spencer, philosopher and critic, was a contemporary of Darwin and someone equally keen to push forward the programme of scientific naturalism. Even before Darwin published his *Origin*, Spencer was attempting to build a philosophy of knowledge on the unifying principle of evolution. In his essay of

1852, 'A Theory of Population Deduced from the General Law of Animal Fertility', he introduced the famous phrase 'survival of the fittest'. It was this phrase that Darwin borrowed from Spencer and used in his fifth edition of the *Origin*.

Spencer's curiosity in evolution was first aroused when he stumbled across fossils in his first career as a civil engineer. In 1840, he read Lyell's *Principles of Geology* (1830–33), a work that contained a discussion of Lamarck's evolutionary ideas, and thereafter became attracted to Lamarckian notions of the evolutionary process. Spencer's thinking on evolution applied to the human psyche came to fruition in 1855 when he published his *Principles of Psychology*. The strain of completing this work led him to suffer a nervous breakdown that left him incapacitated for about 18 months; modern readers who wish to plough through Spencer's writings might be warned of a similar risk. In this book, Spencer reasoned that Lamarckian processes could have led to modern human faculties. The love of liberty, for example, may have originated in the fear that animals show when restrained. This then evolved to a political commitment whereby individuals seek liberty for themselves and others as a principle.

What is more interesting is that Spencer advanced a view of the human mind, a sort of evolutionary Kantism, that he thought would provide a solution to the age-old controversy between the disciples of Kant and Locke. For Locke, the human mind is essentially structured by experience: we are born with a blank slate, a *tabula rasa*, upon which experience scribes. In this empiricist conception of the mind, there are few inherent specific structures or mechanisms. There is a congruence between our mind, our knowledge and the world itself because experience has shaped for us our perceptual categories and modes of perception. Kant, however, held that the human mind is provided at birth with *a priori* categories or inherent specialised mechanisms (such as Euclidean geometry and notions of space and time) that structure the world for us. This philosophical divide between the views of Locke and Kant was revived in the mid-19th century by debates between J. S. Mill and William Whewell, the former advocating an inductive view of knowledge in the manner of Locke, and the latter espousing a more Kantian position.

Spencer's proposed solution to this age-old conundrum was to suggest that Mill was right in proposing that experience shapes our mental operations but wrong in suggesting that each individual has to start the process from scratch when he or she is born. Like Whewell, Spencer believed that the mind is born ready equipped with perceptual categories and dispositions, but with the crucial addition that these Kantian categories were themselves the consequences of mental habits acquired through inheritance. Our mind structures our experience, but the structures used have been laid down during the evolution of the species. In broad outlines, and if we replace Lamarckian inheritance by natural selection, this is a view that is regarded as substantially correct and one that is perhaps Spencer's greatest legacy. It was this view of human knowledge, sometimes called 'critical realism', that was employed by ethologists such as Lorenz and Eibl-Eibesfeldt later in the 20th century (see also Chapter 7). Unfortunately for Spencer's reputation, and probably as a result of the failure of his views in other areas and his association with a discredited social Darwinism, it is Darwin rather than Spencer who usually receives credit for cutting this particular Gordian knot.

Darwin himself took a curious view of Spencer. In a letter to Lankester in 1870, he suggests that he will be looked upon as the greatest philosopher in England, 'perhaps equal to any that have lived' (Darwin, 1887, iii, p. 120). Elsewhere Darwin is less impressed and complained to Lyell that Spencer's essay on population was 'dreadful hypothetical rubbish' (Darwin, 1860). To Romanes in 1874, he confessed that 'Mr Spencer's terms of equilibration etc. always bother me and make everything less clear' (quoted in Cronin, 1991, p. 374). History has sided with the latter two views rather than the former.

1.3.3 *Evolution in America: Morgan, Baldwin and James*

Lloyd Morgan is most frequently remembered today for his canon, designed, as noted earlier, to avoid the traps of anthropomorphising. Another of his achievements was the formulation of a novel theory of the interaction between mental and physical evolution. One of the reasons Morgan rejected Lamarckism was precisely because he thought his own theory rendered it superfluous. While on a lecture tour of America in 1896, he gave a lecture in which he outlined his new theory. By a remarkable coincidence, the man who spoke next on the platform, the evolutionary psychologist James Mark Baldwin (1861–1934), delivered exactly the same idea. The theory became known as 'organic selection' but is usually today known as the 'Baldwin effect' and describes how a learnt adaptation could become fixed in the genome.

Suppose a sudden change in environmental conditions causes stress for a particular group of animals and thus exerts a selective pressure. Some who are unable to adapt their behaviours in their own lives will perish while those who are flexible enough to accommodate to the changed conditions may stand a better chance of surviving. Those individuals which are saved may over time throw up variations that are favourable in their new environment, thus allowing natural selection to fix these variants in the **gene pool**. By this means, what began at the level of phenotype as a non-inherited behavioural reaction to new conditions could over time become fixed by natural selection.

As an illustration, suppose that a change in conditions means that food for some animals is to be found lower down in the soil and that, to reach it, animals now have to dig deeper. Those which have the ability to learn that food lies at a deeper level, or are perhaps less inclined to give up at shallow depths, will survive food shortages. This could then give natural selection enough time to select out genuine genetic differences that enhance food-digging capabilities, such as large claws or an innate tendency to dig deep for food. In short, phenotypic plasticity can give rise to genetic adaptability.

Modern biologists have usually steered well clear of the Baldwin effect and have been reluctant to exploit its explanatory potential. The reason no doubt is that, *prima facie*, it smacks of Lamarckism – although in reality it is consistent with modern Darwinism. Richards (1987) and Dennett (1995) give sympathetic discussions of the Baldwin effect, and Deacon uses the Baldwin effect in his attempt to account for the origins of human language (Deacon, 1997).

Both Morgan and Baldwin had high expectations of the power of evolutionary thinking to explain the development and adaptation of individuals within social groups and even the progress of society itself. So too did the philosopher and psychologist William James (1842–1910), the elder brother of the novelist Henry James. In 1875, James taught a course in psychology at Harvard using Spencer's *Principles* as his text. Later he wrote his own textbook, *Principles of Psychology*, published in 1890. The book, originally intended only as a brief introduction to the subject, took him 11 years to write and became a major landmark in the discipline. Like Darwin and Spencer before him, James looked to animals for the instinctive roots of human behaviour and morality. Parental affection and altruism, for example, were to be seen as evolved traits. Morality as a whole was unashamedly a product of heredity.

In his psychology, James made frequent use of the concept of instinct, ascribing to humans such inherent traits as rivalry, anger, a hunting instinct and so on. He also speculated that some pathological conditions in humans could be remnants of animal instincts possessed by our ancestors that resurface in the human psyche following injury or disease. Many animals, rodents for example, prefer to remain close to cover and only reluctantly rush across open ground. James saw in such behaviour a distant basis for agoraphobia in humans. In all, James listed over 30 classes of instinct possessed by humans – enough to satisfy the keenest ethologist – and regarded humans as being more richly endowed with instincts than any other mammal. Such instincts he saw as being elicited by environmental stimuli. A given stimulus could spark off the emergence of a jumble of potential instinctive reactions, but humans were not mere automata since reason could intervene to select the most appropriate response to the situation.

James' singular contribution was to apply the idea of natural selection to ideas themselves. James raised the philosophical question of how we are to choose the best ideas in a world where humans project a range of competing theories, all claiming to have a correspondence with the truth. To answer this, James advanced an evolutionary epistemology that became known as pragmatism. In this scheme, truth is the idea that works, or more specifically the one that survives best in the intellectual environment into which it is cast.

James' thinking, particularly his use of the concept of human instincts, was enormously influential in American psychology in the first few decades of the 20th century. The historian Hamilton Cravens estimated that, between 1900 and 1920, over 600 books and articles were published in Britain and America that employed the idea of human instincts (Degler, 1991).

1.3.4 *Galton and the rise of the eugenics movement*

William James died in 1911, the same year as saw the demise of Darwin's cousin Francis Galton (1822–1911). Both of these thinkers left an influential legacy of ideas that were to have a major impact on subsequent attempts to apply evolutionary theory to human behaviour. The influence of James was, as noted above, initially highly productive, but it was the association of the ideas of heredity and instinct with the more insidious notions of Galton and the eugenicists that were to have a more lasting, and ultimately damaging, effect on the evolutionary **paradigm**.

The idea that natural science could throw light on pressing social problems was a typical product of Victorian positivism and scientism. It was this overweening ambition of scientists such as Galton that led to an entanglement of evolution with naïve and dangerous political thought. Francis Galton noted that our very humanity served to blunt the edge of natural selection. Care for the sick and the needy led to the procreation of the less fit and could in time, Galton feared, lead to the deterioration of the national character. Whereas Darwin was content to let things be, the remedy advocated by Galton was to propose that the state should intervene to modify human mating choices. Those with heritable disorders or the constitutionally feeble should be discouraged from breeding while the better sort of person should be positively encouraged. Such a programme Galton called **eugenics** – a term he coined in 1883 from Greek roots meaning 'well born'.

The fallout from the eugenics programme was for the most part detrimental and led to tragic consequences. Between 1907 and 1930, 30 states in America passed laws allowing for the compulsory sterilisation of criminals and mental defectives. By the early 1930s, about 12 000 sterilisations had been carried out in the United States. In America, the eugenicists were also particularly active in lobbying for the restriction of immigrants from southern and eastern European countries, whom they regarded as intellectually inferior. Consequently, in 1924, the US Congress passed the infamous Immigration Restriction Act. The way in which eugenic notions, and the idea that there are heritable mental differences between human groups, infected attitudes to race and immigration, and informed IQ testing and sterilisation practices, has been documented on many occasions (see especially Gould, 1981). Some of the issues raised will be considered again in the Epilogue, where it will be shown that eugenics has no logical linkage to the evolutionary programme.

By the 1930s, however, the damage was done. In the mind of the public and many scientists, the evolutionary approach to human nature had become entwined with an odious set of political beliefs. There thus began in the middle of the 20th century an almost total eclipse of serious efforts in the natural sciences to study human behaviour from an evolutionary perspective, a neglect that lasted for several decades, illustrating James' own observation that ideas are selected and thrive if they suit a given social and intellectual milieu. In this respect, it is also the social context that helps to explain the move away from hereditarian and evolutionary thinking in anthropology and the social sciences that led to the triumph of cultural over biological explanations of human nature.

1.4 The triumph of culture

1.4.1 Franz Boas

Darwin, in common with many anthropologists of his day, subscribed to the notion that differences in the level of culture between peoples were based upon biological inequalities. It was reasoning such as this that allowed racism to enter evolutionary thought. The efforts of Franz Boas (1858–1942) helped to turn the tide away from racist ideas in anthropology, and in so doing, Boas effectively enshrined culture as the central concept to be employed in explaining the social behaviour of man. In his youth, Boas worked in Berlin at the Royal Ethnographic Museum, an institution with a tradition of espousing cultural rather than biological explanations for human differences. In 1888, he joined the faculty of Clark University in America, later being appointed professor of anthropology at Columbia University, where he exerted an immense influence over American anthropology.

In 1911, Boas published two works that had a revolutionary impact on the social sciences. One was a book, *The Mind of Primitive Man*, the other a report entitled 'Changes in the Bodily Form of Descendants of Immigrants'. The first work was an assembly of previously published essays and conveyed the central message that the mind of primitive man (people in traditional or tribal cultures) did not differ in mental capability from that of civilised people. In effect, Boas was denying any significant innate differences between indigenous 'savages' and civilised peoples. Boas pointed to history and culture rather than biology as explanations for social differences in behaviour and customs. In place of innate racial differences, Boas substituted a single human nature common to all humans but shaped by culture.

In his report on bodily changes in the form of American immigrants, Boas published a series of remarkable findings that surprised even himself. He looked at whether the cephalic index, the ratio in humans of head length to head width, remained constant after racial groups had emigrated to America. The significance of this unusual statistic was that the cephalic index was widely used in the late 19th and early 20th centuries as a way of classifying human types. It was thought that races could be reliably classified along a spectrum ranging from the wide-headed (brachycephalic) southern Europeans to the long-headed (dolichocephalic) Nordic types. It was generally assumed that this measure was immune from environmental influences and remained stable from generation to generation. In studying the head shapes of immigrants to the United States and their subsequent offspring, Boas found that the head shapes of the children appeared to change if the mother had been in America for 10 years or so before conception. The round heads of the European Jews and the long heads of southern Italians converged towards a common type. So surprising were the results that Boas made further enquiries to rule out illegitimacy. For Boas, such results were proof concrete that culture and environment exert a strong pressure on the most basic of human features. If humans were plastic in their morphology, reasoned Boas, their mental plasticity would also be assured.

Boas' emphasis on the importance of culture did not imply that he was an extreme environmentalist like his contemporary Watson. Boas was primarily concerned with demonstrating the importance of culture and history as explanatory causes for the differences in achievement between (rather than specifically within) racial groups. Boas accepted that heredity played a major role in shaping the traits that a child possessed. His core message was that variations within any racial group are so large that they call into question the concept of race as a useful idea at all, a position largely endorsed by most modern geneticists.

It was the work of Boas and his followers that persuaded many social scientists to abandon Darwinian approaches to human social behaviour. However, relinquishing Darwin was, for many, not such a great sacrifice. Lamarckism always seemed more attractive anyway since it added a hopeful significance to the process of social reform. The notion that striving brought about heritable change also seemed more in keeping with human dignity. Moreover, Lamarckism was able to persist into the early years of the 20th century since biologists were not unanimous in its rejection. When Lamarck was finally ousted from biology in the 1920s, the purposeless process of Darwinian natural selection seemed markedly at odds with the powerful intentionality that emanated from human nature. Better therefore to fly into the arms of environmentalism than to suffer the cold embrace of natural selection. It is ironic that just as scientists such as R. A. Fisher, J. B. S. Haldane and Sewall Wright were, in the early 1930s, publishing their great summary works, which finally demonstrated the consistency between Darwinian natural selection and Mendelian genetics, most social scientists were moving rapidly in the opposite direction.

Social scientists began to erect what Tooby and Cosmides were later to call the standard social science model of human nature (Tooby and Cosmides, 1992). The model, which in its various versions dominated psychology and the social sciences from about the 1930s to the 1970s (and is still uncritically accepted in some quarters), contains a number of components. First, it stresses the insignificance of intergroup variations in genetic endowment. In other words, people at birth are by and large everywhere the same. Second, since adult human behaviour varies widely across and within cultures, it must be culture itself that supplies the form to the adult mind, disposes it to think and behave in culturally specific ways, and shapes adult behaviour. On the first point, evolutionary psychologists and sociobiologists are largely in agreement with the social scientists. It is on the second point that the standard model receives its greatest challenge from evolutionary theory.

Having abandoned Darwin, the social sciences found that it was the unity of human nature rather than its diversity that was remarkable. There remained of course one striking feature of the human condition that posed a challenge to environmentalism, and that was sex. Physical differences between the sexes were undeniable, as was the fact that in virtually every society males and females differed in their behaviour and their social roles. Darwin, with his theory of sexual selection (see Chapter 5), offered a biological approach that promised to be fruitful. More powerful forces than scientific rationality were, however, at work, and this was the next bastion of biology that came under attack from the social sciences in the 1920s and 30s.

One figure in the vanguard was Boas' student Margaret Mead (1901–1978). In her twenties, Mead studied the life of the islanders of Samoa. Her work culminated in a popular and influential book called *Coming of Age in Samoa,* published in 1928. This was followed in 1935 by *Sex and Temperament in Three Primitive Societies.* The central message of both books was that sex roles and sex differences in behaviour were not biologically ordained. Mead concluded that sex roles were interchangeable, so any differences between the personalities of each sex observed in any culture must be socially produced. Human nature, as she put it, is 'almost unbelievably malleable'.

A number of problems were subsequently noted with Mead's interpretation of her data, and this is considered in the last chapter. What is noteworthy here is the enormous influence that Mead's work had at an academic and popular level. The links between biology and human nature were now being severed as fast as they were once made. It was, however, the revulsion against eugenic ideas put into practice that was decisive.

1.4.2 *The revolt against eugenics*

Despite the promise offered by the eugenics movement to provide a scientific programme for social improvement, many sociologists had never swallowed the arguments. Experimentally, it was difficult to demonstrate that mental traits, good or bad, could be inherited. The family genealogies that Galton constructed showing how genius and talent ran in families could always be explained by nurture as well as or instead of nature. Moreover, it was not obvious that the most successful in society possessed the best genes. Boas, despite his insistence on cultural explanations for racial differences, accepted that personal qualities could be inherited from parents but questioned whether anyone would agree on what qualities were desirable and worth fixing through eugenics. He also, probably rightly, believed that individuals would not accept the infringement by the state of something so fundamental as the right to procreate. For social scientists, there was the additional drawback that many pressing social problems, such as inequalities in wealth distribution, poverty, prostitution and so on, did not seem particularly amenable to eugenic solutions. The Great Depression in America in the early 1930s, in which the wealthy, the intelligent and the poor alike were all affected by more powerful factors than their genetic endowment, seemed to confirm to many the irrelevance of the purported link between intelligence and success.

Following World War II, repugnance in the face of Nazi atrocities – many of them committed with eugenic principles in mind – led to a statement in 1951, by the United Nations Educational, Scientific and Cultural Organisation (UNESCO), that there existed no biological justification for the prohibition of mixed race marriages. The statement, endorsed by a number of prominent geneticists, also asserted that scientific knowledge provided no basis for believing that human groups differed in their innate capacity for intellectual and emotional development. This of course was not the same as saying that evolutionary theory had nothing to say about human behaviour, but the impression conveyed was that environmentalism had been given official and political sanction.

With whatever justification, an association began to form between eugenic thinking, right-wing ideologies and an evolutionary approach to human nature. Terms such as 'biological determinism' and 'biology is destiny' were applied, however fallaciously, to attempts to identify a biological basis for human nature and then roundly attacked. Compared with environmentalism, biological approaches to human nature seemed tainted by association and looked distinctly reactionary in their implications.

1.4.3 *Behaviourism as an alternative resting place*

Thus, in the middle years of the 20th century, the insecure foundations and dubious political connotations of the eugenics programme, the rigidity of the concept of instinct, and the paucity of experimental data to substantiate a belief in human instincts, gave pause to those who might otherwise have sought a basis to human psychology in biology. There was also some guilt among psychologists over the way in which hereditarian notions about IQ had been used to justify some racist immigration policies.

It is just possible that psychology would not have so readily abandoned the evolutionary paradigm if it had had nowhere else to go, but there was an alternative in the form of **environmentalism**, or more specifically behaviourism. In 1917, Watson was already claiming that a child is born with just three basic emotions: fear, rage and joy. For some psychologists, this was a positive relief from the dozens of instincts described by James and others, and the various traits claimed by the eugenicists to have a genetic basis. By 1925, Watson was denying the usefulness of the concept of instinct at all in human psychology. Although we should be wary of regarding Watson as typical of behaviourism as a whole, it is fair to say that most psychologists were not too sorry to abandon the theory of human instincts.

There were also methodological considerations. In the 1920s, there was a move away from what were seen to be deductive methods (that is, the deduction, or more strictly the induction, of human characters by extrapolation from animals and our supposed evolutionary past) towards an experimental approach. Behaviourism fitted neatly with the new image of psychology as a discipline based on rigorous experimental methods free from disputatious distractions, which psychologists were keen to cultivate.

In the 1970s, two psychologists, Wispe and Thompson, looking back over the hegemony that behaviourism once exerted over psychology, concluded that the whole idea of instincts never fitted comfortably with the American outlook (Wispe and Thompson, 1976). In America, where even the humble farmer's son could become president, the burgeoning science of psychology more naturally warmed to a position that did not suggest biological predestination. In contrast to 'biologism', behaviourism (in which anybody, with some exaggeration, could be anything) seemed liberal and liberating. Such sociological considerations may indeed provide a handle on the drift away from biological theories of human nature in the 1930s.

1.5 The rise of sociobiology and evolutionary psychology

As we have noted, in the 1960s, behaviourism came under attack at the same time as ethology was looking towards a revision of some of its basic concepts. If anything, it was behaviourism that suffered the most from such radical re-evaluations, while ethology continued to thrive. In the 1970s and 80s, the contribution of ethology to the natural sciences was recognised by the award of Nobel prizes in 1973 to Konrad Lorenz, Niko Tinbergen and Karl von Frisch for their work on animal behaviour. Then in 1981, Roger Sperry, David Hubel and Torsten Wiesel were awarded the prize for their work in neuroethology. The classical approach to ethology associated with Lorenz was continued in Germany by Irenaus Eibl-Eibesfeldt (1989). In the UK, the Netherlands and Scandinavia, where the influence of Tinbergen was more dominant, a more flexible approach to study of animal behaviour was emerging. At Cambridge in the 1950s, for example, under the leadership of William Thorpe, ethologists became particularly interested in behavioural mechanisms and ontogeny, while at Oxford Tinbergen headed a group more concerned with behavioural functions and evolution. In a sense, these two groups divided up Tinbergen's four 'whys' (Durant, 1986).

Meanwhile, aided by some fresh ideas from theoretical biology, a new discipline called sociobiology was emerging that applied evolution to the social behaviour of animals and humans. Sociobiology, like behavioural ecology, is concerned with the functional aspects of behaviour in the sense raised by Tinbergen. It drew its initial inspiration from successful attempts by biologists to account for the troubling problem of altruistic behaviour. Of all the problems faced by Darwin, he considered the emergence of altruism and, more specifi-cally, the existence of sterile castes among the insects as two of the most serious. Darwin provided his own answer in terms of community selection but also came tantalisingly close to the modern perspective when he suggested that if the community were composed of near relatives, the survival value of altruism would be enhanced. Haldane came even closer when, in his book *Causes of Evolution* (1932), he pointed out that altruism could be expected to be selected for by natural selection if it increased the chances of survival of descendants and near relations.

Sociobiologists such as Barash suggested that the new discipline represented a new paradigm in the approach to animal behaviour (Barash, 1982). Others such as Hinde concluded that 'sociobiology' was an 'unnecessary new term' since behavioural ecology covered the same ground (Hinde, 1982, p. 152). Behavioural ecology is an established and uncontroversial epithet for those who study the adaptive significance of the behaviour of animals. It would be a mistake, however, to treat 'sociobiology' and 'behavioural ecology' as interchangeable terms. Behavioural ecologists tend to focus more on non-human animals than humans and have a particular concern with resource issues, game theory and theories of optimality. Sociobiology itself deals with human and non-human animals, although it started out with a particular concern with inclusive fitness, and can be thought of as a hybrid between behavioural ecology, population biology and social ethology. The term 'human behavioural ecology' comes close to that of 'sociobiology'.

A book that served in some ways as a seed crystal for the new approach – at least in the sense that it galvanised its opponents – was *Animal Dispersion in Relation to Social Behaviour*, written by V. C. Wynne-Edwards and published in 1962. In this work, Wynne-Edwards advanced a position that was already present in the literature but had aroused no real opposition, namely that an individual would sacrifice its own (genetic) self-interest for the good of the group. It was attacks on this idea that catalysed the emergence of a more individualistic and gene-centred way of viewing behaviour (see Chapter 3).

One example of the new approach that countered group selectionist thinking came in 1964 when W. D. Hamilton published two ground-breaking and decisive papers on inclusive fitness theory (see Chapter 3). Then in 1966, G. C. Williams published his influential *Adaptation and Natural Selection*, in which he argued that the operation of natural selection must take place at the level of the individual rather than that of the group, and in a similar vein exposed a number of what he thought were common fallacies in the way in which evolutionary theory was being interpreted. In the 1960s and 70s, the British biologist John Maynard Smith pioneered the application of the mathematical theory of games to situations in which the **fitness** that an animal gains from its behaviour is related to the behaviour of others in competition with it. Then in the early 1970s, the American biologist Robert Trivers was instrumental in introducing several new ideas concerning reciprocal altruism and parental investment (see Chapters 3 and 4). It is a sad reflection on the specialisation of academic life and the disunified condition of the social sciences that while this revolution was occurring in the life sciences, psychology initially remained aloof from these new ideas.

The book that encapsulated and synthesised these new ideas more than any other was E. O. Wilson's *Sociobiology: The New Synthesis*, published in 1975. The book became a classic for its adherents and a focus of anger for its critics. In this work, Wilson irritated a number of scientists by forecasting that the disciplines of ethology and comparative psychology would ultimately disappear by a cannibalistic movement of neurophysiology from one side and sociobiology and behavioural ecology from the other. Others were alarmed that Wilson extended biological theory into the field of human behaviour. Although only

Figure 1.3   The American biologist Edward Wilson (born 1929)

Wilson did pioneering work on the social insect before publishing his massive and controversial work *Sociobiology* in 1975

one chapter out of the 27 in Wilson's book was concerned with humans, a fierce debate over the social and political implications of Wilson's approach ensued (see Chapter 12). It was as if Wilson had stormed the citadel that had been fortified from biology by social scientists over the previous 40 years, with predictable repercussions.

It is here that recent history merges with current perspectives. Sadly, there is as yet no satisfactory single name for describing the Darwinian paradigm applied to human behaviour. For many, the term 'sociobiology' is redolent of the painful debates that surrounded Wilson's early work. The description 'sociobiology' is still used but, partly to avoid associations with the vitriolic debates of the 1980s and partly to reflect a change in emphasis and ideas, other terms have sprung up.

It is significant that, following Wilson's book, a number of new journals, such as *Behavioural Ecology and Sociobiology* and *Ethology and Sociobiology*, appeared to cater for this growing area. In recent years, the latter journal has changed its name to *Evolution and Human Behaviour*, reflecting a new diversity of approach to the study of human behaviour.

There is also now a vibrant school of thought in America that calls itself evolutionary psychology. Evolutionary psychologists focus on the adaptive mental mechanisms possessed by humans that were laid down in the distant past – the so-called **environment of evolutionary adaptation (EEA)**. To be fair, sociobiology too had emphasised the importance of adaptive mechanisms forged in the geological period known as the Pleistocene. There are those who insist that evolutionary psychology and sociobiology are really the same thing. Robert Wright, in his *The Moral Animal: Evolutionary Psychology and Everyday Life* (1994), for example, suggests that 'sociobiology' was simply dropped as a name (for political reasons) although its concepts carried on under new labels:

> Whatever happened to sociobiology? The answer is that it went underground, where it has been eating away at the foundations of academic orthodoxy. (Wright, 1994, p. 7)

As one might expect from one of its founders, E. O. Wilson prefers to stick with the name 'sociobiology':

> in the interests of simplicity, clarity, and – on occasion – intellectual courage in the face of ideological hostility, evolutionary psychology is best regarded as identical to human sociobiology. (Wilson, 1998, p. 165)

As will be explained in Chapter 2, this book takes a catholic approach to the project of 'Darwinising man'. It adopts as a fundamental premise the fact that we evolved from ape-like ancestors and that our early environment has shaped our bodies and minds. The first species of the genus *Homo* evolved into being about two and a half million years ago, *Homo sapiens* making an entry about 200 000 years ago. If we date the origin of modern humans to the Neolithic revolution (the origins of agriculture), at about 10 000 years ago, this represents only 5 per cent of the time since the origins of *Homo sapiens* and only 0.5 per cent of that since the first appearance of the hominids. It follows from all this that we largely carry genes hard won by our ancestors in the Palaeolithic period, genes that are not always optimally suited for modern life. Nevertheless, there are features of modern life, such as making friends, finding a mate and raising children, that we experience in ways not totally dissimilar from our predecessors. Moreover, human behaviour is not 'hard wired': our genetically based behavioural mechanisms are moulded by development and modified by learning. Natural selection has also probably endowed us with conditional developmental strategies whereby particular behaviour patterns are triggered by, and are thence adaptive to, specific contexts.

In view of the rather sordid history of some aspects of evolutionary theorising about man, it is necessary, even at the risk of repetition elsewhere, to make some statements about how the Darwinian paradigm now stands on some of the issues raised in this chapter. Nearly all sociobiologists and evolutionary psychologists now assert the psychic unity of mankind; this is done not out of political correctness – a poor foundation for knowledge anyway – but simply because the biological evidence points in that direction. It follows that racial and between-group differences in behaviour are attributable to the influence of culture, or at least the interaction between commonly held genes and their fate in different cultures. In this sense, Darwinians agree with Boas: race is not a predictor of inherent potential and capabilities, and there is no racial hierarchy worth taking seriously. Where Darwinians profoundly disagree with the 'nurturists' is the view of the latter:

1. that our evolutionary past has no bearing on our present condition
2. that human behaviour and the human mind and are moulded and conditioned solely or largely by culture
3. and, crucially, that culture bears no relation to our genetic ancestry and hence can only be explained in terms of more culture.

In contradistinction to this, Darwinism asserts the existence of human universals upon which the essential unity of mankind is predicated. Some cultures may amplify some of these universals and suppress others, and culture itself may in some as yet unclear way reflect the universals lurking in the human gene pool, but the crucial point is that these universals represent phylogenetic adaptations: they have adaptive significance. This is not to suggest of course that there are 'genes for' specific social acts; genes describe proteins rather than behaviour. Human behavioural genetics is still in its infancy, and we still have to rely upon statistical correlations to mediate between adaptationist theories and facts. It will be genes in concert with other genes and environmental influences (meaning both the cellular and extracellular environments) that are seen to shape the neural hardware of the brain that forms the ultimate basis of human universals.

To Darwinise man, we need to begin with what Darwin actually said and how subsequent scientific work has modified or confirmed his conclusions. This is the subject of the next chapter.

SUMMARY

Darwin and some of his immediate followers sought an explanation of animal and human minds based on the principle of psychoneural monism: the idea that the mental and physical life of animals belong to the same sphere of explanation and that both have been subject to the force of natural selection. The essential continuity between animal and human minds was also assumed. This approach stimulated the application of concepts drawn from the emotional life of humans to account for animal behaviour. At its worse, this tradition relied too heavily on anecdotal information and was prone to anthropomorphising animals.

In the 20th century, two distinct approaches to animal behaviour emerged: ethology and comparative psychology. The ethological tradition in Europe was based around the work of such pioneers as Heinroth, Lorenz and Tinbergen, entailing the study of a variety of animal species in their natural habitat using a broad evolutionary framework. Comparative psychology took root particularly in America and built upon the work of Thorndike. Watson and Skinner were extreme representatives of this tradition. With the passage of time, comparative psychology became increasingly concerned with the behaviour of a few species in laboratory conditions, to the neglect of an evolutionary perspective.

Biology became entangled with political movements claiming to seek the improvement of society, or the race, or the national character. Galton was one of the chief founders and exponents of eugenics, the idea that the state should occupy a role in directing the breeding of individuals. Galton and his followers recommended that the able should be encouraged to procreate and the unfit or undesirable discouraged. Eugenic ideas were put into practice in the United States and Germany, with repressive and hideous consequences.

The theory of human instincts begun by Darwin was continued in America by James and in Europe by Heinroth, Lorenz and other ethologists. The experimental problem of demonstrating the action and existence of human instincts, as well as the reactionary associations of biological theories of human nature, led most social scientists and anthropologists in the middle years of the 20th century to reject biological explanations of human behaviour and assert the primacy of culture.

In the 1960s and 70s, a number of fundamental papers and books set in motion the sociobiological approach to animal and human behaviour. Sociobiology is predicated on the view that animals will behave so as to maximise the spread of their genes. Behaviour is therefore examined largely with functional (in the sense of Tinbergen) questions in mind. When applied solely to humans, the approach is also sometimes called evolutionary psychology, some regarding this movement as a new paradigm. This new movement seeks to revive the Darwinian project of demonstrating the evolutionary basis of human behaviour and of many facets of human culture.

FURTHER READING

Degler, C. N. (1991) *In Search of Human Nature: The Decline and Revival of Darwinism in American Social Thought*. Oxford, Oxford University Press.
True to its title, this book examines the period 1900–88. A penetrating sociological analysis of the fate of Darwinian ideas in America.

Dewsbury, D. A. (1984) *Comparative Psychology in the Twentieth Century*. Stroudsburg, PA, Hutchinson Ross.
A book from a leading comparative psychologist that is both a history of the discipline and a defence of its importance.

Gould, S. J. (1981) *The Mismeasure of Man*. London, Penguin.
An impassioned exposé of the flaws in intelligence testing.

Richards, R. J. (1987) *Darwin and the Emergence of Evolutionary Theories of Mind and Behaviour*. Chicago, University of Chicago Press.
A detailed and thorough work. Covers the ideas of Darwin, Spencer, Romanes, Morgan and James. Lays particular emphasis on the evolutionary origins of morality. Contains the authors' contemporary defence of using evolution as a basis for ethics.

Smith, R. (1997) *The Fontana History of the Human Sciences*. London, Fontana.
A panoramic survey of the human sciences from the scientific revolution to the late 20th century. Many chapters are of relevance to the study of evolution.

Thorpe, W. (1979) *The Origins and Rise of Ethology*. New York, Praeger.
An inside account of ethology from someone who helped to shape the discipline. Plenty of anecdotal information.

2

Darwin's Legacy

If I were to give an award for the single best idea
anyone has ever had, I'd give it to Darwin, ahead of
Newton and Einstein and everyone else. In a single stroke,
the idea of evolution by natural selection unifies the realm of
life, meaning and purpose with the realm of space and time,
cause and effect, mechanism and natural law. But it is not just a
wonderful scientific idea. It is a dangerous idea.

(Dennett, 1995, p. 2)

Compiling a league table of great ideas is not the best way to write a history of science, but with Darwinism we can make an exception. The scope and power of Darwinian thought seems to call out for and deserve superlatives. By linking life forms to all other life forms, and life itself to non-living, natural processes, the theory of evolution by natural selection probably represents the most profound of all human insights. Darwinism, in effect, provides an answer to one of the greatest questions: 'What is life?'

This chapter outlines the ideas central to Darwinism and examines some of the difficulties that Darwin faced. Many of the problems that confronted Darwinism in the 19th century have been largely resolved by work over the past 75 years. A successful theory, however, has not only to confront empirical evidence, but also to show that it can do so better than alternative accounts. With this in mind, we will contrast Darwinism with Lamarckism as alternative ways of explaining how organisms become adapted to their environments. The chapter also looks at Darwin's ideas on sexual selection, ideas unpopular in his day but revived over the past 20 years. Finally, there are at present distinct schools of thought on how Darwinism can properly be applied to the behaviour of contemporary humans. The chapter concludes that Darwinian anthropology, sociobiology and evolutionary psychology have too much in common to be called rival approaches, and that all can lay claim to Darwin's legacy.

2.1 The mechanism of Darwinian evolution

The essence of Darwinism can be summarised as a series of statements about the nature of living things and their reproductive tendencies:

- Individuals can be grouped together into species on the basis of such characteristics as shape, anatomy, physiology, behaviour and so on. These groupings are not entirely artificial: members of the same species, if reproducing sexually, can by definition breed with each other to produce fertile offspring

- Within a species, individuals are not all identical. They will differ in physical and behavioural characteristics

- Some of these differences are inherited from the previous generation and may be passed to the next

- Variation is enriched by the occurrence of spontaneous but random novelty. A feature may appear that was not present in previous generations, or may be present to a different degree

- Resources required by organisms to thrive and reproduce are not infinite. Competition must inevitably arise, and some organisms will leave fewer offspring than others

- Some variations will confer an advantage on their possessors in terms of access to these resources and hence in terms of leaving offspring

- Those variants that leave more offspring will tend to be preserved and gradually become the norm. If the departure from the original ancestor is sufficiently radical, new species may form, and natural selection will have brought about evolutionary change

- As a consequence of natural selection, organisms will become adapted to their environments in the broadest sense of being well suited to the essential processes of life such as obtaining food, avoiding predation, finding mates, competing with rivals for limited resources and so on.

One of the crucial points to appreciate about Darwinian thinking is that evolution is not goal directed. Organisms are not getting better in any absolute sense; there is no end towards which organisms aspire. Creatures exist because their ancestors left copies (albeit imperfect ones) of themselves. A useful analogy is suggested by the American biologist Stephen Jay Gould when he compares life on earth to a branching bush that is periodically but ruthlessly pruned.

One of the great triumphs of Darwinism was that it offered an explanation of the structure and behaviour of living things without recourse to any sense of purpose or **teleology**. Teleology (Greek *telos*, end) is the doctrine that things happen for a purpose or are designed with some express end. It was a cast of mind widespread in the early 19th century that natural phenomena and living things were designed with some motive or purpose in mind. Even by the mid-century, the critic Ruskin could still write that the structure of mountains was calculated for the delight of man and that in their contours 'the well-being of man has been chiefly consulted'. The teleological way of thinking was systematised and popularised by Archdeacon William Paley. In his classic *Natural Theology* (1802) – a work which Darwin knew well and once even admired – he advanced the famous watchmaker analogy. Just as a watch could not arise by chance but immediately suggests a designer and maker, so too the intricate organisation of living things suggests a cosmic designer. The whole of creation could then be seen as God's 'Book of Works', a book, moreover, that provided ample evidence for a benign Creator. It followed that creatures appeared to be fitted to their mode of life because they were designed that way by the supreme artificer we call God.

Purpose, teleology and the whole concept of providential design were all to be swept away by Darwin's 'dangerous idea'. For Darwin, there was no grand plan, no evidence that life forms were placed on earth by a Creator, no ultimate purpose or inevitable progress towards some goal. The watchmaker is blind.

2.1.1 *The ghosts of Lamarckism*

One person to suggest the possibility of the transmutation of species before Darwin was the French thinker Jean Baptiste Lamarck (1744–1829). His views are now virtually totally discredited, but they once served as the only serious alternative to Darwinism as a way of explaining the adaptive nature of evolutionary change. Lamarck argued that the characteristics (**phenotype**) acquired by an individual in its lifetime could be passed on to subsequent generations through the germ line (**genotype**). This view is usually parodied by the suggestion that the large muscles of a blacksmith, acquired through use during his lifetime, will result in slightly larger than average muscles in his son (and presumably daughter). Darwin is known to have rejected **Lamarckism**, but what he was really rejecting was the additional notion within Lamarckism that creatures possessed an inherent tendency to strive towards greater complexity. It was this teleological idea of purpose, rather than Lamarck's mechanism of adaptation, that Darwin wisely attacked. Indeed, it comes as a surprise and shock to many to learn that Darwin accepted the possibility of 'the effects of use and disuse' as a mechanism for influencing the characteristics of the next generation.

Lamarckism has at least two failings. Suppose we suggest that the hind legs of a rabbit developed their strength by the continued action of rabbits running away from predators. The first problem, as noted with the blacksmith analogy, is that, as an empirical fact ascertainable through experimentation, acquired characteristics are not passed on to offspring. The other problem is more serious, relating to the difficulty of explaining why exercise should increase the strength of muscles in the first place. Logically, exercise could reduce muscle strength, leave it unchanged or increase it. We know that it tends to increase it, but Lamarckism needs to explain this. To a Darwinian, it is straightforward: animals possess physiological mechanisms whereby exercise promotes stength because they help to increase their survival chances. Lamarckism is hard pushed to explain the direction of the muscular response.

The ghost of Lamarck was virtually (but not quite) laid to rest by August Weismann (1839–1914). Weismann distinguished between the germ line of a creature and its body, or 'soma'. Characteristics acquired by an individual affected the somatic cells (all the cells of the body other than the sperm or eggs) but not the germ line (information in the sperm or eggs). Weismann's essential

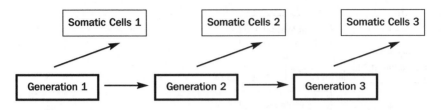

Figure 2.1   The germ line and information flow, according to Weismann

insight was that information could flow along the direction of the germ line and from the germ line to the somatic cells, but not from somatic cells to germ line (Figure 2.1).

In establishing his case, Weismann (largely between 1875 and 1880) cut off the tails of mice over a number of generations and showed that there was no evidence that this mutilation was inherited. This is in fact a rather unfair experiment – to both Lamarck and the mice – since, in strict Lamarckian terms, the mice were not striving in any way to reduce their tail length. It turns out that Weismann performed the experiment as a counterblast to those who argued at the time that it was well known that if a dog's tail was docked, its pups often lacked tails (Maynard Smith, 1982). For an equally forceful experiment, consider the fact that many Jewish male babies are still circumcised despite the fact that they are the products of a long line of circumcised male ancestors.

2.1.2 *The central dogma in a modern form*

The **central dogma** in its modern form can be understood in terms of the flow of information within individuals and between generations carried by macromolecules; this is shown in Figure 2.2. There is small number of cases in which this picture is an oversimplification, but none of these seriously challenges the general application of this schema (Maynard Smith, 1989).

We must still, however, deal with the question of why a two-way flow of information between genotype and phenotype never evolved. After all, male germ cells or gametes in the form of spermatozoa are made continuously by the body of most mammals at an incredible rate, so why couldn't information about the current state of the body be fed back to the germ line in the same way perhaps as an engineer uses his experience of the behaviour of the prototype to alter the blueprint for the next improved version? The answer seems to be that most phenotypic changes (with the exception of learnt ones) are not useful or adaptive. They result typically from disease, injury or ageing. A hereditary mechanism that enabled parents to transmit such changes would not be favoured by natural selection (Maynard Smith, 1989). As is so often the case, alternatives to Darwinism are unlikely, for good Darwinian reasons. For further arguments on the improbability of Lamarckian inheritance, see R. Dawkins (1986).

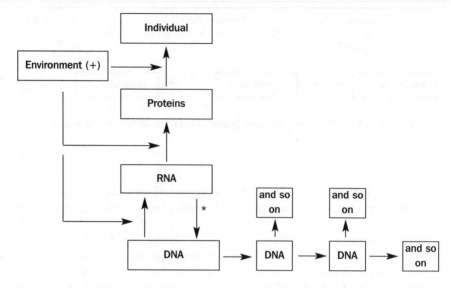

* The reverse flow between DNA (deoxyribonucleic acid) and RNA (ribonucleic acid) is postulated to account for the fact that some viruses carry RNA coding for an enzyme (reverse transcriptase) that can copy base sequences from its own RNA onto the DNA of the host.
(+) Note that here the 'environment' can be considered to be the cellular environment of DNA as well as the environment in which the organism lives.

Figure 2.2   The central dogma of the molecular basis of inheritance

2.2 Darwin's difficulties

There were four notable areas in which Darwin's theory faced serious problems:

1. the mechanism of inheritance
2. the means by which novelty might be introduced into the germ line
3. the existence of altruistic behaviour
4. the fact that some features of animals ostensibly seem to place the animal at a disadvantage in the struggle for existence.

Concerning the first two problems, Darwin knew nothing of course of the schema shown in Figure 2.2. His own theory of inheritance (sometimes called the theory of pangenesis involving 'gemmules' circulating around the body) is so wide of the mark that it is not worth considering. In Chapter 3, we will show how inheritance and novelty can be explained in terms of molecular genetics. With respect to the last two problems, Darwin made more headway, and although they will be tackled in the light of modern ideas more thoroughly in Chapters 3 and 5, it is worth looking at the issues as Darwin saw them.

Figure 2.3   Charles Darwin (1809–1882)

2.2.1 *The problem of altruism*

When the natural theologians of the 19th century looked at the world, they saw a harmonious edifice redolent with signs of design, purpose and God's constant beneficence. Darwin's vision was different: he saw a pointless brutal world populated by creatures engaged in an unrelenting struggle for existence. Darwin was also emphatic that every complex structure and instinct should be useful to the possessor and not useful to another unrelated animal. He noted that natural selection 'will never produce anything in a being injurious to itself, for natural selection acts solely by and for the good of each' (Darwin, 1859, p. 20). Yet Darwin could not fail to notice that nature revealed plenty of examples of co-operation: animals groom one another, share food and expose themselves to risk for the benefit of others.

It seems, however, that the problem of **altruism** did not trouble Darwin as much as it has troubled commentators since (see, for example, Cronin, 1991). Darwin's own response to the problem of self-sacrifice is not entirely clear, and he fluctuates between levels of explanation in the various editions of *Origin*. He sometimes draws close to an individualistic explanation but usually argues that the sterility of some may be of benefit to the larger whole:

> With strictly social animals, natural selection sometimes acts indirectly on the individual, through the preservation of variations which are beneficial only to the community... Many remarkable structures, which are of little or no service to the individual or its offspring, such as the pollen-collecting apparatus, or the sting of the worker bee, or the great jaws of soldier-ants, have been thus acquired. (Darwin, 1871, quoted in Cronin, 1991, p. 303)

Community-level explanations of insect altruism were invariably those resorted to for 100 years after the publication of *Origin*, although they are now regarded as flawed. The ground-breaking work that finally solved the problem of altruism that Darwin neglected came from Hamilton (1964), who extended Darwin's notion of **fitness** by developing the idea of inclusive fitness and kin selection. The concept of kin selection has now become an extremely powerful tool in the unravelling of many aspects of human behaviour (see Chapters 3 and 11).

2.2.2 *Natural selection or sexual selection?: dilemmas over sex*

Darwin's idea of natural selection was that creatures should end up with physical and behavioural characteristics that allow them to out-compete their rivals. Most features of plants and animals should therefore have some adaptive function in the struggle for existence, and nature should allow no extravagance or waste. So what about the peacock's train? It does not help a peacock to fly any faster or better. Neither is it used to fight rivals or deter predators; in fact the main predator of peafowl, the tiger, seems particularly adept at pulling down peacocks by their tails. It would seem to be a magnificent irrelevance, a positive encumbrance that should have been weeded out by natural selection long before now. Such features seem to defy the adaptionist programme, and even Darwin once remarked: 'The sight of a feather in a peacock's tail, whenever I gaze at it makes me sick!' (quoted in Cronin, 1991, p. 113).

Nor is the peacock's tail an exception: many species of animals are charac-terised by one sex (usually the male) possessing some colourful adornment that serves no apparent function or even seems dysfunctional. When males and females differ like this in some physical characteristic, they are said to be sexually dimorphic (literally, occurring in two shapes). Sexual dimorphism is found to varying degrees throughout the animal kingdom. Some species, for example kittiwake gulls, are hardly dimorphic at all; others, such as peacocks (*Pavo cristatus*) and widow-birds (*Euplectes progne*), have males with extremely long and colourful tails. Some of these differences could in principle be the result of natural selection: males and females may exploit different food resources, and female mammals are generally adapted to provide more care to the offspring than are males. But however ingeniously we work to apply the principle of natural selection, we are still confronted by the almost mocking sight of the peacock's tail. With characteristic insight and determination, Darwin had, by 1871, provided the answer to this seeming paradox. In his *Descent of Man and Selection in Relation to Sex* (1871), he gave the explanation that is still accepted (with refinements) today. The force of natural selection is complemented by the force of sexual selection: individuals possess features that make them attractive to members of the opposite sex or help them to compete with members of the same sex for access to mates. Thus, for example, the tail of the peacock has been shaped for the delectation of the peahen, while the antlers of male red deer serve as weapons in male versus male contests to win females.

The theoretical core of *The Descent of Man* failed to take hold, and over the following 100 years, sexual selection was rejected, ignored or relegated to some minor role. The consensus now, however, is that natural selection is not sufficient to explain the diverse forms of appearance and behaviour in sexually reproducing species. It must be complemented by the force of sexual selection, including female choice.

Given the research of the past 20 years, which has revived the theory of sexual selection, it is perhaps fitting to let Darwin have the last word. According to Romanes, a paper by Darwin read for him before a meeting of the Zoological Society in 1882 only a few hours before his death contained the passage:

> I may perhaps be here permitted to say that after having carefully weighed, to the best of my ability, the various arguments which have been advanced against the principle of sexual selection, I remain firmly convinced of its truth. (quoted in Cronin, 1991, p. 249)

2.3 Testing for adaptive significance

To say that a feature or behavioural trait is adaptive is to say that it confers more reproductive success on individuals who possess those features than is seen in those who do not or did not possess it. To demonstrate this effectively, we need to show how a feature in question confers some reproductive advantage. This is by no means easy. Giraffes with long necks may have an advantage over rivals with shorter necks in terms of grazing from tall trees, but they also probably had a better view of approaching predators. They may also have the edge in aggressive disputes with rivals of the same sex. It is all too easy to jump to conclusions: a given trait may be advantageous in a number of different ways in a given species or even in different ways in different species. Rabbits may have large ears to detect predators, but the large ears of an African elephant probably have more to do with heat regulation than sound detection.

It is also easy to find adaptations that are not there. Consider for a moment balding in human males; what adaptive function could it have? You may suggest that it helps exposure to sunlight and the synthesis of vitamin D. It could show that the male has high levels of testosterone and is thus virile. It could be an adaptive response to the need to lose heat on the African savannah plains; this, after all, is probably why humans lost their body hair. Bald men especially will be good at devising flattering and functional explanations. Most of them are probably false and amount to what Gould, after Kipling, has called 'Just so stories'. This particular trap is examined below.

2.3.1 *Pitfalls of the adaptationist paradigm: 'Just so stories' and Panglossianism*

In his *Just So Stories*, Rudyard Kipling gave an amusing account of how animals came to be as they are. The basic structure of the stories is that when the world was new, animals looked very different from today's types. Something then happened to these ancestral species that left them in the form we see now. The elephant, for example, once had a short nose, but after a tussle with a crocodile its nose was pulled and stretched into a trunk (Figure 2.4). In evolutionary biology, the 'Just so story' has become a metaphor for an evolutionary account that is easily constructed to explain the evidence but makes few predictions that are open to testing.

Figure 2.4   How the elephant acquired a long trunk (Kipling, 1967)

A similar trap is what Gould and Lewontin have referred to as 'Panglossianism' (Gould and Lewontin, 1979). In Voltaire's book *Candide*, Dr Pangloss is the eternal optimist who finds this world to be the best of all possible worlds, with everything existing or happening for a purpose, our noses, for example, being made to carry spectacles. In evolutionary thinking, Panglossianism is the attempt to find an adaptive reason for every facet of an animal's morphology, physiology and behaviour. Panglossian explanations are fascinating exercises in the use of the creative imagination. Consider why blood is red. It could help to make wounds visible, it could indicate the difference between fresh and stale meat and so on. Yet blood is red simply as a consequence of its constituent molecules, for example haemoglobin, and has probably never been exposed to any selective force. The evolutionist must be prepared to accept that just as some genetically based traits may no longer be adaptive, so some adaptive features may not be directly genetically based – although learning mechanisms will themselves have a genetic basis (Box 2.1).

It was Williams who, in 1966, helped to clarify what is meant by an **adaptation**. An adaptation is a characteristic that has arisen through and been shaped by natural and/or **sexual selection**. It regularly develops in members of the same species because it helped to solve problems of survival and reproduction in the evolutionary ancestry of the organism. Consequently, it can be expected to have a genetic basis ensuring that the adaptation is passed through the generations. Williams suggested that three criteria in particular should be employed to

BOX 2.1

Behaviour that is genetic but not adaptive,
or adaptive but not genetic

We must be wary of interpreting the basis of all behaviour as being genetic adaptation. There may be non-adaptive or non-genetic explanations for the phenomenon we are investigating. Some such alternative explanations could be:

1. Genetic but not adaptive

- *Phylogenetic inertia*
  Organisms may show signs of an ancestry from which they are unable completely to escape even though the features in question are no longer adaptive. It is not optimal, for example, for a hedgehog to curl into a ball as a defence against oncoming traffic. The human skeletal frame is not an optimum form for vertical posture, as anyone with a bad back will confirm. When a moth circles a candle flame, sometimes ending its revolutions by incinerating itself, it is obeying a genetic rule to non-adaptive ends. The rule is one that helps it to navigate by moonlight (or the sun), but moth genes have not kept up with artificial lights.

- *Genetic drift*
  Some genetic polymorphisms may exist in a population as a result of chance mutations that are neither advantageous nor disadvantageous, or that have not yet had time to be weeded out by natural selection. One special case of genetic drift is the **founder effect**. If a new popula-tion is formed from a few individuals, alleles may be fixed in the population that were once only a partial sample of a larger population. The new populations may look different, not for adaptive reasons but simply because of the effect of the founders being a limited sample from a larger and more diverse gene pool. The fact that the blood group B is virtually absent from North American Indians is probably the result of genetic drift rather than adaptive change.

2. Adaptive but not genetic

- *Phenotype plasticity*
  The phenotype of an organism can often be moulded by external influences during ontogeny to suit the prevailing environmental conditions. Bone, for example, grows in such a way as adequately to resist the pressures applied. The growth of corals and trees is well adapted to the direction of water and air currents. We could say of course that the mechanism to so adapt is genetic and thus heritable, but the adaptation itself is not.

- *Learning*
  Humans in particular have a great capacity to learn from each other, from experience and from their culture. If humans in widely dispersed and different cultures show similar patterns of behaviour that appear well adapted, this may be because of similar shared genes, but it could also be that they have come to the same conclusion on how to behave by parallel social learning.

ascertain whether the feature in question is truly an adaptation, these being: reliability, economy and efficiency (Williams, 1966). The first criterion is satisfied if the feature regularly develops in all members of the species subject to normal environmental conditions. Economy is satisfied if the mechanism of characteristic solves an adaptive problem without a huge cost to the future success of the organism. Finally, the characteristic must also be a good solution to an adaptive problem; it must perform its function well. If these three criteria are satisfied, it looks increasingly unlikely that the feature could have arisen by chance alone.

In searching for adaptations, we must be wary of the pitfalls of Panglossianism and try to avoid them by making precise predictions about how a feature or behavioural pattern under investigation confers a competitive advantage. Some of the specific procedures that can be used to test hypotheses are considered in the next section.

2.3.2 *The testing of hypotheses*

There is no such thing as a single scientific method. Different disciplines have different ways of gathering evidence, performing experiments, constructing models and testing hypotheses. One crucial feature common to all sciences, however, is the rigorous interplay between theory and experience. One of the most successful methods that structure this interplay is the so-called 'hypothetico-deductive method'. It was the philosopher Karl Popper who particularly noted the importance of this approach. The essential idea is that a **hypothesis** is framed to account for a particular phenomenon. The consequences of the hypothesis being correct are deduced and turned into predictions. These predictions are tested by experiment or by the analysis of other evidence and, if they are found not to hold, the original hypothesis from which the predictions were deduced is rejected or at least considerably modified. If a hypothesis successfully predicts an outcome, we can cautiously say that the hypothesis is supported (Figure 2.5).

A useful classification of different methods of testing hypotheses is suggested by Buss (1999), who distinguishes between 'theory-down' and 'observation-driven' approaches. The theory-down approach can be used to derive specific hypotheses from higher-level theories. The theory of sperm competition, for example, is one such high-level theory that can be used to derive subsidiary hypotheses. The theory suggests that aspects of the physiology and mating behaviour of males can be understood by the fact that, in some species, sperm from more than one male are likely to be present at the same time in the reproductive tract of a female. From this, we could derive the hypothesis that in conditions where the risk of sperm competition is high, males will tend to produce and/or ejaculate more sperm. This can then be tested either within a species in variable conditions or between species with different mating habits.

The observation-driven strategy is a sort of bottom-up approach. Humans are curious creatures prone (probably through inherited psychological mechanisms) to finding patterns amid sensory data. Our pre-existing concepts and expectations in part structure the patterns we find. To take a simple example, we may observe that, in Western cultures, women tend to spend more on cosmetics than men do. From this, we could derive a hypothesis that cosmetics enhance the signalling of reproductive fitness to males and hence play upon a male's evolved perceptual apparatus for evaluating attractiveness. To test this, we would then need to examine data on the types of transformation brought about by cosmetics and whether these characteristics in the natural state are correlated with fitness. The fact that we noticed the original pattern is in part conditioned by the fact that, as students of evolutionary psychology, we are alert to sex differences in behaviour, but our expectations do not prejudge the outcome of the tests that may follow.

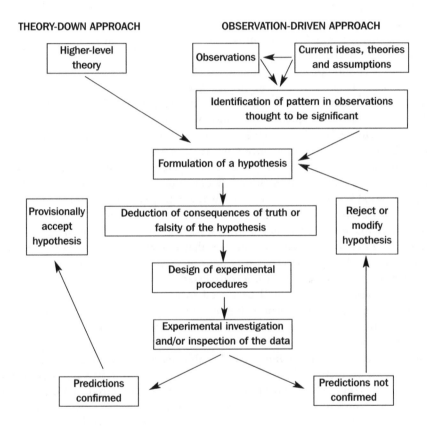

Figure 2.5   The hypothetico-deductive method applied to evolutionary hypotheses – an idealised view

Note that current ideas and assumptions (the prevailing paradigm) will influence the observations that are made and the types of pattern that are thought to be interesting, as well as the types of hypotheses framed to account for them

This second technique is in fact a very useful one, partly because we may notice many patterns before we have a scientific explanation for them. It is crucial to realise that it is not simply forcing the ideas to fit the facts. In the physical sciences, a similar technique is sometimes known as 'retroduction'. As an example, astronomers knew of Kepler's laws of the motion of planets around the sun before a theory could explain them. Newton and others partly used the method of retroduction to decide what a higher-level explanation would have to look like in order to generate the known laws. Newton's answer was the inverse square formula for gravitational attraction. In evolutionary theory, the method is sometimes called 'reverse engineering': the features of an organism can be used to infer backwards to the function for which it was designed.

Popper and others pointed out that to suppose that the hypothesis is proved by a successful prediction is to commit the fallacy of affirming the consequent. The fallacy arises because it is conceivable that a false hypothesis could give rise to a successful prediction. Where we are more certain is in the conclusion that if the prediction is not observed, there must be something wrong with the hypothesis. This distinction was Popper's most valuable contribution to the problem of establishing criteria for what counts as science and what is non-science – the so-called demarcation problem. The essence of science, according to Popper, is that it formulates hypotheses that are in principle falsifiable. Non-sciences (astrology may be an example here) make do with vague general statements that conveniently avoid an open confrontation with facts (Popper, 1959).

Some critics of modern evolutionary theory have suggested that evolutionary hypotheses are non-scientific since they are *ad hoc* and, being specific only to the trait in question (like the *Just So Stories*), lack generality to allow testing elsewhere. It has also been suggested that evolutionary theory is not truly predictive (as good science should be) since it refers to past events rather than future occurrences. These points, as a general critique of evolutionary reasoning, are in fact easily dismissed. Evolutionary hypotheses are often of necessity *post hoc* in that they refer to evolutionary processes that operated long ago when we were not there to observe them, but they are not inevitably *ad hoc*. In addition, the prediction of future events is not a necessary condition for a discipline to be scientific; if it were, we would be forced to re-evaluate the status of geology, palaeontology and other sciences dealing with the past. It is of course important that scientific theories predict unknown findings, but these could be in the past, present or future. We should also note that evolutionary theory is in principle falsifiable, although it has not yet been falsified. Scientists would be forced to radically revise their notions if, for example, an adaptation was discovered that functioned solely for the benefit of another species or conferred an advantage on a same-sex competitor with no benefit in return.

2.4 Adaptations and fitness: then and now

The terms 'adaptive' and 'fitness' are troublesome ones. An adaptive character should obviously palpably assist the survival and reproductive chances of its bearer, but this criterion is not always easy to apply. Debate still exists, for example, on whether language is adaptive and has been the subject of natural selection. 'Fitness' is a term with its own difficulties. It is sometimes complained that at the core of Darwinism lies the tautology of the 'survival of the fittest', for what is fittest if not best able to survive? So all we have is the 'survival of those best able to survive' – which tells us nothing. The eight points at the start of this chapter show this to be false in that they make specific and testable claims about the nature of inheritance and reproduction. But the word 'fittest' is misleading. As Badcock (1991) points out, human males would probably be fitter in the sense of living longer and enjoying a reduced susceptibility to disease if they were castrated or had evolved lower levels of testosterone, but this may not be the best way to produce offspring. Work by Westendorp and Kirkwood (1998) has also shown that life expectancy may be increased for couples who do not have children, but again this is not a good strategy for increasing Darwinian fitness. Darwinism is ultimately about the differential survival of genes rather than about the fitness of the gene carriers.

We must also not be tempted to hybridise Darwin with Pangloss and expect every adaptation to be perfect for the task in hand. In some cases, the environment can change more rapidly than natural selection can keep up with, and an adaptation is left high and dry, looking imperfectly designed. Some features may also be caught in an adaptive trough such that a large change would take the organism to a higher (better-adapted) peak but small changes would decrease its reproductive fitness. In these cases, the organism will be stuck with a less than perfect adaptive feature.

There are also developmental constraints acting on behaviour. An interesting example is to be found in brood parasitism. When the female cuckoo (*Cuculus canorus*) lays its egg in the nest of a host, the host sometimes rejects the egg. The very survival of the cuckoo, however, shows that this is not done unfailingly. Once a cuckoo is hatched, it is hardly ever rejected. The behaviour of the host is hardly optimal, so why has it not caught on and learnt to recognise its own eggs? Lotem *et al.* (1995) have shown that the problem lies in the fact that the hosts learn the characteristics of their eggs from those they produce when they breed for the first time. Eggs that differ from this learned set are rejected. It follows that any host parasitised during its first breeding attempt will learn the cuckoo's egg as its own and thus accept it in this and future breeding cycles. The rule is not perfect, but it works well for most birds.

Adaptation must always represent a trade-off between different survival and reproductive needs. A big body may be helpful in fighting off predators, but big bodies need lots of fuel and time to grow. It is also important to consider how behaviour leads to fitness gains over the whole lifespan of the animal. An animal must devote resources to growth, repair and reproduction. Over a single year behaviour may not seem to be optimal, whereas over a lifespan a different picture may emerge. This selective allocation of resources is often known as a 'life history strategy'. It explains why, for example, many mammals forego reproduction until they have reached a suitable size. It is a concept that can also be used to explain ageing and death (see Chapter 9 and also Kirkwood, 1977). The perfect organism would live forever, have no predators to fear and reproduce constantly. Fortunately, compromises have to be made, and the world is a more interesting place as a result.

2.4.1 *Evolutionary psychology or Darwinian anthropology?*

Suppose we identify some physical or behavioural trait of humans, such as hairlessness (compared at least with other primates) or the specific mating preferences of either sex, and attempt to demonstrate that they have some adaptive significance and have been the subject of natural or sexual selection. Among other problems, one serious question to address is 'Adaptation to what?'; to current conditions, or to conditions in the past? The problem is the same for non-human animals: a trait that we study now may have been shaped for some adaptive purpose long ago. The environment may have changed so that the **adaptive significance** of the trait under study is now not at all obvious. Indeed, it may now even appear maladaptive. When human babies are born, they have a strong clutching instinct and will grab fingers and other objects with remarkable strength. This may be a leftover from when grabbing a mother's fur helped to reduce the number of accidents from falling. It is not, however, clear that it helps the newborn in contemporary culture.

The problem is especially acute for humans since, over the last 10 000 years, we have radically transformed the environment in which we live. We now encounter daily conditions and problems that were simply absent during the period when the human genome was laid down. It could be expected that we have adaptations for running, throwing things, weighing up rivals and making babies but not specifically for reading, writing, playing tennis or coping with jet lag. One crucial question is whether the human psyche was designed to cope with specific problems found in the environment of our evolutionary adaptation (EEA) before the invention of culture (roughly the period between 2 million and 40 thousand years before the present) or whether our psyche is now flexible enough to give rise to behaviours that still maximise reproductive fitness in current environments. The problem is so serious that it has led to two basic schools of thought in the application of evolutionary theory to human behaviour: that of the evolutionary psychologists, who argue for the former model of the mind, and that of the Darwinian anthropologists, sometimes called human sociobiologists or biological anthropologists, who argue for the latter. A conjectured relationship between ancestral adaptation and current behaviour is shown in Figure 2.6.

Evolutionary psychologists would argue that human behaviour as we observe it today is a product of contemporary environmental influences acting upon ancestrally designed mental hardware. The behaviour that results may not be adaptive in contemporary contexts. We should focus then on elucidating mental mechanisms rather than measuring reproductive behaviour. We should expect to find mind mechanisms that were shaped by the selection pressures acting on our distant ancestors. An analogy is often drawn with the human stomach. We cannot digest everything we put in our mouths; the human stomach is not an all-purpose digester. Similarly, the mind is not a blank slate designed to solve general mental problems because there were no general mental problems in the Pleistocene age, only specific ones concerning, hunting, mating, travelling and so forth.

An example of this approach is discussed in Chapter 7, where we examine the work of Tooby and Cosmides on logical reasoning. They show that humans make mistakes when reasoning through a problem at an abstract level, but that performance improves when the problem is couched in a form in which cheaters would be detected. They suggest that the detection of cheaters would have been important in ancestral social environments where altruism was reciprocally exchanged. A more mundane example concerns our food preferences. As humans, we are strongly attracted to salty and fatty foods high in calories and sugars. Our taste buds were probably a fine piece of engineering for the Old Stone Age when such foods were in short supply and when to receive a lot of pleasure from their taste was a useful way to motivate us to search out more. Such tastes are now far from adaptive in an environment in developed countries where fast food high in salt, fat and processed carbohydrates can be bought cheaply, with deleterious health consequences such as arteriosclerosis and tooth decay.

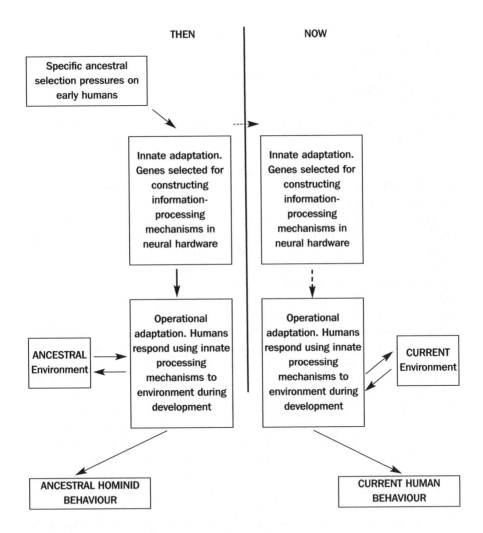

Figure 2.6   The relation between ancestral adaptations and current behaviour

Evolutionary psychologists argue that contemporary humans carry genes for innate mental hardware or mind modules designed in response to ancestral selection pressures. Current behaviour is a product of environmental influences on ancestral and innate mental mechanisms (adapted from Crawford, 1993)

At this point, it is worth stressing the difference between behavioural adaptations and cognitive adaptations. A pattern of behaviour may be adaptive without any cognitive component. The greylag goose noted in Chapter 1 that rolls its egg back to the nest is a case in point. The behaviour is highly adaptive – millions of eggs must have been saved by this device – but the fact that the behaviour continues when the egg is removed shows that is it rather simply constructed, inflexible and not 'thought about'. In contrast (Tomasello and Call, 1997), a cognitive adaptation:

- involves decision-making among a variety of possible courses of action
- takes place directed towards goals or outcomes
- probably involves some sort of mental representation that goes beyond the information immediately presented to the senses.

In a sense, cognitive adaptations are the result of evolutionary processes that have relinquished hard-wired solutions to directing behaviour optimally in favour of judgements made by an individual organism. In such cases, an organism (although not necessarily consciously) has goals, makes decisions and calculations according to context and its own life experiences, and hence chooses an appropriate **strategy**.

Daly and Wilson (1999) argue that the use of the term 'evolutionary psychology' only for humans introduces an unnecessary species divide. The divide is unwarranted by the history of the subject since many ideas from animal behaviourists have been incorporated into human behavioural studies and, moreover, the principles that apply to the human animal must also apply to non-human animals. Consequently, they prefer the term 'human evolutionary psychology' when referring only to humans.

For the sake of convenience, we will include human sociobiology, human behavioural ecology and human ethology under the heading of Darwinian anthropology. Darwinian anthropologists argue that ancestral adaptation was not so specific and that we possess 'domain-general' mechanisms that enable individuals to maximise their fitness even in the different environment of today. They suggest that different contemporary environments will give rise to different fitness maximisation strategies. The way to look for adaptations is not to try to find ancient mental mechanisms but to look at current behaviour in relation to local environmental conditions. In this sense, their approach is similar to that of the behavioural ecologists who study non-human animals. The differences between Darwinian anthropology and evolutionary psychology are shown in Table 2.1.

Objections to the approach of evolutionary psychology have focused on the mysterious EEA. Much clearly depends on the EEA, so what was it like, and how far do we go back? Betzig (1998) makes the point that, for the past 65 million years, our ancestors existed as some sort of primate, so do we look to the selective pressure on primates over this period as a clue to the human psyche? We could narrow it down by suggesting that we have spent the last 6 million years as one species or other of hominid (*Homo habilis, Homo erectus* and so on; see Chapter 6), so should we consider those environmental conditions? Archaic *Homo sapiens* and the subspecies *Homo sapiens sapiens* (that is, us) have been around for about 200 000 years, mostly spent in a hunter-gatherer lifestyle – or rather lifestyles, for even among contemporary hunter-gatherers, there are large differences in mating behaviour, paternal investment and diet – so perhaps we should focus on this period.

Table 2.1 Contrast between the methods and assumptions of Darwinian anthropology and of evolutionary psychology

Darwinian anthropology (Human sociobiology, Darwinian social science, human behavioural ecology, human ethology)	Evolutionary psychology
Behaviourist approaches	*Cognitivist approaches*
Culture should be viewed as part of a fitness maximisation programme. Humans are flexible opportunists and so **optimality** models (for example foraging and birth intervals) can be used. Game theory can help in investigating decision-making	Fitness maximisation of current behaviour is not a reliable guide to the human mind since current environments differ enormously from ancestral ones. Design (by natural selection) is manifested at the psychological rather than the behavioural level
Concentration on behavioural outcomes rather than beliefs, values, emotion and so on	Studies should look for mental mechanisms that evolved to solve problems of the Pleistocene, the environment of evolutionary adaptation
Measures lifetime reproductive success of individuals in relation to their environment. Counts current babies. Methods typical of those of behavioural ecologists	Needs to focus on the conditions and selective pressure of ancestral rather than contemporary environments
Ancestral adaptations have given rise to 'domain-general' mechanisms	Ancestral adaptations have given rise to 'domain-specific' modules designed to solve specific problems. The mind is like a Swiss army knife, consisting of discrete tools or problem-solving algorithms. Such modules may now function in maladaptive ways
Genetic variability can still exist and is particularly influential in choice of mate	The evolved mental mechanisms we now possess show little genetic variability; they point to a universal human nature

Over the past 2 million years, there have been a series of EEAs rather than just one. The EEA has acquired an almost fabled status, but Tooby and Cosmides point out that there was not one single EEA. They argue that it is 'a statistical composite of the adaptation-relevant properties of the ancestral environments encountered by members of ancestral populations, weighted by their frequency and their fitness consequences' (Tooby and Cosmides, 1990, p. 386). This is a fine definition in theory, but integrating such factors across time for human evolution is extremely difficult.

If we are only allowed to speculate about adaptations to an EEA, and given that establishing the features of EEA will be difficult enough, we are restricted in how far we can understand human nature with any precision. Moreover, it has been pointed out that, as humans, we know the probable responses of our own species to questionnaires and other measures of thinking, so there arises the temptation to select features from what we imagine to be the EEA to predict correctly outcomes that are already foreseen (see Crawford, 1993). It would be better if palaeontologists and palaeogeographers supplied the conditions of EEA and an intelligent chimp made the predictions about human behaviour.

Betzig takes the approach that we should look at the contemporary behaviour of humans in all its cultural manifestations through Darwinian spectacles, and that the behaviour of modern humans is still governed by the iron logic of fitness maximisation. We must also be wary of treating the EEA as some sort of 'land of lost content' when the human genome was in total harmony with its surroundings. In reality, no organism is perfectly adapted to its environment. Adaptations are compromises between the different requirements of an animal's life. What may appear maladaptive today, such as an infection of the appendix, may also have been maladaptive in the Pleistocene era.

One fundamental criticism of the approach of Darwinian anthropology is that if it is stated that the reproductive advantages of current behaviour reveal adaptations, this is tantamount to arguing that the present determines the past, for how can current conditions give rise to adaptations that preceded them? This argument is, presumably, not applied to non-human animals, in which it can easily be shown that the reproductive advantages of current behaviour reveal the existence of adaptations, because current behaviour and past behaviour were similar and took place in a similar context. It is not such a serious problem as it may appear for human behaviour, however, if there are features of past and present environments that are similar. Suppose we find that young men who are rich and undertake risky forms of behaviour, such a flying light aircraft, are more sexually successful (measured in, say, the number of sexual partners) than poorer or more cautious types. Financial wealth and aeroplanes are products of fairly recent culture, but the ability to command resources and the willingness to undertake risks are forms of behaviour that could have taken place at any time over the past 200 000 years. We may find that the number of offspring of wealthy high status males in today's culture is no greater than average, but this merely shows that sexual success has become uncoupled from offspring production. Contraception has in this instance severed the link between status and its reproductive consequences, and we need to measure success in other terms, such as number of sexual partners, which would once have been proportional to lifetime reproductive success.

Darwinian anthropologists focus on 'adaptiveness' to different environments, implying that humans will tend to maximise their reproductive output in the circumstances in which they find themselves. In contrast, evolutionary psychologists focus on 'adaptation' – discrete mechanisms and traits that an organism carries as a result of past selective pressures. Some have claimed that the study of adaptation is the essence of Darwinism; as Symons (1992, p. 154) remarked, 'since Darwin's is a theory of adaptation, not adaptiveness, the Darwinian Social Science hypothesis is not clearly derived from the theory of evolution'. This seems unnecessarily restrictive. Darwinism is an evolving set of ideas, and there should be no need to be bound by every last refinement of the ideas of the master: we reject without compunction, for example, Darwin's acceptance of Lamarckism. A more important point, however, is that behavioural ecologists have shown that non-human animals do display adaptiveness: they have a range of strategies that are evoked in different environmental conditions. The sex ratio of opossums, for example, varies according to environmental conditions in a manner predicted to increase reproductive success (see Chapter 4). Baker and Bellis (1989) provide evidence that human males adjust the number of sperm in their ejaculate according to the likelihood of sperm competition (see Chapter 5), and bush crickets alter their sexual strategy according to the availability of food resources (see Chapter 5). It would be a simple or a foolish and very 'un-Darwinian' animal that only had a single behavioural strategy to be expressed in all circumstances.

It is also important to note that natural selection can shape how development and learning occur in relation to local environments, as implied in Figure 2.6 above. It follows that behaviour does not have to be forced into the category of 'hard-wired mental modules'. Natural selection could have shaped our minds to respond to what is fitness-enhancing under prevailing conditions, and to behave accordingly. As an example of this, Bobbi Low at the University of Michigan has provided evidence that the training of children is related to the type of society in a way that tends to maximise fitness. Low found that the more polygynous the society, the more that sons were reared to be aggressive and ambitious. The logic here is that, in polygynous societies, successful males stand to secure more matings. If Low is right, the behaviour of adult males is strongly influenced by their childhood training, but such training is a response to local social and ecological conditions. It shows that humans may have a repertoire of strategies activated by environmental cues (Low, 1989).

There are problems, however, in choosing measures of fitness. Consider how a male could maximise his fitness in the late 20th century. A scenario envisaged by Symons (1992, p.155) is both amusing and instructive:

> In a world in which people actually wanted to maximise inclusive fitness, opportunities to make deposits in sperm banks would be immensely competitive, a subject of endless public scrutiny and debate, with the possibility of reverse embezzlement by male sperm bank officers an ever-present problem.

The answer of course is that natural selection did not provide us with a vague fitness-maximisation drive. The genes made sure that fitness maximisation was an unconscious urge; like the heart beat, it is too valuable to be placed under conscious control. Instead, males and females were provided with sexual drives that could be moderated according to local circumstances. Counting the size of the queue outside a sperm bank would be a fruitless way of assessing a male's fitness – maximising sex drive and counting partners and real sexual opportunities might be better. If sperm banks were set up to allow males to deposit sperm more naturally (in vivo), I suspect that the queues would be longer.

In an important article, Crawford (1993) suggests that we should look for similarities rather than differences between ancient and current environments. It is easy to draw sharp contrasts between ancestral and modern environments by choosing features that have drastically changed, such as population densities, jet travel, computers and so on, but such a comparison is pointless unless we specify the nature of the adaptations that are supposed to be out of place as a result. Moreover, if the world is so fundamentally different, why do humans seem to be thriving? We live surrounded by space-age technology, but the fundamental patterns of life go on: couples meet and have babies; people make friends and enemies, argue and settle arguments, gossip intensely about each other and so on. One of the ironies of the modern condition is that we launch high-tech satellites into orbit around the planet to beam down soap operas and pornography.

It must also be remembered that we have built modern culture around ourselves. We visit or live near to our relatives, houses are designed for the nuclear family, we work in groups with hierarchies – all features probably not far removed from those of the ancestral condition. Crawford advises that we should assume a basic similarity between ancient and contemporary environments with respect to particular adaptations, unless there are signs of stress and malfunction in humans, or the behaviour is rare in the ethnographic record, or unusual reproductive consequences are observed. Polyandry (the sharing of one wife by several men), for example, is rare in human society, and there are strong reasons for suspecting that we are not well-adapted to this way of life.

In this book, the approach taken is that there is room for both methodologies (Sherman and Reeve, 1997). The methodological issues are serious, and the interested reader is referred to texts at the end of the chapter, but the human brain is complex enough and powerful enough to accommodate behaviours that are learned or unlearned, behaviours that are soft wired and hard wired, behaviours that adjust to local conditions and behaviours that are invariant. In subsequent chapters, we offer significant findings and successful predictions from both perspectives (for a review, see Daly and Wilson, 1999).

2.4.2 *Orders of explanation in evolutionary thinking*

One of the great strengths of evolutionary thinking is that it enables us to answer 'why'-style questions in a scientific and non-metaphysical fashion. Consider the questions posed in Table 2.2. There are at least three types of answer to these questions: **teleological, proximate** and **ultimate**.

If we answer that the reason the fur of a stoat turns white in winter is to help with camouflage, we are, strictly speaking, reasoning **teleologically**. Camouflage is a consequence of the fur turning white; it is an effect of the change of colour, and an effect cannot be a cause. To avoid this, we might resort to identifying prior causes that triggered the change in colour, such as a hormonal response to falling temperatures and reduced daylight. This may be a correct response physiologically but is somewhat unsatisfying; all we have done is to identify a **proximate** causal mechanism. We have provided the 'how' of the process but have not explained why such processes exist. In the language of Tinbergen we have identified a causal mechanism (see Chapter 1).

The **ultimate** causal explanation rests in the third column of Table 2.2: genes that code for a change in coat colour exist because they conferred a survival value on the stoats that possessed them. Natural selection cannot think ahead like the teleologist and plan a set of genes to achieve some purpose. We sweat because a chance mutation in our ancestral genes conferred some fertility advantage on our predecessors; sweating is an adaptive or functional response. One of the remarkable features of Darwinism is that, for the first time in the life sciences, it provided satisfactory answers to 'why'-type questions. Without Darwin, nothing in life really makes sense. Consider the question 'Why are we here?', which carries a miasma of spurious profundity. To a committed Darwinian, we are here because we carry genes that were successful at self-replication. Similar genes that were less successful are not here for us to observe: none of us is descended from sterile ancestors. The nature of genes is to make copies of themselves not with any grand plan or purpose in mind but simply because that is what they do. In a sense that is non-tautological, we are here because we are here. To paraphrase Wittgenstein, a cloud of metaphysics is thereby condensed into a drop of Darwinism.

If we compare the terms 'proximate' and 'ultimate' to the four 'whys' of Tinbergen considered in Chapter 1 (causation, development, evolution and function), we can note that proximate translates to causal and ultimate to functional. The distinction is important as a means to an intellectual understanding of evolution, but, practically speaking, the four questions of Tinbergen should not be treated as isolated areas of inquiry but instead as concerns that are interdependent. Natural selection has shaped behaviour to serve its present function in ensuring the survival of the genes responsible, but we must also acknowledge that natural selection has determined the way in which the causal mechanisms that initiate the behaviour begin their work. Ontogeny (development) may also be linked to function in that the precise course of development in an individual may be sensitive to local conditions in order to achieve the best adaptive fit to current circumstances.

Examples of this are discussed in Chapter 3, where it is shown that the mating behaviour shown by an individual is sensitive to variables such as the behaviour of others and the abundance of resources. Individuals may have several strategies for mating that are triggered by different environmental events. Another example concerns the Westermarck effect discussed in Chapter 4, in which the ontogeny of sexual desire is influenced by members of the opposite sex associated with during childhood. If Westermarck was right, we develop to experience no sexual desire for those whom we grew up with in close proximity. The adaptive significance of this, and hence of the incest taboo that proscribes mating between kin, is that sibling mating can produce congenital defects in the newborn. Thus in this case, we see an integration of function, causal mechanism and ontogeny.

In the study of sperm competition examined in Chapter 8, we are often dealing with morphology, causal mechanisms and their ontogeny, and adaptive significance as a series of integrated ideas. In sexually reproducing species in which a female is likely to mate with more than one male, one strategy for a male to increase his chances of fathering an offspring is to adjust the number of sperm (gametes) produced in relation to the likelihood of the female mating with other males. Studies on the genitalia of primates (see Chapter 8) seem to confirm these predictions in terms of behaviour and morphology. One of the most neglected questions of Tinbergen in behavioural studies has often been that of evolutionary history – the very question that Lorenz thought most crucial. Recent studies on phylogeny are redressing this imbalance and promise to throw light on functional questions. The evolution of concealed ovulation in human females, for example, could help to elucidate the function it served and serves (see Chapter 8).

Table 2.2   Types of explanations in evolutionary thinking

Question	Teleological	Proximate	Ultimate
Why does the fur of stoats (*Mustella erminea*) turn white in the winter?	To become better camouflaged	Hormonally mediated response to day length and ambient temperature	Advantages once (and still) conferred: differential survival of genes
Why do humans sweat when hot?	To lose heat by evaporative cooling	Response of sweat glands to high temperature	Advantages once (and still) conferred: differential survival of genes

SUMMARY

■ The theory of evolution provides a naturalistic account of the variety, forms and behaviour of living organisms. In constructing his theory, Darwin jettisoned the idea of purpose in nature (teleology) and the notion that organisms conform to some abstract and pre-existing blueprint or archetype.

■ Evolution occurs through the differential reproductive success of genes. To understand the mechanism of natural selection, we need to distinguish between genotype and phenotype. The genotype consists of the genes that carry the information needed to build organisms. The phenotype is the result of the interaction between genes and the environment within an individual. Whereas environmental factors may strongly influence the way in which genes are expressed within an organism, the outcome of this interaction cannot be communicated to the genotype. Characteristics acquired in the lifetime of an individual are thus not inherited by the offspring.

■ Darwin was unable satisfactorily to explain the mechanism of inheritance and how novel and spontaneous differences between offspring and parents (which form the raw material for evolution) could arise.

■ The existence of altruistic behaviour posed a problem for Darwin that he did not satisfactorily resolve.

▨ Darwin and Wallace both appreciated the need for a supplementary theory of sexual selection to explain the physical and behavioural attributes of animals.

▨ We should expect the behaviour of animals to be adaptive in the sense that it has been selected by natural and sexual selection to help confer reproductive success on individuals. There are various ways in which the adaptive significance of behaviour can be demonstrated and investigated. Some of these involve experimental manipulation of the natural state, and some involve looking for correlations between behaviour and environmental factors. In such studies, we must be constantly wary of finding convenient but spurious explanations designed *post hoc* to fit the facts.

▨ A debate exists over the correct way to apply Darwinian reasoning to human behaviour. Darwinian anthropologists, sometimes called human sociobiologists, human behavioural ecologists or human ethologists, suggest that current human behaviour measured in terms of reproductive success shows signs of adaptiveness. Evolutionary psychologists argue that the correct Darwinian approach is to look for adaptations to ancestral environments that can now be identified with discrete problem-solving modules in the brain. It is suggested here that there is room for both interpretations.

KEY WORDS

Adaptation ■ Adaptive significance
Altruism ■ Central dogma ■ Fitness
Founder effect ■ Function ■ Genotype ■ Hypothesis
Lamarckism ■ Optimality ■ Phenotype ■ Proximate cause
Sexual selection ■ Strategy ■ Teleology ■ Ultimate cause

FURTHER READING

Barkow, J. H., Cosmides, L. and Tooby, J. (1992) *The Adapted Mind*. Oxford, Oxford University Press.
See Chapters 1 and 2 for discussion of the approach of evolutionary psychology. Discusses some complex methodological issues.

Buss, D. M. (1999) *Evolutionary Psychology*. Needham Heights, MA, Allyn & Bacon.
An essential book for any undergraduate studying this field. One of the best textbooks written so far specifically dealing with evolutionary psychology.

Crawford, C. and Krebs, D. L. (1998) *Handbook of Evolutionary Psychology*. Mahwah, NJ, Lawrence Erlbaum.
Numerous chapters by leading authorities on the whole field of evolutionary psychology. See Chapters 1, 8 and 9 for a discussion of methodological issues.

Cronin, H. (1991) *The Ant and the Peacock*. Cambridge, Cambridge University Press.
Excellent historical account of the theories of kin selection (the ant) and sexual
selection (the peacock). Closely argued and packed with references. A book for the
serious historian of ideas.

Ridley, M. (1993) *Evolution*. Oxford, Blackwell Scientific.
A good overview of the whole theory of evolution.

3

The Selfish Gene

...do not look for them floating loose in the sea; they gave up that cavalier freedom long ago. Now they swarm in huge colonies, safe inside gigantic lumbering robots, sealed off from the outside world, communicating with it by tortuous indirect routes, manipulating it by remote control. They are in you and me; they created us, body and mind; and their preservation is the ultimate rationale for our existence. They have come a long way, those replicators. Now they go by the name of genes, and we are their survival machines.

(Dawkins, 1976, p. 21)

Applying Darwinian thinking to animal behaviour is based on the understanding that genes will influence behaviour in ways that tend to ensure their own reproductive success. Obviously, the environment in which an organism grows, lives and learns from also influences its behaviour. It is a truism worth repeating, however, that environmental factors need something to act upon. Without an environment, genes would have nothing to do; without genes, the environment could have no influence. Individuals are products not of genes or environment, nature or nurture, but of both.

In this chapter, we will examine the nature and properties of these genes at the molecular and cellular levels. Our understanding of the genetic basis of behaviour and the whole operation of natural selection is also greatly assisted by a clarification of what can properly be considered to be the unit of natural selection, and by the crucial distinction that must be made between the replicators (genes) and the vehicles of these replicators (bodies). The concept of the selfish gene has been much maligned, but properly understood it does not suggest that all individuals must behave selfishly in the pejorative sense. This chapter will demonstrate how altruism can possibly arise in a world of selfish replicators. Two essential concepts that emerge by the end of the chapter are kin selection and reciprocal altruism. These concepts offer ways of understanding the biological basis of human altruism. The application of these ideas to humans is then more fully explored in subsequent chapters.

3.1 Some basic principles of genetics

3.1.1 The genetic code

In Chapter 1, we saw how Weismann argued that inheritance is best thought of as a flow of information along the germ line. With this in mind, there are four basic questions that we need to answer in order to appreciate the genetic basis of this information in the context of evolution:

1. How is genetic information stored and preserved?
2. How is this information used to build organisms?
3. How is the information passed on to new organisms in the act of reproduction?
4. How do novelty and change enter the information and thus provide the raw material for natural selection to act upon?

The storage and preservation of information: the language of the genes

The language of **genes** is written on an extremely long molecule called deoxyribonucleic acid, or **DNA** (Figure 3.1). The molecule consists of two strands, each strand having a backbone of alternating ribose sugar and phosphate groups, and each sugar group having one of four **bases** attached to it. This sequence of bases is the genetic code, which prescribes the development of each individual. Since each base can only chemically bond with one of the other three (its complementary partner), the sequence of bases on one strand uniquely defines the sequence on the other strand. The base pairs are said to be complementary. There are four types of bases: cytosine (C), thymine (T), adenine (A) and guanine (G). The complementary pairings of the bases is as follows:

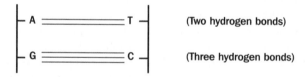

It is estimated that the human **genome** contains about 3×10^9 base pairs, of which, at the time of writing, 60×10^6 have been mapped by the human genome project. Each strand of the DNA molecule is wound into a helical shape, two strands thus giving us the double helix (Figure 3.2). It is the sequence of bases on any one strand that contains the information necessary for the development and functioning of each cell. Since each base determines the base on the opposite strand, this information is therefore contained within just four characters. At first sight, it may appear unlikely that just four characters could carry the information necessary to define the development of complex organisms – even the English alphabet has 26 letters. We must remember, however, that molecules are small and, relatively speaking, the DNA molecule is long. The enormous storage capacity achieved by modern computers is achieved using just two characters: a zero (0) and a one (1), corresponding to the states of transistors in the microcircuitry.

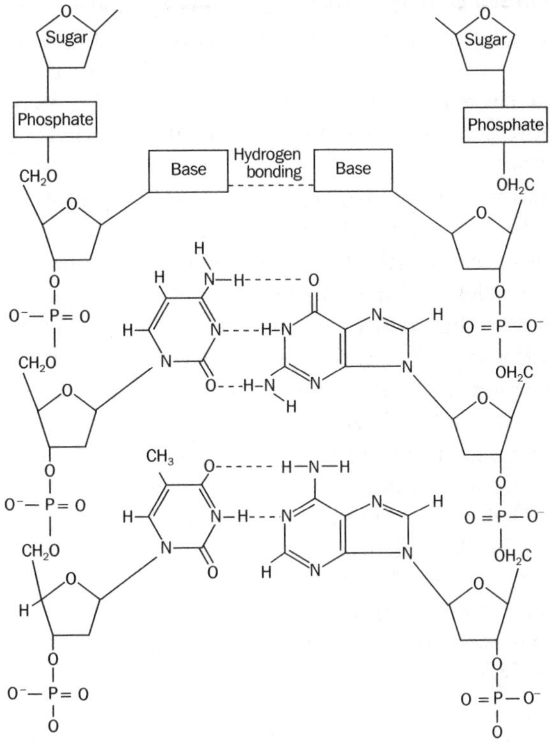

Figure 3.1   Structure of DNA, composed of sugar, phosphate and base groups

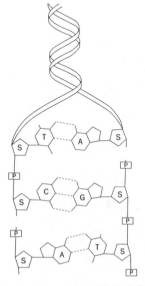

S = sugar; P = phosphate; A, C, T, G = bases

Figure 3.2   DNA as a double helix of complementary polynucleotide chains

If the sequence of nucleotides in the entire human genome were represented on the page of a book by symbols (so that a sequence would look like, for example, AGTCGAATTGCC...), a gene would on this scale spread over about three pages, an average **chromosome** would take up about 50 books of the size of this one, and the entire genome (present in just one of your cells) would spread over about 1000 books.

The preservation during growth and reproduction of the information contained in the base sequences of DNA raises two related questions. First, when a cell divides during the growth of an organism, how is information passed to the daughter cells in order that they develop and perform appropriately? Second, when organisms reproduce either sexually or asexually, how is information passed from parent to offspring? The process, at least in the early stages, is essentially the same. An examination of the structure of DNA in Figure 3.3 shows that if the two strands were divided and each retained its sequence of bases, each strand could serve as a template to create another double helix. This potential was immediately obvious to Watson and Crick when they established the structure of DNA in 1953. Their own suggestion in their paper submitted to *Nature* has become a classic of understatement:

> It has not escaped our notice that the specific pairing we have postulated immediately suggests a possible copying mechanism for the genetic material. (Watson and Crick, 1953, p. 737)

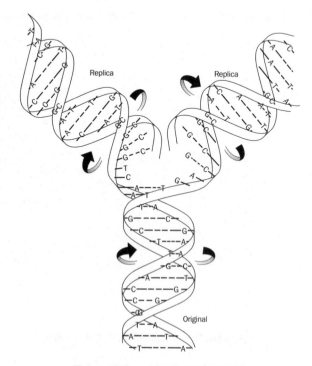

Figure 3.3   Replication of DNA

Figure 3.4 James Watson (b. 1928), on the left, and Francis Crick (b. 1916), on the right, shown with their model of DNA at the Cavendish Laboratory, Cambridge, England in 1953. Their discovery of the helical structure of DNA ranks as one of the most important scientific findings of the 20th century

How genetic information is used to build organisms: the translation of the language

Following the rediscovery of Mendel's work by de Vries in 1901, and subsequent work by de Vries and Morgan, it became generally accepted that heritable information was carried on units called genes. Even as long ago as 1909, it was suspected that genes influenced the **phenotype** through the production of enzymes. Archibald Garrod, an English physician, proposed that inherited diseases, so-called inborn errors of metabolism, were caused by defective enzymes, which reflected a defective encoding of the genetic information. Later work supported Garrod's insights and led to the 'one gene, one enzyme' hypothesis. In this view, each unit of information (the gene) specified one enzyme. All enzymes belong to the class of molecules called proteins, and this 'one gene, one enzyme hypothesis' was later modified to 'one gene, one protein'. When it was realised that some proteins consisted of several different chains coded for by different genes and assembled after each chain was produced chemically, the view was modified again to the idea of 'one gene, one polypeptide'.

A gene can thus be regarded as a stretch of DNA that in some way contains the information necessary for the synthesis of a polypeptide. There are about 80 000–100 000 genes per human cell. Not all of the DNA in a cell codes for proteins, and there are some rather mysterious sections that seem to do nothing at all, sometimes called, to the great irritation of geneticists, 'junk DNA' or more respectfully 'introns' (Figure 3.5). It is estimated that about 95 per cent of human DNA is of the non-coding or 'junk' sort, leaving only 5 per cent with a function that we currently understand (Sudbury, 1998).

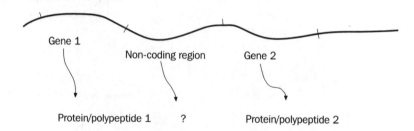

Figure 3.5   DNA as sections of genes and non-coding sections

Proteins are made up of long chains of chemical units called **amino acids**. There are thousands of different kinds of protein but only 20 types of amino acid common to all organisms. We have seen how the information along the DNA is in the form of a four-character language: A, C, T, G. Now, if each base coded simply for one type of amino acid, we would only have four possible acids that could be encoded. If a pair of bases such as AT or CG coded for one amino acid, there would still be only 16 (4^2) permutations possible. But if triplets of bases such as AAA, CCG, GCA and so on coded for each amino acid, there are 64 (4^3) possibilities. This is the minimum number of base combinations that are required to encode the 20 or so amino acids used to build organisms. Moreover, this triplet code would allow spare information that could convey instructions such as 'start', 'stop' and so on. It turns out that the triplet code is the one used by DNA.

Proteins are not assembled directly from the DNA template. The fine detail of the biochemistry is beyond the scope of this book, but, in brief, the information along the DNA is first transcribed to a very similar long molecule called messenger ribonucleic acid (mRNA) in the nucleus of the cell. The RNA then carries this information to the cytoplasm, where it is translated into polypeptides according to the triplet language just described (Figure 3.6).

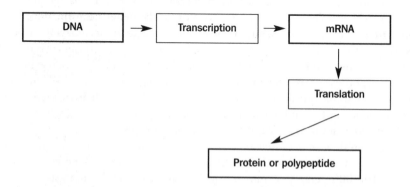

Figure 3.6   Transcription and translation of DNA and RNA

Deciphering the triplet code began in 1961 when the codon TTT was established as coding for the amino acid phenylalanine. The remarkable fact about the 64 codons that have now been deciphered is that they have the same meaning in virtually all organisms. It is as if all living things share one universal language. Hence in laboratories, bacterial cells can translate genetic messages from human cells and vice versa. This consistency of the genetic vocabulary is what makes genetic engineering possible, and it also implies that the code must have been established early in evolution. By examining the similarity of amino acid sequences in proteins in different species, we gain some insight into similarities between their DNA and hence, by making a few assumptions, into how closely related they are in evolutionary time. Haemoglobin, for example, is a molecule found in monkeys, chickens and frogs. The precise arrangement of haemoglobin amino acids in a typical monkey differs by only 5 per cent (8 being different in a chain of 125) from that in humans, whereas that of a chicken differs by about 35 per cent. Studies such as these fit almost perfectly with what one would expect concerning the ancestry of creatures from fossil evidence and morphological similarities. They have led to the oft-quoted remark that we differ from common chimps by only 1.6 per cent, which all sounds rather depressing. We can be assured, however, that differences in phenotype are not a simple linear function of DNA differences (see also Chapter 8).

In summary, the genotype of an organism is determined by the sequence of **codons** (base triplets) on its DNA. This sequence in turn commands the types of protein that will be synthesised. The assembly of proteins into cellular structures and their action in metabolic pathways, in combination with environmental influences during development, determines to a large degree the form, functions and behavioural patterns that together comprise the phenotype. Now, there is obviously a huge leap between this well-established mechanism of how genes make proteins and the final assembly of an organism in all its multicellular complexity. Embryonic development is a vast subject and one still not properly understood. It is almost certain, however, that its processes are under the control of that master blueprint, the DNA in our cells.

How the information is passed on in reproduction – the flow of information through the germ line

The idea that all the information we need to build an organism is found in the DNA has now entered popular culture. Novels and films have explored the idea that an individual can be cloned from a small tissue sample, an idea relying upon the fact that every diploid cell in the human body contains a full complement of DNA. These cells would be invisible to the naked eye, yet each one contains about 3 metres of DNA. If the cell were enlarged to the size of a full stop on this page, the DNA would stretch for 150 metres, although it would be too thin to be visible. On this length are to be found about 3 billion base pairs and somewhere between 80 000 and 100 000 genes. The capacity of cells to store information along stretches of DNA would be the envy of any computer engineer.

The DNA of each cell nucleus is usually bound to groups of proteins and exists as long thin diffuse fibres that are difficult to see. As a cell prepares to divide, however, the fibres coil into visible structures called chromosomes. It is convenient to consider the reproduction of cells in terms of the fate of chromosomes. Chromosomes consist of DNA coiled around protein bodies and then coiled on itself twice again.

We have already seen how the structure of DNA lends itself to replication: the DNA molecule can act as a template for the synthesis of more identical DNA either for cell division within an organism or to form the basis of a new organism. Cell division that contributes to growth and repair in an organism is called **mitosis,** and each new cell is simply a replicate of the original. By the time you have finished reading this sentence, several thousand of your cells will have divided by mitosis.

In order to transmit DNA in the process of sexual reproduction, the cells responsible divide in a different way, called **meiosis**. It is worth examining the process of meiosis in some detail since it helps us to understand why sexual reproduction should exist at all – a topic covered in Chapter 4 – as well as throwing some light on the essential differences between males and females.

To understand the process of meiosis, and hence the implications for sexual reproduction, we need to tackle the arrangement of chromosomes in cells. The position of a particular gene along a length of DNA, and hence on the chromosome, is called the **locus** of that gene. All human beings belong to the same species and thus have some obvious similarities that must have a genetic basis. In features such as hair type, eye colour and so on, there are also differences, and it follows that there must be different forms of the genes that determine these characters. These different forms are called allelomorphs, or more commonly **alleles**. Now, most cells in a typical animal contain chromosomes in matching pairs. In human cells, there are to be found 46 chromosomes made up of 23 pairs, each member of a pair being very similar except for the two that are called **sex chromosomes**. The sex chromosomes are called X or Y, which refers to their appearance under an optical microscope. It could be said then that humans have 22 matching pairs (of **autosomes**) and then either one X and one Y chromosome (XY) if they are males, or one X and another X chromosome (XX) if they are female.

Figure 3.7 shows a pair of chromosomes aligned side by side. The loci of the genes are given by the letters Aa, BB and Cc. In the case of the allele B, the two forms of the gene on each member of the chromosome pair are the same, so the **genotype** is said to be **homozygous**. If that allele coded for spots on the fur of an animal, the phenotype of the organism would be spotted. In the case of allele A, it is not as simple. Both genes refer to the same trait but exist in different forms; the genotype at this locus is thus **heterozygous**. The final outcome that is expressed depends a lot on the type of gene. If this were the locus for eye colour, a coding for blue eyes and A for brown, the phenotype would show brown eyes. We say that brown is the **dominant allele** and blue the recessive. In other cases, the outcome is intermediate between the homozygous condition for each allele.

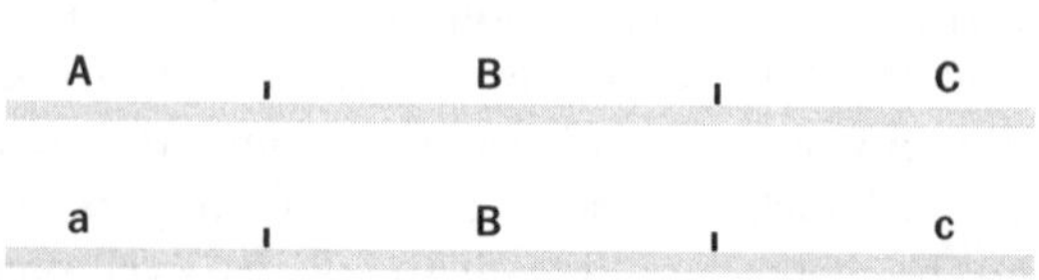

Figure 3.7   Complementary pair of chromosomes
showing homozygosity for allele B and heterozygosity at A and C

Why chromosomes exist in pairs when it seems that one set would do relates to the phenomenon of sexual reproduction. Of the 46 chromosomes in each of your cells, 23 are provided by your mother and 23 by your father. You share 50 per cent of your genome with your biological mother and 50 per cent with your biological father. When cell division occurs in your body to replace damaged cells, or simply as part of growth, each new cell has the same 46 chromosomes as the one from which it grew. When animals produce sex cells or **gametes** by the process of meiosis, however, the procedure is different. In each human sperm and egg, there are to be found only 23 chromosomes, that is, half the number in normal 'somatic' cells. Fertilisation brings about the **recombination** of these to 46 in 23 pairs, and so the life of a new organism begins.

Figure 3.8 shows a simplified account of meiosis for a simple organism with only one pair of chromosomes (noting that even fruit flies have four). The loci for eye colour and fur colour are shown. Figure 3.8 focuses on meiosis leading to the production of sperm (spermatogenesis), but the same steps occur in the formation of the eggs in the female (**oogenesis**). Figure 3.9 shows the fusion of two gametes (sperm and egg) to produce a fertile zygote.

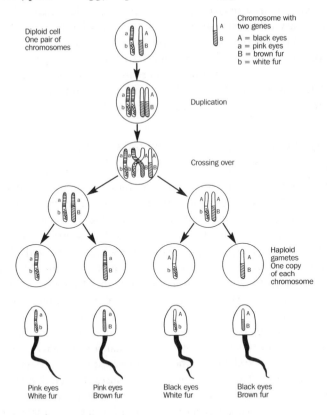

Figure 3.8   Simplified picture of meiosis and spermatogenesis

Notice that, in Figure 3.8, one pair of chromosomes with just two genes on each gives rise to four different gametes. The number of possible gametes in human spermatogenesis or oogenesis from 23 pairs of chromosomes with 100 000 genes is vast.

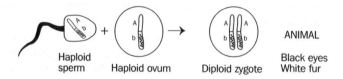

Figure 3.9   Fertilisation of an ovum

In Figure 3.8 above, we can also see the process of **crossing over**. During meiosis, chunks of DNA from each complementary chromosome are swapped. The effect of this process is extremely important and goes to the heart of why the gene must be considered to be the unit of natural selection (see below), as well as perhaps to why sex exists at all. The consequences of crossing over can be understood by comparing gametes with ordinary cells. Looking down an imaginary high-powered microscope at a pair of complementary chromosomes inside an ordinary cell, it would in principle be possible to see one as coming from your mother and one as originating from your father. Looking along the single chromosome inside one of your **haploid** gametes, we would see a patchwork of genes from your mother alternating with genes from your father. Meiosis has thus stirred up the genes from your father and mother, reconfigured them and presented a whole new array to the world.

How novelty enters the information – mutations and meiosis

Modern genetics provides an answer to the problem that so troubled Darwin of how spontaneous novelty arises and allows natural selection to take effect. Much of the apparent novelty that arises in any one generation comes from the shuffling of genes during meiosis and sexual reproduction, as discussed above. Fundamental changes must, however, be caused by changes in the base sequence of the DNA. We now know that chemical and physical agents such as high-energy radiation can have a mutagenic effect on DNA and cause alterations in its structure. Mutagenesis can also occur spontaneously by errors in replication. In Chapter 4, we will examine how sexual reproduction may have begun as a way of reducing the number of these spontaneous errors.

Even a change to one base pair can have a profound effect. A change in a single base pair is thought to bring about sickle-cell anaemia when altering the base pair from TA to AT causes one change to the amino acid sequence in haemoglobin, glutamine becoming valine. This simple substitution causes an alteration in the shape of the red blood cells (they appear sickle shaped) and a reduction in the oxygen-carrying capacity of the blood. Sickle-cell anaemia provides an example of **pleiotropy** (Greek *pleion*, more), a case in which one gene influences many traits. The sickle-shaped cells produced when the defective gene is present on both chromosomes (that is, when the chromosomes are homozygous) are quickly broken down by the body. Blood does not flow smoothly and parts of the body are deprived of oxygen.

The physical symptoms range from anaemia and physical weakness to damage to the brain and other major organs, and heart failure. There is no cure for the condition, which causes the death of about 100 000 people worldwide each year.

Sickle-cell anaemia is by far the most common inherited disorder among African-Americans, affecting 1 in 500 of all African-American children born in the United States. Its high frequency in the population, and the fact that natural selection has not eliminated it (many sufferers dying before they can reproduce), is probably caused by the fact that, in Africa, the possession of one copy of the sickle-cell gene confers some resistance against malaria. If Hb is taken to be the normal haemoglobin gene and Hbs the sickle-cell gene, people who inherit both sickle-cell genes (and who are therefore Hbs Hbs) suffer from sickle-cell anaemia. People who inherit only one copy of the sickle-cell gene and are Hb Hbs are said to have sickle-cell trait, only some of their red blood cells being oddly shaped. It is in this latter condition that the gene gives an advantage in protecting against malaria since the malarial parasite (*Plasmodium*) cannot complete its life cycle in the mutant cells.

Box 3.1 gives another example of the practical consequence of a genetic alteration.

BOX 3.1

Queen Victoria's gene: haemophilia and the breaking of nations

A particularly fascinating example of how a small change in one gene can have profound effects concerns the inheritance of **haemophilia** in Queen Victoria's family.

Of the 23 pairs of chromosomes in human cells, all are homologous except for the pair of sex chromosomes. In females, this pair is referred to as XX (from the shape they appear under an optical microscope), and that in males XY. When male cells divide to produce haploid gametes, it follows that half will carry an X chromosome and half a Y chromosome. Females produce eggs carrying only X chromosomes. When a Y male gamete fertilises a female egg, a zygote (XY) is produced that will grow into a male. When an X gamete from the male meets an X gamete from the female, an XX zygote is produced, which grows into a female. Hence the sex of a child is determined by sperm from the male, and on average an equal number of males and females is produced.

It seems that Queen Victoria must have been a carrier on her X chromosome of a defective allele leading to haemophilia. Haemophilia is caused by an allele producing a defective version of a protein involved in blood clotting. We can represent Queen Victoria as XHXh, meaning that she 'carried' the mutant allele (h) on one chromosome but had the normal allele on the other (H). Consequently, Victoria herself had no symptoms of the disease but passed the defective gene down the generations.

BOX 3.1 (cont'd)

One way of consolidating power and strengthening alliances is through marriage, and this tendency led to the defective gene being passed through the courts of Spain and Russia. The effect on the Russian monarchy is shown in Figure 3.10. Potts (1995) charted the fate of this gene. He argues that Tsar Nicholas' desire to keep secret the fact that his only son and male heir, Aleksei, suffered from haemophilia was a factor in the downfall of the Romanov regime. The Serbian peasant Rasputin seemed able to exert a hypnotic affect over the Tsarevitch Aleksei and enabled his internal bleeding to subside (restriction of movement enabling some healing to occur). When the February revolution began, the new government offered to make Aleksei a constitutional monarch. Instead, the Tsar, reluctant to allow his haemophilic son to succeed to the throne, abdicated. In 1918 the family was tracked down by revolutionaries and shot in Siberia. In July 1998 the bones of the Romanovs were finally given a state funeral in Moscow. They were identified as the authentic remains by DNA fingerprinting.

A complete account of the way in which Queen Victoria acquired the defective gene has not yet been written. There is no sign of haemophilia in her mother's family since it fails to appear in the numerous descendants of her first marriage. It is possible that a **mutation** occurred in Queen Victoria herself or in her father, the Duke of Kent. Strangely, the Duke's own genetic defect (porphyria) was not passed on to Victoria or her family. Potts raises the intriguing possibility that because of the pressure on Queen Victoria's mother (Mary Louise Victoria, Duchess of Kent) to produce an heir, she may have engaged in 'extrapair copulation'.

3.1.2 *From genes to behaviour: some warnings*

Our current knowledge of the molecular basis of inheritance, gene expression and the inheritance of genetic disorders is considerable. The conditions discussed above – haemophilia and sickle-cell anaemia – are, however, physiological rather than behavioural conditions, and it is important to point out that a simple one-to-one correspondence between genes and behaviour is hard to find. Part of the reason for this is that most behavioural characteristics are polygenic, that is, they are the result of the expression of many genes rather than just one (as in the case of sickle-cell anaemia). Research has, however, demonstrated that some behaviour is monogenic, and even for complex polygenic effects, the behaviour of some organisms has been clearly shown to be a product of genotype. A monogenic determination of behaviour was demonstrated by Walter Rothen-buhler (1964) in his studies of the behaviour of honeybee colonies. He was able to show that the behaviours 'uncap a comb cell' (uncap) and 'remove dead larvae' (remove) were both under the control of single separate genes.

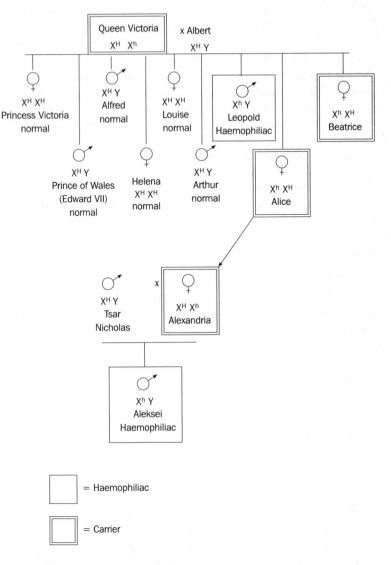

Figure 3.10   Haemophilia in the monarchies of Europe

Most behaviour patterns are not this simple. It is possible, for example, to breed rats that are good at travelling through mazes and remembering the path they took. Strains can be developed that are 'maze bright' and 'maze dull'. It can be conclusively shown that this ability reflects genotype in that the separate **populations** reared under identical conditions perform differently at navigating through mazes (Tryon, 1940). If however, both populations are reared in an 'enriched environment' (given interesting toys and play objects), the difference between the two phenotypes disappears: the effect of genotype can be masked by environmental influences (Cooper and Zubek, 1958).

In this whole area, there remains a great deal of heated discussion about the idea that behaviour is 'innate' or has a genetic basis. Marian Dawkins (1986) suggests that much of the dispute can be related to a lack of precision in terminology. It should in fact be pointed out that when we say 'gene for', we are not implying a simple one-to-one correspondence. The trait in question, be it morphological or behavioural, could be the result of many genes interacting with the environment. It would perhaps be better to say that the behaviour in question has some genetic basis and that genetic differences exist between individuals displaying this trait. A spectrum clearly exists in the genetic 'determination' of behaviour. We can identify at least four positions on this spectrum:

1. *Unlearnt patterns of behaviour invariant with respect to environment*
   Clearly, all genes need an environment in which to thrive and replicate. If *Teleogryllus* crickets are deprived of oxygen (an environmental commodity), they will die and not sing. In a wide band of normal environmental conditions, however, they sing a song that is so characteristic of the species that even if individuals are reared in isolation, have never heard the song or indeed are subjected to a cacophony of non-song-like sounds, they still grow up to sing the species-typical song of the cricket. Male sticklebacks act aggressively towards the red bellies of other males when confronted with them for the first time, even if they have never seen another fish or even their own reflection. Both these behaviour patterns are clearly something with which the creatures are born; they must have some genetic foundation. An environment is needed only in the trivial sense of providing a platform for the actions.

2. *Unlearnt but modified by learning*
   Laughing gull (*Larus atricilla*) chicks have an innate tendency to peck at their parents' bills to persuade them to regurgitate food. Very young chicks will even peck at long, thin red knitting needles. As they mature, however, they become more discriminating and gradually learn to recognise the finer details of a gull's bill.

3. *Selectively learnt or selectively expressed*
   To sing properly in a chaffinch-like way, young chaffinches must learn the song from other birds, but they are not simple mimics. If exposed to a wide variety of song sounds, they will tend to pick out those sounds characteristic of their species. One cannot make a chaffinch sing or call like a duck. It is clear that even simple creatures such as scorpion flies or bush crickets have a range of behavioural strategies that are contingently expressed in relation to environmental conditions (see Chapter 4). Different environments trigger different responses. In these cases, genes have coded for the mental machinery that solves a problem in a particular environment. Language acquisition is probably another example of learning that has a genetic basis. We have an innate disposition to learn a language, but we need cues from the environment to trigger the process and shape the outcome.

4. *Learnt behaviour*

Perhaps the classic case of this is the ability of tits to open the foil tops of milk bottles to extract the cream. It is not something with which they are born, and milk bottles have not formed part of their 'environment of evolutionary adaptiveness', yet the behaviour has spread, each tit learning it from another. The disposition to peck and explore and the shape of beak are all obviously laid down by genes, but this particular behavioural phenotype has to be learnt from scratch.

Despite these qualifications and notes of caution, we may begin tentatively to approach human behaviour. A great deal of human behaviour must necessarily fall into the learnt category: there are no genes for playing tennis, watching television or playing the violin. But given that behaviour is under the control of a nervous system, and that every nerve cell has a full set of genes that regulate the chemistry of the cell, it would be surprising if there were to be found no genetic influences on even complex aspects of human behaviour. Studies on identical twins are increasingly confirming this.

Dizygotic or fraternal (non-identical) twins result from the separate fertilisation of two eggs by two sperm, but identical twins are monozygotic; that is, a single female cell is fertilised by a single sperm, but as the zygote grows, for reasons that are unclear, it separates into two. An inspection of the process of meiosis reveals that, on average, fraternal twins share half their genome, whereas truly zygotic twins have their entire genome in common. It is now possible to use DNA fingerprinting to establish zygosity with some confidence. Research has shown that even though environmental influences have an effect on the behaviour of these people, zygotic twins are more alike both physically and behaviourally than fraternal twins. Moreover, zygotic twins, even when separated and raised in different environments, reveal bizarre similarities (Plomin, 1990).

It is worth repeating here, however, the point made in Chapter 1 concerning **heritability**. Heritability estimates refer to the variation between individuals that can be accounted for by differences in the genome. For the evolutionary psychologist, it is features that have low heritability that are interesting since these point to similarities in anatomy and behaviour that are common to humans and that may be explained in adaptive terms. This is because, through time, natural selection will tend to favour advantageous genes, and a population will become increasingly homogeneous. In other words, the genetic variation will be used up. The crucial point is that heritability is not a property of a gene or a trait but instead reflects the distribution of alleles in a population and the state of the environment at any time. In highly genetically homogeneous populations, such as may be the case for many psychological adaptations common to humans, phenotypic variation will largely be caused by environmental influences, so heritability will be low. Suppose, however, that the environmental influences to which people are exposed become more similar. Differences will then largely be due to genetic differences, so heritability will increase. The interpretation of heritability estimates must be carried out with great caution (see Bailey, 1998).

3.2 The unit of natural selection

3.2.1 *The lure and failure of group selection*

In Chapter 1, we saw how genes serve as the units of inheritance. It remains now to establish what the unit of selection is. Does natural selection operate on genes, on individuals or on populations? To address this question, it is instructive to revisit one of Darwin's difficulties touched on in Chapter 2. In a world where the 'fittest survive', why do some individuals forego their own reproductive interests in favour of others? How could neuter wasps and bees have evolved that leave no offspring but instead slave devotedly to raise the offspring of their queens? Why does the honeybee die when it stings? What advantage is there to an insect that abandons camouflage and garishly advertises its unpleasant taste when the first to do so died in the process. Our grasp of evolutionary processes suddenly seems threatened by swarms of difficult questions.

It is tempting to respond to this difficulty by reference to what Cronin has called 'greater goodism' – the idea that individuals will serve the greater good at a cost to themselves. The **altruism** of insects described above could thereby be explained by suggesting that such behaviour serves the interests of the hive, the group or even the species. This line of argument is often associated with V. C. Wynne-Edwards (1962), who argued that the dispersal of animals in relation to food supply is such that the final population density reached is optimal for the group. In this view, groups possess some mechanism whereby the selfish inclinations of individuals to overgraze are restrained in favour of the longer-term interests of the group. Wynne-Edwards' book, *Animal Dispersion in Relation to Social Behaviour*, had the major effect not of converting biologists but of rousing several prominent Darwinians, chief among whom were George Walden and John Maynard Smith, to the attack. The net result of the debates that followed is that **group selection** is now regarded by many as an untenable heresy or at best an unlikely scenario. Others suggest that sociobiologists may have been overhasty in rejecting group selection, particularly in relation to human groups (Wilson and Sober, 1994).

The problem with Wynne-Edward's theory is that groups ('demes') are also composed of individuals that evolve. Selection can, in principle, take place between individuals in a group and between groups, and predicting the effect of selection at two levels is not intuitively obvious. Using mathematical models of population genetics, Maynard Smith concluded that the restrictive conditions needed for group selection to take place are hardly ever met. In effect, selection on individuals in a group will virtually always swamp any group selection effect; individuals will not restrain their selfish interests in favour of the greater good of the group (Maynard Smith, 1989). It turns out that, in order to explain the apparently sacrificial behaviour of bees wasps and other insects, we need to delve even more deeply to find the unit of natural selection.

Group selection ideas were largely dismissed in the 1970s and are still an anathema in many quarters. There are a sizeable number of evolutionists, however, who consider that natural selection could still operate at the level of the group. The debate is complex, and opinion is likely to remain divided for many years (see Wilson, 1992a).

3.2.2 *The unit of selection: replicators and vehicles*

It comes as a surprise to find that, despite a considerable knowledge of the structure of DNA and the processes of transcription and translation, there still remain fundamental questions about what exactly is evolving and being selected. The issue is often called the 'unit of selection' question. Dawkins takes the view that if, for the unit of selection, we are looking for an irreducible entity that persists through time, makes copies of itself and on whose slight changes natural selection can act, the individual organism will not do. The reason is that organisms do not make facsimiles of themselves: in sexual reproduction, offspring often differ markedly from their parents. The process of meiosis and recombination found in sexual reproduction serves as a method of shuffling genes every generation. It is as if a Premier League football team is forced to split and swap players with another team before every new match. (In this sense, natural selection favours 'teams of champions' rather than 'champion teams'.) Moreover, accidental changes to an organism (acquired characteristics) are not inherited by the offspring. We are looking for a unit in which changes that have an effect on reproduction are inherited. Even organisms that reproduce asexually are not the replicators. An aphid that has lost a bit of leg does not reproduce by **partheno-genesis** to make copies of 'itself' complete with shortened leg. It 'strives' (apparently) to make copies of its genome. Consequently, the unit of selection is neither the group nor the individual organism but the gene itself (Dawkins, 1976; Hull, 1981).

The question is somewhat clarified by making a distinction between units that reproduce and entities that expose themselves to natural selection. This distinction is important because genes are not exposed directly to selective forces: the environment 'sees' not the genes but only the phenotypic expression of many genes working together. Dawkins introduced the terms 'replicators' and 'vehicles' for this purpose. In this view, organisms become vehicles for the replicators. The properties of these organisms are conferred on them by genes (in concert with environmental influences), and these properties influence the survival of the organism and ultimately the replicators.

In a sense, it does not matter in principle where the properties that are adaptive are manifested: they could be in the individual organism, the group or the population. When bees cluster together in winter to form a tight ball, this has the effect of reducing the surface area to volume ratio of the colony and thus reducing the heat loss, which has obvious survival value. This geometric property is a product of a colony of animals even though the gene(s) responsible are carried by individuals and probably amount to some simple instruction about moving close to their neighbours. In most cases, it is far more probable that these properties are expressed at the level of individual organisms. This is largely because the different genes in any body have the same 'hoped-for route' into future generations, and same-body genes must collaborate to pursue a common purpose. This of course has the effect of keeping at bay any conflict of interests between genes found in one body.

Interestingly, as genes pursue their reproductive 'goals', there does arise the possibility that they will care more for themselves than for the body that harbours them. If DNA can survive and be replicated without phenotypic expression, then so be it. This could explain why many organisms have large amounts of repetitive or 'junk' DNA whose function, if any, is not yet known. This may look non-adaptive and puzzling until we take a gene-centred view. This 'parasitic' DNA (assuming that it carries no benefit for its vehicle and that its replication is a net drain on resources) is highly adaptive for itself (Doolittle and Sapienza, 1980).

The whole issue is, however, still not resolved. If we accept the gene as the unit of selection (which is compelling for the reasons outlined above), we still face the problem of **linkage disequilibrium**. Some genes do not behave as separate units during reproduction but remain linked with other genes. So, how large or small a fragment of genome should count at the fundamental unit? This debate is also likely to continue.

A great deal of science relies upon metaphors, and evolutionary theory is no exception. If we speak of DNA as the 'blueprint for life', this conveys a misleading impression of how individual genes affect the phenotype. On an architect's blueprint, it is possible to estimate how the lines on the paper will translate into the finished building. If we add or remove a few features, we can make a good guess on the effect on the final outcome. But genes are not like this. A better analogy might be a recipe for a cake. From the list of ingredients and instructions for assembly, it is not easy to predict the appearance of the product and even more difficult to forecast the taste: adding and removing ingredients could have profound effects on both. We could say that the level of selection for cakes is the cake itself – its appearance and taste. Cakes are exposed to humans and are judged accordingly. Individual cakes do not survive, but recipes may persist if they are successful. By analogy, genes are selected according to the success of the vehicles that they influence. This gene-centred view of selection helps to explain some aspects of altruism, as we shall see in the next section.

3.3 Kin selection and altruism

To ensure its survival, a gene normally impresses itself on its vehicle in ways that enhance the chance of the vehicle, and ultimately itself, reproducing. It is because of this strong association between genotype and phenotype that the behaviour of the Hymenoptera (the group that includes ants, wasps and bees) appears extremely puzzling. In some of these species, individuals care for the offspring of the queen, defend and clean the colony and, in short, devote their lives to the survival and reproduction of other individuals in the colony while they themselves remain sterile. For this and other reasons, Darwin declared that the insects posed a 'special difficulty, which at first appeared to me insuperable, and actually fatal to my whole theory' (Darwin, 1859b, p. 236) (see Chapter 2). In terms of the language of vehicles and replicators, the problem becomes how to explain the existence of a replicator that instructs an individual to behave altruistically if the individual that carries it does not reproduce to pass it on. The answer was provided by Hamilton's (1964) theory of **kin selection**.

3.3.1 *Hamilton's rule*

In 1963, Hamilton laid down the conditions required for gene coding for social or altruistic actions to spread. This theory is also known as the **inclusive fitness** theory. The mathematics of Hamilton's original papers is extremely complex. It was West-Eberhard (1975) who showed how Hamilton's rule could be simplified, and it is the simplified form that is now commonly encountered.

Consider two individuals X and Y, which are related in some way, and that X helps Y. An altruistic act can be defined as one that increases the reproductive success of the beneficiary (Y) at the expense of the donor (X).

Let b = benefit to recipient
 c = cost to donor
 r = the **coefficient of relatedness** of the recipient to the donor. This is the same as the probability that the gene for helpful behaviour is found in both the recipient *and* the donor.

The condition for assistance to be given and thus for the gene to spread is: 'help if $rb - c > 0$', which is the same as $rb > c$. Figure 3.11 shows an example of this. It should be clear that although the reproductive success of the helping gene in X is reduced, this is more than compensated for by the potential increased success of the gene appearing in the offspring of Y.

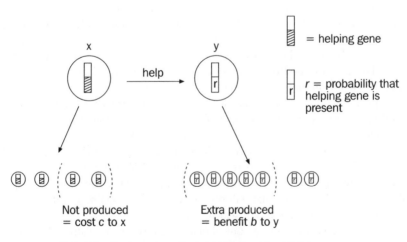

Notice that the focus is on the helping gene and not just any gene shared in common since it is precisely the spread of this helping gene that has to be explained

Hamilton's rule is: 'help if $rb - c > 0$', or $rb > c$. Let $b = 5$, $r = 0.5$ and $c = 2$. If X loses two offspring (produced in this case by asexual reproduction), then since $5 \times 0.2 > 2$, it 'pays' the gene in X to give assistance to Y

Figure 3.11   Conditions for the spread of a helping gene

3.3.2 *Coefficient of relatedness*

The situation presented in Figure 3.11 is highly simplified. Figure 3.12 shows a more detailed example of two **diploid** individuals I and J, with alleles aA and mM respectively, that mate sexually to produce four types of offspring: am, aM, Am and AM.

The coefficient of relatedness can be understood in a number of ways. The important thing to remember is that we are concerned with how closely individuals are related to each other rather than how similar they are. There are three ways of conceptualising this:

1. *r* is a measure of the probability over and above the average probability (which is determined by the gene's average population frequency) that a gene in one individual is shared by another
2. *r* is the probability that a random gene selected from I is identical by descent to one present in J
3. *r* is the proportion of genes present in one individual that are identical by descent with those present in another.

The terms 'identical by descent' and 'probability above average' are there to take into account the fact that a gene may be present in two individuals because it is common in the population (that is, it has a high average probability) or because they are siblings and the gene came from the father *or* mother.

Figure 3.12 shows how two unrelated parents produce siblings with an average value of *r* = 0.5. It follows that it may pay siblings to help each other if the gene for helping thereby increases in frequency. This is, in effect, what the distinguished biologist Haldane was suggesting when he said (reputedly in the Orange Tree Pub on the Euston Road in London) that he would lay down his life for at least two brothers or eight cousins.

It is probably kin selection that led to the bright warning colours found in caterpillars. A bright colour must have first appeared as a suicidal advertisement. The problem is to explain how could it have started and spread if the caterpillar were instantly eaten and destroyed when the gene appeared. If the gene for bright colour, however, ensured that it was not destroyed again because it was found in related individuals, the bargain would have been a good one. A bird that swallows a distasteful caterpillar will probably not do so again, and if the gene is found in siblings (for example), it will survive.

Violent acts would seem to be out of place in a discussion of altruism, but if we take a gene-centred view, the killing of relatives and violent acts such as cannibalism also become understandable. Hamilton's equations suggest that a gene for a particular activity will thrive if $rb > c$. If the action 'eat your brother' causes more copies of this gene to prosper, as it may do if the environment is such that many siblings would not survive, this action will be favoured. This seems to happen with herons and egrets, in which siblicide can be common; the gain to the siblicide gene apparently outweighing the cost (Mock, 1984). Similarly, some rodent species will ingest their offspring if there is little hope of survival since the 'ingest offspring' gene may survive in future reproductive acts by the parents.

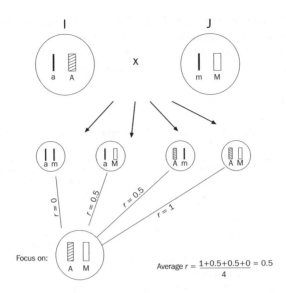

r = coefficient of relatedness, which can be thought of as the probability that a gene sampled at random in one individual is also present in another. In this case, r between siblings is 0.5

Figure 3.12   Coefficient of relatedness between siblings

3.3.3 *Application of Hamilton's rule and kin selection*

The social insects provide plenty of examples of altruistic behaviour to test Hamilton's insights. The fact that worker bees have barbed stings that, when used to attack predators, remain embedded in the victim, causing the death of the worker, is understandable when we realise that other members of the nest are close relatives. This extreme form of self-sacrifice is maintained as a behavioural and morphological trait in the face of natural selection because the survival of the genes responsible is ensured by their presence in the rest of the colony.

The Hymenoptera are said to be **eusocial**: female workers forego reproduction to help the queen raise more sisters rather than producing offspring of their own. The genetics of the Hymenoptera is unusual and is known as haplodiploidy. The males are haploid (that is, possess one set of chromosomes) and develop from unfertilised eggs, whereas the females are diploid and develop from haploid eggs laid by the queen and fertilised by a haploid male.

Table 3.1 shows the r values associated with diploidy and haplodiploidy. Hamilton originally suggested that the fact that sisters are related to each other by an r value of 0.75, compared with one of 0.5 for a potential daughter, predisposes sisters to help each other (or, more strictly, to help the queens to produce more sisters) rather than flying off and producing daughters of their own. Note also in Table 3.1 that male drones are only related to the rest of their sisters in the hive by 0.25, and this could explain their notorious laziness. It pays the sisters to invest in each other, rather than allow the queen to produce more sons, since sisters are related to each other by *r* values of 0.75 while they are related to their brothers by only 0.25. Hence female workers, who are far more numerous than the drones, dominate the hives. When winter arrives and food resources must be used wisely, the drones pay a terrible price for their 'laziness' and the low value of *r* they bear to their sisters; they are mercilessly driven from the hive to die. These are presumably highly moral acts in the world of bees.

Table 3.1   Coefficients of relatedness *r* between
kin pairs in humans and Hymenoptera

Actor	Recipient	*r* value (probability of recipient holding any one gene identified in the actor)
Diploid (for example humans)		
Parent	Offspring	0.5
Individual	Full sibling	0.5
Individual	Half-sibling	0.25
Individual	Identical twins (monozygotic)	1.0
Grandparent	Grandchild	0.25
Haplodiploid (for example Hymenoptera, assuming queen mates with only one male)		
Queen	Daughter	0.5
Queen	Son	0.5
Son	Queen	1
Sister	Sister	0.75
Brother	Sister	0.5
Sister	Brother	0.25

The haplodiploidy hypothesis generated an intense amount of interest in that it seemed to provide an elegant confirmation of kin selection theory. In reality, the evolution of eusociality is probably not a simple function of haplodiploidy. Many Hymenoptera taxa that are haplodiploid are not eusocial, and some eusocial insects such as termites are not haplodiploid but diploid. It follows that haplodiploidy is neither a sufficient nor a necessary condition for the evolution of eusociality. The balance of evidence does, however, suggest that along with ecological influences, it has probably been a strong contributing factor. Bourke and Franks (1995) and Bourke (1997) provide excellent reviews of recent thinking.

Other acts of altruism

Co-operative breeding, as illustrated by the Hymenoptera, is just one example of altruistic behaviour. More generally, numerous studies on co-operative breeding illustrate the application of kin selection theory (see Metcalf and Whitt, 1977; Emlen, 1995), but altruism can take many other forms: sharing food, grooming, raising alarm calls and so on. For animals that live in social groups, the raising of an alarm call by one individual at the approach of a predator is often considered to be altruistic behaviour since the call itself could enable the predator to locate the caller. It would presumably pay the first observer of a predator to skulk silently away, leaving others to their fate. Numerous studies have been carried out on such calls, most reporting that animals are much more likely to give alarm calls when the beneficiaries are close relatives. In experiments reported by Hoogland (1983), it was found that black-tailed prairie dogs (*Cynomys ludovicianus*) were much more likely to give an alarm call at the sight of a predator if close relatives were present in the social group.

Although human altruism is examined more thoroughly in Chapter 11, it is worth noting here that humans are more inclined to behave altruistically towards kin than unrelated individuals. Most people choose to live near their relatives, sizeable gifts are exchanged between relatives, wills are nearly always drawn up to favour relatives in proportion to their genetic relatedness and so on. In an intriguing direct experimental investigation of this effect, Dunbar and others asked volunteers to sit with their back against the wall in a skiing posture but with no other means of support. After a while, it becomes painful to maintain this position. The subjects were given 75 pence for every 20 seconds they could maintain this posture, with the proviso that the money must go to one of three categories of people: themselves, relatives of varying degrees of genetic related-ness or a major children's charity. The results were unambiguous – subjects tended to endure more discomfort and thus earn more money for themselves or close relatives than distant relatives or a charity (Dunbar, 1996a).

Human kin make powerful claims on our emotional life. It is significant that movements that extol the 'brotherhood of man' or group solidarity, such as the world's great religions or organised labour, often resort to kin-laden language. A trade union leader (of the Old Left) may typically use the term 'brothers and sisters' when addressing fellow members. The New Testament is full of terms such as 'father' and 'son' applied to individuals who are not related. The cohesiveness of human groups is assisted by the fact that any group in the early evolutionary history of hominids probably contained a large number or relatives.

The problem with groups, however, especially modern ones, is that they are vulnerable to free riders – individuals who claim membership but who are not genetically related to anyone in the group or who will extract favours from the group without any repayment in kind. What would be desirable in these situations is some sort of cultural badge, a mark of belonging to the group that indicates commitment. It could be that language dialect serves as some sort of badge of membership; as George Bernard Shaw suggested, no sooner does one Englishman open his mouth to speak than another despises him. Groups break down into smaller groups, and there begins an 'in group' and 'out group' morality. Dunbar has explored the suggestion that language dialects may serve a marker function. He found, using computer simulations, that as long as dialects change slightly over a few generations, cheaters (that is, outsiders who drift into the group to extract benefits and then drift out) find it difficult to gain a foothold. Dunbar thinks it likely that 'dialects arose as an attempt to control the depredations of those who would exploit people's natural co-operativeness' (Dunbar, 1996a, p. 169). This raises the further interesting possibility that the diversification of dialects into different languages is related to the need for group cohesiveness.

3.4 **Kin recognition**

3.4.1 *Kin recognition and discrimination*

For kin selection to operate, it requires animals to be able to recognise or at least discriminate between kin and non-kin. The evidence for kin recognition or 'reading *r* values' is necessarily indirect since it is an internal process, but if we observe animals treating kin differently from non-kin, this **kin discrimination** could be used as evidence for kin recognition.

There are probably two basic reasons why it is in the interests of an animal to recognise kin. First, kin selection requires that acts of altruism be directed according to Hamilton's formula $rb > c$. This requires some assessment of costs, benefits and *r* values. Second, it is important for sexually reproducing organisms not to mate with close relatives, otherwise deleterious gene combinations may result. It follows that evidence for kin discrimination does not automatically imply the existence of an altruistic gene. We will outline some kin recognition mechanisms and evidence in their favour and then return to this second and important point.

There are probably at least four mechanisms that are available to animal species: location, familiarity, phenotype matching and recognition alleles ('green beards'). We will examine each in turn.

Location

If animals live in family-based groups, in for example burrows or groups of nests, there is a good chance that neighbours will be kin. In these circumstances, a simple mechanism such as 'treat anyone at home as kin' may suffice. Cuckoos of course exploit this. It is a remarkable sight to observe a reed warbler (*Acrocephalus scirpaceus*) feeding a young cuckoo many times its own size under the delusion that it is a reed warbler chick, whereas adult cuckoos (*Cuculus canorus*) outside the nest will be aggressively attacked.

Familiarity

Work by Holmes and Sherman (1982) showed that non-sibling ground squirrels raised together were no more aggressive to each other than were siblings reared together, both these groups being far less aggressive than non-sibling young squirrels reared apart. The mechanisms here may be one of association rather than location. If you have spent some of your early life in close proximity with another individual, the chances are that you are related. Similar evidence comes from work on mice. It has been shown (Porter *et al.*, 1981) that, in the case of the spiny mouse (*Acomys cahirinus*), siblings prefer to huddle together if reared together but not if reared apart. Non-siblings reared together will display the same preference for each other's company as siblings reared together and will shun siblings reared apart. The proximate mechanism appears to be olfactory.

Phenotype matching

Family members often resemble each other because of either similarities in genotype or features such as nest odour derived from the environment. Some animals seem capable of assessing the similarity of genotype using some characteristic of the phenotype, possibly odour. Many creatures probably carry such labels that allow kin to establish their closeness. The sniffing behaviour on greeting, exhibited by many rodents, seems to be designed as a kin recognition system. One compelling example of this general effect concerns the behaviour of sweat bees, some of which act as sentries or 'bouncers' to the hive to monitor the suitability of incomers. Greenberg found a linear relationship between the likelihood that a sentry would admit a bee to the hive and its coefficient of genetic relatedness to the rest of the hive (Greenberg, 1979).

Recognition alleles and 'green beards'

Two factors ensure a high value for r, which is needed if altruism is to spread when costs are significant: kinship and recognition. There is nothing intrinsically special about the location of the altruistic gene in kin; it is merely that kin have a reliable probability that the gene for helping will be present. But it is just a probability. In Figure 3.12 above, if the individual AM were to help am, which it would if the gene A simply instructed it to help siblings, its actions would be wasted.

The importance of recognition is illustrated by the 'green beard' effect, named after a thought experiment of Dawkins. The idea was in fact first proposed by Hamilton in 1964 and given its memorable title by Dawkins (1976). Dawkins considered a gene for helping that would also cause its vehicles to sprout a green beard. This would be an ideal way to focus altruistic efforts. Kin would be ignored if they did not possess the green beard, but anyone with a green beard would be helped without regard to how closely related they were. It is often pointed out that a gene for helping is unlikely to be able also to command the ability to recognise a label and the ability to produce a label, but if two or more genes are closely linked such that they tend to occur together (linkage disequilibrium), this does become a theoretical possibility. The reason why 'green beard' effects are not common probably results from meiotic crossover and recombination (Haig, 1997).

Strictly speaking, a 'green beard' effect is not necessarily kin recognition since the altruistic gene would behave favourably towards any other recognisable 'green beard' irrespective of its kinship. Some evidence for this effect is observed in the congregating behaviour of the larvae of the sea squirt *Botryllus schlosseri*. If individuals share a particular allele found on the histocompatibility region of the chromosome, they will associate whether kin or not (Grosberg and Quinn, 1986).

Another more recent example, and one of the best documented so far, is reported by Keller and Ross in their work on red fire ants (*Solenopsis invicta*). This species exists in two subpopulations. In one, a single queen presides over the colony (monogyne), while the other is polygyne and has several queens. At the locus of the genome known as Gp-9, there are two alleles, B and b. In the polygyne groups, all ants, queens and workers, are Bb. Following sexual reproduction, individuals will be produced that are BB and bb. The bb ants (queens or workers) die young from physiological causes. The interesting fact, however, is that BB queens are killed by the rest of the colony, especially by Bb workers. The BB queens are identified by a chemical coating on the cuticle, but queens that are Bb are spared. So here we have a near-perfect illustration of a 'green beard' effect. The allele b induces the ants to bear a chemical signal advertising this fact as well as acting favourably towards its holders. The b allele, by targeting BB individuals for extermination, ensures that it is itself reproduced in preference to the B allele (Keller and Ross, 1998).

The real importance of 'green beards' is, however, a thought experiment. If you understand why altruism towards 'green beards' should spread faster than that simply towards relatives, you have understood the force of Hamilton's equations and are a long way down the road to a gene-centred view of natural selection.

3.4.2 *Outbreeding: incest taboos and the Westermarck effect*

It is important for many animals to avoid inbreeding since close relatives may be homozygous for deleterious **recessive alleles**. Numerous experiments show the adverse effects on future reproductive success of excessive inbreeding (Slater, 1994). It is estimated that each human probably carries between three and five lethal recessive alleles. Mating with close relatives increases the likelihood that the chromosome will be homozygous at the loci for these recessive and defective alleles. This could be the genetic basis for the strong taboos in incest found in numerous human societies. Animals can avoid inbreeding by a simple dispersal mechanism that forces individuals to leave the group to find a mate. In mammals it is usually males who disperse from the birth group, whereas in birds it is females. (Greenwood, 1980). Such a mechanism does not require kin recognition.

Interestingly, excessive outbreeding also brings danger. Animals could mate with individuals of similar but not identical species, leading to hybrid infertility. In addition, 'winning genes' suited to a local environment could be broken up and dispersed by outbreeding. There is some evidence for 'optimal outbreeding', whereby animals choose mates that are neither too closely nor too distantly related (Bateson, 1982).

In respect of human mating, recent work by Claus Wedekind and his colleagues in Switzerland has shown that human females actually prefer the smell of males who are different from them in terms of the major histocompatibility complex (MHC) region of their genome. This region is deeply involved in self-recognition and the immune response. Differences in this region between individuals can be tested by measuring the antigens produced in their body fluids. It is important for a female to choose a mate who will differ in the MHC region since this provides a cue for genetic relatedness: close relatives will be similar in this region. Moreover, differences in the MHC between a woman and her partner may allow females to produce offspring with a more flexible response to parasites. Evidence has been provided that male odours are related to an individual's MHC. In controlled conditions, women were more likely to find the odours produced by males pleasant if they differed in their MHC. Significantly, the effect was reversed if the women subjects were taking oral contraceptives (Wedekind *et al.*, 1995).

Although human mating will be examined in Chapter 8, it is worth considering here the near-universality of incest taboos in human culture. Very few men have sex with their sisters or mothers. The sexual abuse of daughters by their fathers is more common but still relatively rare compared with heterosexual sex between unrelated individuals. We can consider two explanations for these facts. One is that related individuals secretly desire incest but that culture imposes strict taboos to prevent its occurrence. The other is that humans possess some inherited mechanism that causes them not to find close relatives sexually attractive.

The first of these explanations came from Sigmund Freud. Freud's theory suggested that people have inherent incestuous desires; they are not observed in action very often, partly because they are 'repressed' and partly because society has (presumably for the benefit of the health of its members) imposed strict taboos. Suggesting that incestuous urges are repressed, and thus difficult to observe, makes it difficult of course to refute the idea that we have them in the first place. A further difficulty is that Freud is essentially suggesting that evolution has not only failed to generate a mechanism to suppress incest, but also somehow led to a positive preference for it.

A rival theory was proposed by the Finnish anthropologist Edward Westermarck in 1891. He suggested that men do not mate with their mothers and sisters because they are disposed not to find them sexually attractive. Westermarck suggested that humans use a simple rule for deciding whether or not another individual is related. If humans avoid mating with individuals they have been reared with during childhood, there is a good chance that they will also avoid mating with close relatives. Freud was naturally disposed to reject Westermarck's hypothesis, which he regarded as preposterous, since it flew in the face of his Oedipus complex, something that Freud saw as core to his whole psycho-analytical framework. Strangely enough, Freud could still regard the incest taboo as hereditary. Freud remained a convinced Lamarckian until his death and thought it plausible that the taboo could become fixed in the psyche by what he called 'organic heredity'. It is a pity that when Freud read Darwin, he failed to appreciate fully the power of natural selection.

Evidence in favour of Westermarck's hypothesis comes from Israeli kibbutzim where unrelated children are reared together in crèches. This often results in close friendships, but marriages between kibbutz children are rare (Parker, 1976). Further support for the effect comes from a study on 'minor marriages', or *simpua,* in Taiwan by Arthur Wolf of Stanford University. Minor marriages occur when a genetically unrelated infant girl is adopted by family and raised with the biological sons in the family. The motive seems to be to ensure that a son finds a partner, since the girl is eventually married to a son. Wolf studied the histories of thousands of Taiwanese women and found their experiences to favour the Westermarck hypothesis. Compared with other arranged marriages, minor marriages were much more likely to fail: the women usually resisted the marriage, the divorce rate was three times that of other marriages, couples produced 40 per cent fewer children and extramarital affairs were more common (Wolf, 1970).

The Westermarck effect is an instructive illustration of the relationship between genes and environment. The instruction to avoid sex with others who shared your childhood (a Darwinian algorithm) is a genetic disposition. The target group of individuals to whom this applied is socially determined. The effect is also informative at a deeper level, pointing to a possible model for the relationship between biology and ethics. The standard view from the social sciences has been that morality is something to do with custom, tradition, obligations and contracts, in other words, that it is a cultural phenomenon. If Westermarck is right, the incest taboo is an ethical code of conduct built upon and derived from primal and functional instincts.

There are many unanswered questions concerning kin recognition systems. In addition, we must be wary of expecting to find simple systems in place. From Hamilton's equation $rb > c$, r is only one factor; an animal must also be capable of weighing up b and c if it is to direct its efforts successfully. Parental care or altruism towards offspring is more often observed than sibling care, even though the r values are the same (0.5), because parental care will bring more benefit than sibling care. In other words, the reproductive value of donor and recipient must be considered.

3.5 Reciprocal altruism

3.5.1 *Altruism and selfishness*

So many words in science have a precise meaning different from the looser meaning of their everyday usage. In the main, this does not matter too much, but in the case of the concept of altruism, it has led to many unfortunate misunderstandings. The problem arises because any act that seems to imply that animals are not behaving selfishly has been called altruistic. In most cases, however, the acts are not altruistic at all because the pay-offs are either not so obvious or are delayed. Similarly, the phrase 'selfish gene' coined by Dawkins has led to the criticism that biologists are simply being anthropomorphic. To rescue these useful terms, some clarification is sorely needed.

So far, we have adopted a working definition of altruism as 'an act which enhances the reproductive fitness of the recipient at some expense to that of the donor'. We will now attempt some clarification of situations in which altruism appears to operate and then return to a deeper consideration of this definition. Altruism and co-operation are often observed among symbionts. Symbionts are species that have close ecological relationships with other species. They may be classed as parasites, commensalists or mutualists.

Parasites are organisms that benefit from a relationship with their host at the expense of the host organism. We often find the genes of one organism manipulating the behaviour of the other. The cold virus, for example, not only invades your system and subverts its functions to produce copies of itself, but also manipulates you into helping it spread further by persuading your lungs to expel at a great velocity an aerosol of droplets containing the virus. Dawkins referred to such effects as 'the extended phenotype'. In nest parasitism, the cuckoo manipulates the builder of the nest into giving aid to a completely different species. In these cases, the 'altruism' of the donor is extracted by manipulation, and the donor gains nothing. We could view this as cuckoo genes reaching out beyond the cuckoo vehicle into the behaviour of the nest owner. Dawkins summed this up in his 'central theorem of the extended phenotype':

> An animal's behaviour tends to maximise the survival of the genes 'for' that behaviour whether or not those genes happen to be in the body of the particular animal performing it. (Dawkins, 1982, p. 233)

Another group of symbiotic organisms are those that are said to be mutualistic. **Mutualism** is in fact becoming the preferred term to **symbiosis**. We can distinguish two types of mutualistic behaviour: interspecific (between two or more species) and intraspecific (within a single species). Some species form interspecific mutualistic partnerships because the individuals of each have specialised skills that can be used by the other. Aphids have highly specialised mouths for sucking sap from plants. In some species, this is so effective that droplets of nutrient-rich liquid pass out of the rear end of the aphids undigested. Some species of ant take advantage of this by 'milking' the aphids in the same way as a farmer keeps a herd of cattle. The aphids are protected from their natural enemies by the ants, which look after their eggs, feed the young aphids and then carry them to the grazing area. The ants 'milk' the aphids by stroking their rears to stimulate the flow of sugar-rich fluid. Both sides gain: the ants could not extract sap as quickly without the aphids; the aphids are cosseted and protected from their natural predators by the ants.

In the case of intraspecific mutualism, two or more individuals of the same species co-operate, each gaining a net benefit. If two lionesses co-operate, their chance of capturing a prey is probably more than twice that of each individual. Sharing half the meat then becomes a net benefit to each. In fact, lionesses in a pride are related, as are the male lions that co-operate in taking over a pride, so co-operation is favoured by kin selection and mutualism.

Table 3.2   Matrix of relations defining mutualism, altruism, selfishness and spite

		Recipient	
		Gains	**Loses**
Initiator	**Gains**	Mutualism or **reciprocity**	Selfishness, for example parasitism
	Loses	Altruism	Spite

(adapted from Barash, 1982)

Finally, there are the commensalists. In these cases, one organism benefits while the other is neither harmed nor helped. Pilot fish that follow sharks feed on what is left over that the shark would anyway ignore.

The essential differences between mutualism and parasitism can also be described in a matrix (Table 3.2). It is interesting to note that spite has not yet been observed in the natural world (other than in humans). This absence is entirely what we would expect from a natural selectionist point of view. Genes have been selected to enhance their own fitness and could not be selected to damage themselves at the same time as they damage others, with no resulting benefit. Its presence in humans is therefore a puzzle. We could invoke the power of memes (see Chapter 11) to infect our decision-making circuits and direct behaviour that is contrary to the natural selection of genes but ensures the survival of memes. Or we could simply say that some human actions have cut themselves free from the logic of reproductive fitness and have achieved some autonomy. A third alternative is that spite is maladaptive because it represents a miscalculation of the effects of certain courses of action. The threat of spite could bring rewards: 'If you don't do what I want, we will both suffer' may sometimes work, but if the person threatened calls the bluff of the aggressor, it becomes maladaptive.

3.5.2 *Reciprocal altruism, or time-delayed discrete mutualism*

Acts can often appear altruistic in the sense of Table 3.2 when in fact we have simply not taken a sufficiently long-term view of the situation. Trivers (1971) was one of the first to argue that altruism could occur between unrelated individuals through a process he termed 'reciprocal altruism', which is really a more refined version of the maxim 'You scratch my back now and I'll scratch yours later.' We are really looking for genes that, by co-operating with each other, enhance their own survival and reproductive success through their own vehicles. In kin selection, an individual may help another in the belief that the helping gene is present in the recipient. In the case of reciprocal altruism, aid is given to another in the hope that it will be returned. For it to work, we again need an asymmetry between the value of the gift to the donor and that to the recipient. To use a human analogy, it would be pointless to give away £5 only to receive it back again the next instant; the time would be not well spent. But if the £5 represented a small sum to you but helped to save the life of an unrelated individual, it might be worth it if there were a probability of finding yourself in a similar life-threatening situation.

We can now see that there is a rather fine line between our definition of mutualism and that of reciprocal altruism. The most useful distinction is that mutualism involves a series of constant reliances. An extreme form of this is lichen, which are composed of an alga and fungus in an inextricable symbiosis. Similarly, the bacteria in one's gut that help in the digestion of food are mutualistic in that the exchange of food products is virtually constant. We can think of reciprocal altruism as a sort of time-delayed mutualism. An exchange takes place in which it seems that the beneficiary gains and the initiator loses according to Table 3.2. What we expect, however, is that in cases of reciprocal altruism, the favour is returned at a later date. Game theory has also been used to distinguish between reciprocal altruism and mutualism (see Chapter 9), with the suggestion that rewards and punishments (for cheating) have different values.

3.5.3 *Conditions for the existence of reciprocal altruism*

We would expect the following conditions to obtain if reciprocal altruism were to be found:

1. An animal performing an altruistic act must have a reasonable chance of meeting the recipient again to receive reciprocation. This would imply that the animals should be reasonably long lived and live in stable groups in order to meet each other repeatedly
2. Reciprocal altruists must be able to recognise each other and detect cheats who receive the benefits of altruism but give nothing back in turn. If defectors cannot be detected, a group of reciprocal altruists would be extremely vulner-

able to a take-over by cheaters. Codes of membership for many human groups, such as the right accent and the right clothes, as well as the initiation rituals and signals of secret societies, could serve this function

3. The ratio 'cost to donor/benefit to receiver' must be low. The higher this ratio, the greater must be the certainty of reciprocation.

Although humans spring to mind as obvious candidates, species practising reciprocal altruism need not be highly intelligent.

Examples of reciprocal altruism

One of the best-documented examples concerns vampire bats (*Desmodus rotundus*). These were studied by Wilkinson (1984, 1990), who found that vampire bats, on returning to their roost, often regurgitate blood into the mouths of roost-mates. Such bats live in stable groups of related and unrelated individuals. A blood meal is not always easy to find: on a typical night, about 7 per cent of adults and 33 per cent of juveniles under 2 years of age fails to find a meal. After about 2–3 days, the bats reach starvation point. It might be thought that regurgitation is an example of kin selection, and some of this is undoubtedly occurring, but the exponential decay of loss of body weight prior to starvation suggests that the conditions for reciprocal altruism could be present.

Figure 3.13 shows weight loss against time. In essence, the time lost by the donor is less than the time gained by the benefactor. The bats' mode of life also

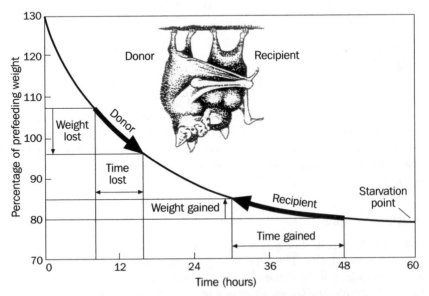

Figure 3.13   Food-sharing in vampire bats shown
in terms of a cost-benefit analysis (adapted from Wilkinson, 1990)

In this diagram, the donor loses about 12 per cent of its prefeeding body mass and
about 6 hours of time before starvation. The donor gains, however, 18 hours of time
and 5 per cent of its prefeeding weight. The fact that the time gained is so much
greater than the time lost favours reciprocal altruism

means that they constantly encounter each other. Wilkinson conducted experiments whereby a group a bats was formed from two natural clusters, nearly all of the bats being unrelated. They were fed nightly from plastic bottles. Each night, one bat was removed at random and deprived of food. Wilkinson noticed that, on returning to the cage, it was fed by other bats from its original natural group. Reciprocal partnerships between pairs of bats were also noticed.

Reciprocal altruism has also been documented in gelada baboons. Dunbar (1980) found a positive correlation between support given by one female gelada baboon to another and the likelihood that it would be returned. Evidence that reciprocal altruism may be at work in chimpanzee groups has been forthcoming from the work of de Waal on captive chimpanzees (de Waal, 1997). De Waal found that if chimp A had groomed chimp B up to 2 hours before feeding, B was far more likely to share its food with A than if it had not been groomed. Interestingly, B was equally likely to get food from A whether or not A had groomed it. De Waal's results suggest that grooming serves as a sort of service that is repaid later. Grooming may in fact have a very significant role to play in human and primate evolution, and we shall return to this in Chapter 7.

3.5.4 *True altruism and pseudo-altruism: genes and vehicles revisited*

In our working definition of altruism, espoused earlier as action that has some cost to the donor but also benefit to recipient, we did not specify whether the donor was a gene or a vehicle. We now also see that although reciprocal altruism involves some initial cost, it is later recouped with credit. So what is behaving altruistically?

Gene-level altruism

If we consider our definition at the level of the gene, we are looking for a gene that helps another non-identical gene. In kin selection or kin altruism, the gene responsible is merely increasing its own multiplication rate; it is not assisting another gene. The fact that the gene is present in another vehicle is irrelevant. It becomes clear that there can be no truly altruistic genes. A gene that helped others at expense to itself would quickly become extinct when confronted with a variant that simply helped itself.

Vehicle-level altruism

Following this line of thought, it becomes clear that kin altruism is true altruism at the level of vehicles. The gene(s) responsible may be increasing its fitness in another vehicle, but, by definition, the host vehicle (since we are at this level) is making a sacrifice in favour of genes in other host vehicles. Mutualism and reciprocal altruism are not true altruism at the level of vehicles in the sense that the vehicles co-operate to increase (albeit sometimes in the long term) their chances of both survival and propagation. If we take a long enough timescale, a net sacrifice is not made by either party; instead a net gain flows to each.

Induced altruism

Some authors have suggested that this should be regarded as a distinct form of altruism (Badcock, 1991). It is possible that one organism can induce another to commit an altruist act that benefits only the recipient that induced it. This occurs in the case of nest parasitism. A female cuckoo will dump an egg into the nest of another. This act in effect manipulates the nest owner to care for offspring that are not its own. Another example concerns ants that are captured and induced to act as 'slaves' for their new nest. The ants are in fact usually of a different species from their captors, being captured while young and manipulated to work for the good of their captors.

Meme-led altruism

Whereas a good deal of human behaviour is open to analysis in terms of reciprocal altruism and kin selection, there are undoubtedly many things that humans do that seem impossible to interpret in these terms. Humans display a remarkable ability to help others who are unrelated to themselves, and they do this at great personal cost. Short of saying that there is no biological or Darwinian explanation, and handing over this task to the social sciences, we could speculate that genuinely self-sacrificing human behaviour might be 'meme driven'. A meme is an idea or thought that occupies brains in the same way as a gene occupies bodies. Dawkins (1976) speculated about the possibility of Darwinian rules applying to such ideas (see Chapter 11). Hence the idea that we should help one another at great sacrifice to our biological selves may survive (in brains) even though the vehicles perish in the process. One could argue that we are induced to behave altruistically by memes that have infected us. The concept of memes has split the academic community. Some regard them as empty metaphors, others as useful aids to analyse culture (Blackmore, 1999b).

SUMMARY

▨ The information that instructs the behaviour of each cell in any organism and that passes to offspring along the germ line is contained in the sequence of base pairs found on DNA molecules.

▨ DNA is found in the nucleus of cells, packed with proteins in the form of chromosomes. In all mammals, the somatic cells contain complementary (homologous) pairs of chromosomes (diploidy), one copy in each pair being inherited from each parent. Reproductive cells (gametes) contain only one copy of each chromosome (haploidy). Fertilisation occurs when a sperm meets an egg to produce a zygote. A zygote has the potential to develop into a new individual.

▨ A phenotype is the result of environmental factors acting upon the genotype. The determination of behaviour may be interpreted along a spectrum of weak to strong genetic control.

▨ Although environmental stresses and selection pressures act upon groups and individuals, it is logically more consistent to consider the gene as the fundamental unit of natural selection.

▨ Altruism can be understood biologically in terms of kin selection and reciprocal altruism. In the former, genes direct individuals to help other individuals that are related and thus share genes in common. In the latter, individuals help unrelated individuals in the expectation of favours returned. Kin selection and reciprocal altruism offer insights into the biological basis of altruism in humans.

▨ Kin recognition is important both to ensure that altruism is effectively directed and also to avoid inbreeding. The Westermarck effect first proposed in 1891 has recently received further experimental support and provides a model for how the adaptive significance of a pattern of behaviour (avoiding incest) can be understood at the genetic level. It also illustrates how a functional outcome can be achieved by a series of rules of development enacted during infancy, and furthermore how such development rules are echoed by cultural mores, that is, the incest taboo.

KEY WORDS

Allele ■ Altruism ■ Amino acid ■ Autosome ■ Base ■ Chromosome
Codon ■ Coefficient of relatedness ■ Crossing over ■ Diploid ■ DNA
Dominant allele ■ Eusocial ■ Gamete ■ Gene ■ Genome ■ Genotype
Group selection ■ Haemophilia ■ Haploid ■ Heritability ■ Heterozygous
Homozygous ■ Inclusive fitness ■ Kin discrimination ■ Kin selection
Linkage disequilibrium ■ Locus ■ Meiosis ■ Mitosis ■ Mutation
Mutualism ■ Oogenesis ■ Parthenogenesis ■ Phenotype ■ Pleiotropy
Population ■ Recessive allele ■ Reciprocity ■ Recombination ■ RNA
Sex chromosome ■ Symbiosis

FURTHER READING

Dawkins, R. (1976) *The Selfish Gene*. Oxford, Oxford University Press.
Now a classic, this is probably the best account of gene-centred thinking ever written. A more recent edition was published in 1989.

Dawkins, R. (1982) *The Extended Phenotype*. Oxford, W. H. Freeman.
Shows how the effects of genes reach outside their vehicles.

Dugatin, L. A. (1997) *Co-operation Among Animals*. Oxford, Oxford University Press.
A thorough work, with numerous empirical examples. Dugatin accepts a role for a model of group selection as proposed by D. S. Wilson. An excellent book for the review of altruism among non-human animals, but contains no discussion of humans.

4

Mating Behaviour: From Systems to Strategies

And nothing gainst Time's scythe can make defense,
Save breed to brave him, when he takes thee hence.

(Shakespeare, 'Sonnet 12')

Each generation is a filter, a sieve; good genes tend to fall through the sieve
into the next generation; bad genes tend to end up in bodies that die young or
without reproducing.

(Dawkins, 1995, p. 3)

For individuals of sexually reproducing species, finding a mate is imperative. It is through mating, essentially the fusion of gametes, that genes secure their passage to the next generation; without it, the 'immortal replicators' are no longer immortal. It is hardly surprising then that sex is an enormously powerful driving force in the lives of animals and is attended to with a sometimes irrational and desperate urgency. At a fundamental level, sex is basically simple – a sperm meets an egg – but it is in the varied forms of behaviour leading to this event that complexity is to be found and needs to be understood. In order to understand human sexuality, we need to raise some basic questions concerning the causes, consequences and manifestations of sexual activity in animals as a whole. This chapter begins this task by looking at some current theories of the origin and maintenance of sexual reproduction. It also addresses some fundamental questions, such as why female gametes (eggs) are usually at least 100 times larger than male gametes (sperm) – a phenomenon known as anisogamy – or why the male to female ratio remains so close to 1:1, albeit with some slight but significant variations.

It was once thought convenient to classify sexual behaviour in terms of mating systems, and the terminology of such systems is introduced here. It will be argued, however, that a better approach is to focus on the strategies of individuals rather than the putative behaviour of whole groups. This individualistic approach will reveal that sex is as much about conflict as about co-operation, each sex employing strategies that best serve its own interests.

4.1 Why sex?

In recent years, it has almost become *de rigeur* for books on evolutionary biology to contain a section headed 'Why sex?' Yet for at least 100 years after the publication of Darwin's *On the Origin of Species* (1859b), the existence and function of sex was not really seen as a problem. Sex was viewed as a co-operative venture between two individuals to produce variable offspring. Variation was required to secure an adaptive fit to a changing environment and constant variation was needed to ensure that species did not become too specialised and face extinction if the environment changed.

From a modern, gene-centred perspective, these arguments now appear fatally flawed. Variation and selection cannot act for the good of the species; genes only care for themselves. As soon as we examine the costs and benefits of sexual reproduction, the very existence and maintenance of sex seems all too problematic (Table 4.1).

Table 4.1   Comparison of the costs and benefits of sexual reproduction

Costs	Benefits
Time and effort is spent attracting, defending and copulating with mates. Such effort could have been directed into reproduction	Where parental care is found, two individuals may be able to raise more than twice the number of offspring that one alone could. Consider birds: a single female would find it difficult to incubate and defend her eggs as well as feed herself and her offspring
Individuals may be vulnerable to predation during mating, especially during intercourse or courtship displays	New combinations of genes are created that can exploit variations in environmental conditions (see text)
There is a risk of damage during the physical act of mating	New combination of genes arise to cope with biotic interactions from predators, prey and parasites (see text)
A risk of disease transmission from one individual to another exists	Sex enables deleterious mutations in DNA to be repaired during meiosis and be masked by outcrossing (see text)
The recombination of genes that follows sex may throw up a homozygous condition for a dangerous recessive allele	
Sex introduces same-sex competition. Where polygamous mating is common, an individual may not find a mate at all	
Sex breaks up what might have been a highly successful combination of genes. If it 'isn't broke', sex still 'fixes it'	
Sexual reproduction introduces sibling rivalry since sibs will now only be related by $r = 0.5$ or less (see Chapter 3). Identical offspring would have a greater common interest	
Parthenogenetic females (females who are virgins and produce offspring without copulation) produce offspring faster than sexual females (see Figure 4.1 below)	

4.1.1 *The costs of sex*

'The expense of spirit in a waste of shame' was how Shakespeare summed up the effects of sexual lust. Table 4.1 does seem to show that, compared with the benefits it provides, sex is expensive in terms of cost. This of course begs the question of how big those few benefits are. The first cost on the list goes a long way towards explaining why there are discrete species at all. Imagine a world in which the number of species doubled at a stroke, but the number of individuals, being limited by the carrying capacity of the biosphere, remained the same. Individuals of any one species would on average incur twice the cost of finding a mate – there would simply be fewer of the opposite sex. As we increase the number of species (which can by definition only breed with **conspecifics**, that is, members of the same species) so that there is a smooth gradation between one species and the next instead of discrete jumps, so it becomes more difficult for individuals to mate at all, and some will become extinct.

Why do males exist?

The question 'Why sex?' resolves itself into the question 'Why do males exist?' All organisms need to reproduce, but some manage this asexually: females simply make copies of themselves by a sort of cloning process. This form of reproduction is known as **parthenogenesis** ('virgin birth'), and although it is not found among mammals and birds, it is not uncommon in fish, lizards, frogs and plants. Males are a problem because, in the absence of male care (which is very common), a mutation that made a sexually reproducing organism switch to parthenogenesis (which some organisms can do anyway) should be favoured since it would produce more copies of itself and rapidly spread throughout the population. Put another way, with a given set of environmental limitations, females should be able to produce twice as many grandchildren by asexual compared with sexual reproduction (Figure 4.1).

Almost as if to mock the doubts of biologists concerning the functions of sex, the natural world teems with sexual activity. Intriguingly, asexual species seem to be of fairly recent origin; they comprise the 'twigs' of the phylogenetic tree rather than its trunk or main branches. Some asexual species still betray their sexual ancestry. In the case of the Jamaican whiptail lizard, for example, the female will lay a fertilised egg only when physically 'groped' by a male. The male provides nothing in the way of genetic material, but its physical presence seems to trigger self-fertilisation. In some frog species, the male provides sperm for the activation of the development of the female's eggs, but again no genetic material is transmitted. As Sigmund (1993) has observed, it is a case for the male of 'love's labours lost'. This behaviour is probably of fairly recent evolutionary origin or else the males would have caught on and such time-wasting would be selected against – another caution against always interpreting animal behaviour as optimal. Nearer to home, everyone's back garden probably contains a few dandelions. The gaudy yellow flowers at first sight appear to be made like any other flower to attract pollinators, but dandelions are entirely self-fertilising; their flowers are leftovers from their sexual past when cross pollination did occur.

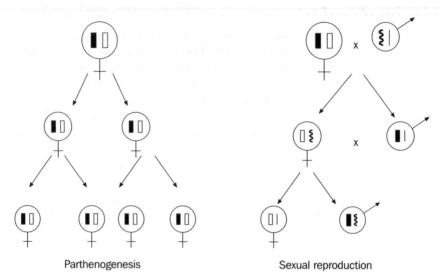

Parthenogenesis   Sexual reproduction

By the second generation, the parthenogenetic female has produced four asexual
daughters, whereas the sexual female has only produced one sexual daughter. This
assumes that males contribute nothing other than gametes to female fecundity

Figure 4.1   Comparison of the fecundity of females breeding
asexually and sexually (assuming that each female can
bear two offspring per generation or season)

In many sections of this book, we will flip between the level of the gene and
that of the individual in developing an argument. Although the fundamental unit
of selection must be the gene, operating at the level of the individual for explanatory purposes is less tedious and not as pedantic as having constantly to refer to
changing gene frequencies. As Dawkins (1976) points out, although the
immediate manifestation of natural selection occurs at the level of the individual,
the differential reproductive success of individuals is ultimately a matter of
changing gene frequencies in the gene pool. In this view, the gene pool of a given
species can be thought of as being akin to the primeval soup of the first replicators. In forming and breaking temporary alliances of genes, sex and the process of
crossing over serves to keep the gene pool stirred.

An argument is sometimes more easily expressed at the individual level but, as
long as it can also be convincingly expressed in an equivalent way at the level of
the gene, this should not cause a problem. Where the levels conflict, the view
taken here is that the individual-level argument should give way to gene logic.
One case of this conflict is the putative 'dilution' cost of sex. Many texts list as a
cost of sex the fact that an individual only passes on half of its genome to each
offspring, whereas it passes all of its genome to offspring by **asexual reproduction**; this is the so-called 'meiotic cost' of sex.

This argument is not compelling because it fails at the level of the gene. The
whole genome is not the unit of selection. What sex does is to force individual
genes to sit alongside genes from another individual – but so what? From a gene-
centred point of view, the gene is indifferent as to whether its neighbour came

from its parent by asexual reproduction or from another parent by sexual recombination. As Dawkins and Treisman (1976, p. 480) point out, an individual is a 'temporary federation of genes each intent on maximising its representation in the next generation'. If we consider sex from a 'sex-inducing' gene's point of view, sex enables the sex-inducing gene to thrive. It follows that, at the locus for sexual reproduction, a female is just as related to her offspring as is an asexual female. The females sacrifice half their genes for sexuality but receive them back again from the male, who must also carry genes for sexuality. Dawkins and Treisman show that at the locus at which the mode of reproduction is determined, the part of the mother's genome determining this is just as well represented in the next generation for sexual as for asexual mothers.

In this respect, the cost of meiosis disappears from a gene-centred point of view. We are left instead with the real cost that parthenogenetic females produce more offspring, and this is a consequence of the fact that males cannot produce babies by themselves. The issue resolves to the problem that a parthenogenesis-inducing mutation in a sexual population should spread rapidly. In fact, given a sex ratio of 1:1, it should spread twice as rapidly as the remaining sexual gene; this is often called the 'twofold cost' of sex (see Figure 4.1 above). If a male is able to help a female to raise the young rather than simply to provide gametes but nothing more, this twofold cost may be lessened. Nevertheless, something about sex must confer an advantage on sexual organisms to drive against this reduction in fecundity.

We will now review some of the major theories advanced to explain the persistence of sexual modes of reproduction in the face of all the apparent disadvantages.

4.1.2 *The lottery principle*

The American biologist George Williams was one of the first to suggest that sex introduced genetic variety in order to enable genes to survive in changing or novel environments. He used the lottery analogy: breeding asexually is like buying many tickets for a national lottery but giving them all the same number; sexual reproduction is like making do with fewer tickets but having different numbers (Williams, 1975). The essential idea behind the lottery principle is that since sex introduces variability, organisms have better chance of producing offspring that survive if they produce a range of types rather than more of the same. On the positive side for this theory, it may help to explain why creatures such as aphids, which can breed both sexually and asexually, choose to multiply asexually when environmental conditions are stable but switch to sexual reproduction when facing an uncertain future. In the steady months of summer, aphids multiply at a fast rate on rose bushes by parthenogenesis, but as winter approaches they have bouts of sex to produce numerous and variable cysts that survive the winter and wait for the return of warmer conditions.

Williams also noticed that when organisms disperse seed beyond their local habitat, they choose sex as a precursor. This he termed the 'elm–oyster model'. When organisms wish to colonise their local area, which, since they are there, they must already be reasonably successful in, they send out runners or vegetative shoots. This is done asexually and is observed in numerous grasses, strawberry plants and coral reefs. But oysters and elms are sexual and produce thousands of tiny seeds that waft on currents of water and air to considerable distances away from the parents. Why do they do this? The answer, according to Williams, is that oyster beds and an elm forest are already saturated and that there is likely to be intense competition for any new living space that the seeds might find. Success goes to the parents who produce a few exceptionally suited seeds rather than to the parents who produce many average ones.

Parasites also provide an illustration of this principle. When a host is first invaded, parasites typically reproduce asexually to fill the host as rapidly as possible. When this niche is filled, new offspring have to leave and infect other hosts. At this stage, the parasite typically switches to sexual reproduction to take advantage of the fact that sex produces variation that may be useful for success in the next round of infecting unknown hosts – some of which may be resistant to genotype of the parent parasites. In short, sex precedes dispersal.

4.1.3 *The tangled bank hypothesis, or spatial heterogeneity*

The lottery principle idea of Williams was developed to form the 'tangled bank' theory of Michael Ghiselin. This term is taken from the last paragraph of Darwin's *Origin*, where he referred to a wide assortment of creatures all competing for light and food on a tangled bank. According to this theory, in environments where there is an intense competition for space, light and other resources, a premium is placed on diversification (Ghiselin, 1974). From a gene-centred point of view, a gene will have an interest in teaming up with a wide variety of other genes in the hope that at least one such combination will do well in a competitive environment. An analogy for the tangled bank theory (even though the tangled bank is already a metaphor) is that of the button-maker. Imagine a button-maker who has made enough identical buttons for everyone's needs in the local area. What is he to do? One answer is to diversify in the hope that he may tap into a latent demand for a slightly different type of button. Thus, in crowded conditions, we would expect to find sex as a means of exploiting tiny variations in the local environment.

Although once popular, the tangled bank theory now seems to face many problems, and former adherents are falling away. The theory would predict a greater interest in sex among animals that produce lots of small offspring (so-called r selection) that compete with each other. In fact, sex is invariably associated with organisms that produce a few large offspring (K selection), whereas organisms producing smaller offspring frequently engage in parthenogenesis. In addition, the evidence from fossils suggests that species go for vast periods of time without changing much. The tangled bank theory would predict a gradual change as types drift through the adaptive landscape. It is only really in the special conditions found on small islands, where populations are tiny and inbred, that we can observe fairly rapid changes.

Another line of evidence used to test the various theories of sex has been that of crossover frequencies in chromosomes. Crossing over during meiosis increases the variability of **gametes**. We might expect then short-lived organisms with high fecundity, which could quickly saturate an area, to have high crossover rates. The opposite seems, however, to be the case. Crossover frequency bears little relationship to the number of young and to body size but is strongly correlated with longevity and age at sexual maturity. Thus, humans have about 30 crossover sites per chromosome, rabbits 10 and mice 3.

4.1.4 *The Red Queen hypothesis*

The Red Queen hypothesis, which now offers one of the most promising explanations of sex, was first suggested by Leigh Van Valen in 1973. Van Valen discovered from his study on marine fossils that the probability of a family of marine organisms becoming extinct at any one time bears no relation to how long it has already survived. It is a sobering thought that the struggle for existence never gets any easier: however well adapted an animal may become, it still has the same chance of extinction as a newly formed species. Van Valen was reminded of the Red Queen in Alice in Wonderland, who ran fast with Alice only to stand still.

The application of this theory to the problem of the maintenance of sex is captured by the phrase 'genetics arms race'. A typical animal must constantly run the genetic gauntlet of being able to chase its prey, run away from predators and resist infection by parasites. Parasite infection in particular means that that parasite and host are locked in a deadly 'evolutionary embrace' (Ridley, 1993). Each reproduces sexually in the desperate hope that some combination will gain a tactical advantage in attack or defence. William Hamilton summed this up in a memorable fashion when he compared sexual species to 'guilds of genotypes committed to free fair exchange of biochemical technology for parasite exclusion' (quoted in Trivers, 1985, p. 324).

The Red Queen hypothesis also gains support from the comparative approach to sexual reproduction developed by Graham Bell in Montreal. Bell (1982) found that sex is most commonly practised in environments that are stable and not subject to sudden change. Asexual species, on the other hand, are often highly fecund small creatures that inhabit changing environments. Even the suggestion that aphids turn to sex when the prospect of hard times looms has been challenged. It turns out that a better predictor of sexuality is overcrowding: aphids will turn to sex in laboratory conditions if they are overcrowded.

The lottery principle suggests that sex is favoured by a variable environment, yet an inspection of the global distribution of sex shows that where environments are stable but biotic interactions are intense, such as in the tropics, sexual reproduction is rife. In contrast, in areas where the environment is subject to sudden change, such as high latitudes or small bodies of water, it seems that the best way to fill up a niche that has suddenly appeared is by asexual reproduction. If your food supply is already dead, it cannot run away, so the best policy if you are an organism feeding on dead matter (a decomposer) is to propagate your kind quickly to exploit the food resource and forego the time-wasting business of sex. In the world of the Red Queen, organisms have to run fast to stay still. A female always reproducing asexually is 'a sitting duck for exploiters from parasitic species' (Sigmund, 1993, p. 153).

Further support for the parasite exclusion theory comes from the fact that genes that code for the immune response – the **major histocompatibility complex** (MHC) – are incredibly variable. This is consistent with the idea that variability is needed to keep an advantage over parasites. Moreover, we have already noted that human females may be choosy about their prospective partners in relation to their MHC genes, genes that are different from their own being preferred (see Chapter 3).

4.1.5 *The DNA repair hypothesis*

Why are babies born young? The question at first sight appears to be a rather stupid one; surely babies are young by definition? But the question we are really asking is how, despite the ageing of somatic cells in, for example, the skin and nervous tissue of the parents, the cells of the newly born have their clocks set back to zero. Somatic cells die, but the germ line appears to be potentially immortal. Bernstein *et al.* lay claim to a solution to this problem:

> We argue that the lack of ageing of the germ line results mainly from repair of the genetic material by meiotic recombination during the formation of germ cells. Thus our basic hypothesis is that the primary function of sex is to repair the genetic material of the germ line. (Bernstein *et al.*, 1989, p. 4)

As we have already noted (Chapter 3) the primary features of sex from a gene's point of view are meiotic recombination and outcrossing. Bernstein *et al.* interpret both these events as responses to the need for repair.

DNA faces two types of disruption. It can be damaged in situ by ionising radiation or mutagenic chemicals, or a mutation can occur through errors of replication, which are best thought of as change rather than damage. Damage to the DNA can take a number of forms, repair mechanisms often being suited to each type. Single-strand damage can be made good by enzymes using the template provided by the other strand, but double-strand damage is more serious: the cell may die or possibly make use of the spare copy in haploid cells. During crossing over in meiosis (see Figure 3.8), the chromosomes line up and the spare copy is used to repair double-strand breaks.

If damage were the only problem faced by DNA, there would not be an automatic need for males. Asexually reproducing females could still be diploid and then produce haploid gametes to fuse with each other to produce offspring by self-fertilisation. In fact, about 17 per cent of plants do just this. Such a process would appear to have all the strengths of keeping a spare copy of vital genes without incurring the cost of sex. Not all damage, however, can be detected by self-inspection. Errors of replication can occur whereby a wrong base is inserted into the strand of DNA. These 'mutations' cannot be detected by enzymes since the strand does not look damaged (comparing one gene with its complementary copy would not help greatly since there is the problem of determining which is the 'correct version').

Most mutations are deleterious, but fortunately they are recessive and their effects consequently swamped by viable alleles on the complementary chromosome. As cell division proceeds, however, the burden of mutation steadily increases, and there will come a time when a genome becomes homozygous for a dangerous recessive allele. This is an example of an effect called 'Muller's ratchet': as time passes, mutations accumulate in an irreversible fashion like the clicks of a ratchet. With the outcrossing brought about by sex, these mutations can be masked in the heterozygous state.

In asexual reproduction, any mutation in one generation must necessarily be passed to the next. Ridley (1993) likened this to photocopying: as a document is copied, and copies made from the copies and so on, the quality gradually deteriorates. In accumulating mutations at a steady rate, asexual organisms face the prospect that they may eventually not be viable. In sexually reproducing species on the other hand, some individuals will have a few mutations while some will have many. This arises from meiosis and outcrossing. Sexual reproduction involves the shuffling of alleles; some individuals will be 'unlucky' and have a greater share than average of deleterious mutations in their genome, and some will be 'lucky', with a smaller share. The unlucky ones will be selected out. This in the long term has the effect of constantly weeding out harmful mutations through the death of those that bear them (Crow, 1997). Eyre-Walker and Keightley (1999) have reported a mutation rate in humans of about 1.6 deleterious mutations per person per 25 years. This would have devastating consequences if it were not for sexual reproduction.

The DNA repair hypothesis will receive vital evidence from the fate of Dolly the sheep. Dolly was produced when, in 1996, scientists at the Roslin Institute in Scotland produced a **clone** of a sheep by introducing DNA from a mature 6-year-old sheep into a developing embryo. The sheep, since it was cloned from cells of the mammary gland of its 'mother', was called Dolly – allegedly after the American singer Dolly Parton. The life history of Dolly will be immensely interesting. The DNA repair theory would predict that she will not live as long as a sheep from a normal birth. The problem for Dolly is that her chromosomes were already old and worn at birth; when Dolly's genome was created it was denied the rejuvenating power of sex.

This theory is not without its problems and critics. Perhaps the best conclusion so far is that it is extremely probable that sex evolved out of the genetic mechanisms for DNA repair and that repair may have been the original function of sex for early organisms. The case that sex is now maintained to check constantly for damage and supervise repair is more controversial.

In summary, we have four major types of theory to account for the origin and maintenance of sex:

1. sex produces variable offspring to thrive as environments change through time
2. sex produces variation to exploit subtle spatial variations in environmental conditions
3. sex enables organisms to remain competitive in a world where other organisms are poised to take advantage of any weakness
4. sex serves to keep at bay the effects of damage wreaked daily on our DNA and thus weed out deleterious mutations.

There is perhaps no one single explanation for the maintenance of sex in the face of severe cost. Genes that promote sexual reproduction could flourish for a variety of reasons. In this respect, we should note that the models are not mutually exclusive: all rely upon sex to maintain genetic variability.

4.2 Sex and anisogamy

Individuals in sexually reproducing species exist in two forms: males and females. The question is, how do we define 'maleness' and 'femaleness'? In most higher animals, the distinction is pretty clear. Even if males and females are morphologically different, we could say that males inject sperm into females. However, to cover cases of external fertilisation, as practised by many fish species, we need a better definition than this. A more comprehensive definition would be that males produce small mobile gametes (sperm) that seek out the larger, less mobile gametes (eggs) produced by the female.

Yet the ancestral state of life on earth must have been that of primitive, single-celled asexual organisms. Now a further problem confronts us: since the first sexually reproducing organisms probably produced gametes from males or females of equal size (**isogamy**), how have we arrived at the situation where, for virtually all cases of sexual reproduction, the size of the gametes from males and females is vastly different? Figure 4.2 shows how great the discrepancy is.

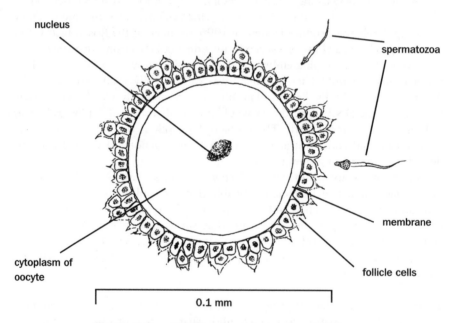

Figure 4.2 Relative dimensions of an egg from a human female and a sperm from a male

Parker *et al.* (1972) suggest one probable scenario. Their argument is essentially that an ancestral state of equally sized gametes quickly breaks down into two strategies: providers and seekers (see Figure 4.3). Parker *et al.* were also able to show that these two strategies are stable in the sense that they can resist invasion from other strategies such as that in which both males and females produce large gametes prior to fusion and thus give them a head start over the smaller zygotes from anisogamy. In this case, the problem for a male once **anisogamy** is established is that any larger gametes, produced to confer an advantage on the zygote, would easily be out-competed by the larger number of rival small gametes. Males and females become locked into their separate strategies.

Other suggestions have been made to account for the origin of anisogamy. One such is the idea that the small size of sperm reduces the likelihood of transmitting cytoplasmic parasites from the male to the zygote (Hurst, 1990). In this respect, it is also significant that humans, in common with many animals, only inherit their mitochondrial DNA (which is different from the DNA of the chromosomes) from the maternal line. Mitochondria are small bodies within cells that serve as energy supply units. They convert the chemical energy of molecules such as sugars to other molecules that can serve as fuel for cellular processes. The origin of the mitochondria in our cells is thought to derive from an invasion of bacteria into cells that then became symbiotic. This peculiar inheritance of mitochondrial DNA has led to some interesting empirical analyses of the lines of descent of modern human populations, leading some to suggest that all modern humans can trace their descent to one female in Africa.

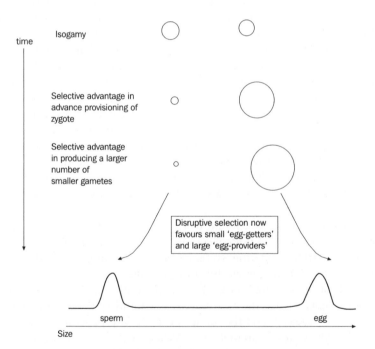

Figure 4.3   The breakdown of anisogamy by disruptive selection

4.3 Describing mating behaviour: systems and strategies

Throughout the animal kingdom, there is one common denominator to sex: the fusion of a large gamete supplied by the female with a smaller one supplied by the male. Yet beyond this, there is a huge diversity and striking contrasts within genera and families in terms of the social systems and behavioural arrangements that facilitate this fusion. In this section, we will discuss the terminology of mating systems. After that, behavioural imperatives that drive these systems will be examined.

There is as yet no universally accepted precise system to classify patterns of mating behaviour. This largely results from the fact that a variety of criteria can be employed to define a mating system. Commonly used criteria usually fall into two groups: mating exclusivity and pair bond characteristics. In the former, a count is made of the number of individuals of one sex with which an individual of the other sex mates. In the latter, it is the formation and duration of the social 'pair bond' formed between individuals for co-operative breeding that is described.

Both sets of criteria have their problems. In the case of mating exclusivity, the crucial act of copulation is not always easy for field workers to observe. With humans we can issue questionnaires and hope for an honest response, but with other animals copulation may be underground, in mid-air, at night or generally difficult to see. Such are the problems here that copulatory activity has often been inferred from the more visible non-copulatory social forms of behaviour such as parental care and nest cohabitation. It is obvious that this approach is open to errors of interpretation. Bird species such as the dunnock (*Prunella modularis*) that were once classified as monogamous on the basis of shared nest-building and parental care of offspring have turned out to be more varied in their mating habits (Davies, 1992). Terms such as 'extrapair copulations' and 'sneak copulations' have consequently entered the repertoire of behavioural terminology.

The problem with the term 'pair bond', as with other theory-laden terms such as 'partner', is that it suggests a degree of harmony between the two sexes that may not exist in reality. Such terms are probably hangovers from pre-1970 ethology when mating was viewed as a co-operative venture between two sexes to perpetuate the species. A more individually focused approach reveals as much conflict as co-operation between the sexes, and the pair bond for many species could be seen with equal validity as a sort of 'grudging truce'. Bearing in mind these reservations, Table 4.2 shows a simplified classificatory scheme combining both sets of criteria.

The fact that naturalists have found it difficult to devise hard and fast definitions for mating systems need not concern us too much. The attempt to match a species with a particular mating system faces a more fundamental array of problems than mere observational difficulty. The first point to note is that the word 'system' can hide a great deal of diversity. Mating involves a number of discrete components, such as the means of mate acquisition, the number of mates or of copulations achieved, and the nature of the 'bond' between the sexes, for example in the division of parental care. Although these components can be interdependent, they can also vary independently. To refer to two species exhibiting monogamy could suggest an underlying similarity greater than is warranted.

A further problem is that the concept of a system does not really capture the diversity of mating behaviour within a single species or even a population. We may find that, in a largely monogamous population, males and females may sneak extrapair copulations. It must also follow that, since the **sex ratio** remains close to 1:1 (see below) in groups where some males or females practise **polygamy**, some

Table 4.2   Simple classification of mating systems

System	Mating exclusivity and/or pair bond character
Monogamy	Copulation with only one partner
Annual	Pair bond formed anew each year
Perennial	Pair bond formed for life
Polygamy	One sex copulates with more than one member of the other sex
Polygynandry	Males and females mate several times with each other and with different partners; for example, a stable group of two males and two females
Polygyny	Males mate with several females, females with only one male
Successive polygyny	Males bond with several females in breeding system but only one at a time
Simultaneous polygyny	Males bond simultaneously with several females
Polyandry	One female mates with several males, males with only one female
Successive	Females bond with several males but one at a time
Simultaneous	Females bond with several males at the same time

individuals may not mate at all. In polygamy, we also have the fact that the two sexes behave differently even though they are of the same species. The classification in terms of polygyny or polyandry is usually based on the sex that 'does best'. Hence elephant seals are often referred to as polygynous because a few males have large harems. The females must be thought of as monogamous.

If individuals are behaving differently like this, it suggests that it may be more productive to examine the rationale behind the behaviour of individuals. After all, ascribing a system to a group of animals does not explain why such a system is found. By analogy, we can say that a collection of atoms in the gaseous state in a closed container exerts a pressure, but we need to look deeper into the impact of individual atoms with the walls of the container before we understand the origin of the pressure.

The essential point to grasp is that species in themselves do not behave as a single entity: it is the behaviour of individuals that is the raw material for evolution. Any system that we care to project onto groups of individuals is at best an emergent property resulting from individual actions, and it becomes hard to justify the application of the term 'system' to an emergent property. A better approach then would be to focus on strategies pursued by individuals in their attempt to optimise their inclusive fitness under within the conditions that prevail at any one time. The most common behaviour of individuals may then allow us loosely to apply the label 'system' as a matter of descriptive convenience.

4.4 Factors affecting expressed mating strategies

4.4.1 A generalised model of mating behaviour

The **lineage** of any given animal will obviously have been exposed for a long time to various biotic and abiotic influences. Selection will thus have led to a set of phylogenetic constraints on the strategies that an animal can employ. Figure 4.4 shows how we might conceptualise the influence of past (phylogenetic) and present (ecological and biotic) factors on the mating behaviour of an individual organism. It is suggested that animals have a range of potential strategies. The strategy that is expressed is a product of local conditions, the learning experiences of the animal and its phylogenetic inheritance. We will now examine each of these factors in turn.

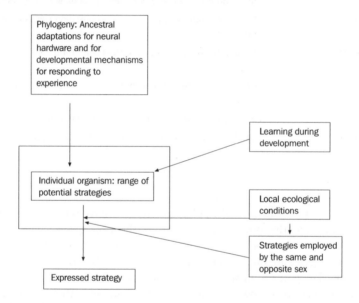

Figure 4.4   Model of factors influencing the mating strategy of an individual

4.4.2 Phylogeny

The physiological apparatus of mating (hormones, genital organs, lactatory devices and so on) will of course be closely tied to the repertoire of behaviour that an organism can evoke. Desertion immediately after fertilisation is not an option for human females since they carry the fertilised **zygote**, but it is for fish. Men cannot lactate to feed babies, but women can. Individuals are thus pre-adapted by their evolutionary lineage to certain modes of behaviour. Hence, varied as taxa are, taxonomy is not totally worthless as a predictor of social behaviour. We can illustrate this point by a comparison of mammals and birds.

Mammals, unlike birds, are mostly live-bearing, and the offspring must consequently be nourished before birth. This is usually done by a placental food delivery system. Once the young have been born, nurture is then given by female lactation. It follows that females make a huge investment in parenting, a good part of which cannot be supplied by the male. In the case of carnivores, the male could in principle share food with the female, but for mammals feeding on low calorie foodstuffs, such as herbivores, the potential for male assistance is extremely limited. If the male is to optimise his reproductive success, his best strategy would be to divert more effort into mating than to parenting, in other words to pursue polygyny. A female will be more disposed towards monogamy (or monandry – having only one male partner) since the bottleneck to her reproductive success is not the number of impregnations she can solicit but the resources she is able to accumulate for gestation and nurture.

In the case of birds, the investment of the female in provisioning the fertilised egg is initially greater than that of the male in that the embryo is laid complete with a packaged food supply. Once hatched, however, the nestlings can be fed by both parents, so both sexes can increase their reproductive success by staying together to help at the nest. Roughly speaking, two parents can feed twice as many young as can one parent. It benefits the male to remain in a monogamous social bond if, as a consequence of his desertion, his offspring have a considerably reduced expectation of survival. If the environment is particularly rich in resources and one parent could cope, the male may be tempted to desert. In principle, the female could desert shortly after laying her eggs.

There has been much discussion (Dawkins and Carlisle, 1976) on why, in these conditions, the male is more inclined to desert than the female. The simplest explanation probably lies in the fact that desertion entails some risk that the eggs may fail and that the consequences of failure are more crippling to the female than to the male. One clutch may represent a significant part of one season's reproductive labours for a female, and consequently she has fewer future opportunities to make good than the male does. Given the physiological differences between birds and mammals, it is no surprise then that most mammalian species are polygynous but that about 90 per cent of all bird species are monogamous.

The genetic legacy of ancient environments may help us to understand current behaviour when simple physiology or comparative socioecology is of little use. For primates, there have been numerous attempts to understand mating behaviour in terms of the ecological conditions faced by a species, with much success. One puzzle, however, that the concept of phylogenetic inertia may help to solve concerns the mangabeys, which are mostly arboreal but still for some reason exhibit the multimale groups typical of terrestrial primates such as baboons and macaques, rather than the uni-male groups typical of tree-dwelling primates. We find, however, that the nearest relatives of mangabeys are in fact terrestrial baboons, from which mangabeys probably evolved. Struhsaker (1969) suggested that phylogenetic inertia was constraining the behaviour of mangabeys. Such a possibility serves as a warning, before we examine the role of ecological conditions, that behaviour may not always be optimally adapted to current environments. This is an ever-present problem when interpreting human behaviour: what may seem maladaptive in current contexts could have been adaptive a few hundred thousand years ago.

4.4.2 *Ecological conditions*

Roughly speaking, for mammals at least, a male's reproductive potential is constrained by the number of females he can impregnate (and of course the inter-male competition that follows from the fact that other males have reached the same conclusion). For females, however, the primary restraint on her reproductive output is not the availability of willing males but ecological factors such as food supply. As noted earlier, it follows that males will be more inclined to pursue polygyny than females will polyandry. One approach then is to start from the perspective of the male and consider how wider conditions favour or militate against his predilection for polygyny.

The ability of a male to achieve polygyny is strongly influenced by the distribution of females. Put simply, if females are widely dispersed, opportunities to practise polygyny look bleak. If females congregate, for whatever reason, the prospects look better. The main influences on female distribution are predation and food (Figure 4.5).

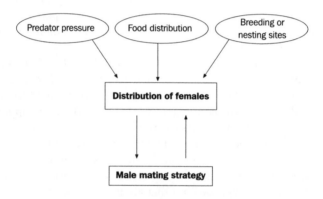

Figure 4.5   Factors determining the distribution of females and the impact on male mating strategy

If a male can command and defend a resource desired by females, the male may secure multiple matings. A good illustration of this is the orange-rumped honey-guide (*Indicator xanthonotus*), the males defending the nests of bees. When females visit the nests to feed on the beeswax, the male copulates with her in what is in effect an exchange of access to food for sex. The correlation between the degree of polygyny achieved by a male and the quantity or quality of the resources at his disposal can be illustrated with numerous examples from birds, fish and mammals (Andersson, 1994). The resources controlled can be food, territory or breeding sites favoured by the female. The phenomenon is sometimes called resource defence polygyny.

We must, however, be wary of examining mating solely from the male perspective. Females have their own reproductive interests at stake and are not simply passive receptacles for male success. If we remember that, in polygyny, some males will have many mates and some none, we could ask what makes a female agree to polygyny when she could presumably mate monogamously with one of the males left over. In making such a decision, the female has a set of costs and benefits to assess. The costs of mating with an already-mated male compared with an unmated male might include sharing resources offered by the male with other females, sharing help (if any) offered by the male with other females, and rivalry from other females. The benefits could be that the female collects a set of successful genes, and if her male offspring inherit such genes, the number of her grandchildren is increased. An additional benefit might be that the female acquires access to high-quality resources. The balance of costs and benefits is explored in the polygyny threshold model. Figure 4.6 shows the usual form in which this model is expressed.

This model has stimulated much field work on resource defence polygyny. Some work, such as that by Pleszczynska (1978) on the lark bunting, has yielded results in keeping with the predictions of the model. Support for the model from human behaviour comes from the work of Borgerhoff-Mulder (1990) on the Kipsigis people of Kenya. Borgerhoff-Mulder found a strong correlation between the area of land owned by a man and his number of wives.

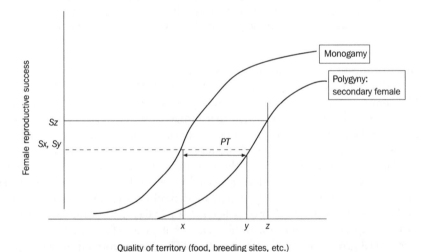

A female mating monogamously on territory of quality x would achieve the same success (Sx = Sy) as a female sharing a male on territory y. PT is the polygyny threshold. If the female increases the quality of the territory to z, her success is greater than with the monogamous mating strategy (Sz > Sx)

Figure 4.6   The polygyny threshold model (after Orians, 1969)

Most studies on resource defence polygyny have reached the conclusion that there is a positive relationship between the number of mates held by a male and environmental factors thought to be indicative of the quality of his territory. The problem comes, however, when measuring the costs and benefits to the female,

Figure 4.7   Polygyny is a common mating system for humans in many
parts of the world. Here a village chief of the Koko village in
Ghana stands proudly before his 5 wives and 38 children

and this means that results are often difficult to interpret. Studies on yellow-
bellied marmots (Downhower and Armitage, 1971) and pied flycatchers
(Askenmo, 1984) have shown that the reproductive success of females sharing a
male partner is actually lower than that achieved by not sharing. This calls into
question the predictions of the model that polygynous females do as well or (if
they cross the polygyny threshold) better than monogamous females.

Males cannot always control the resources that attract females. Resources
may, for example, be too widely dispersed for a single male to control. A male
can satisfy his propensity for polygyny by defending a group of females against
rivals – so-called female defence polygyny. If females spend time in pre-existing
social groups, evolution towards this type of polygyny is facilitated. Females may
form social groups for a number of reasons. The groups formed by many species
of primates probably evolved as a defence against predators, while female African
elephants (*Loxodonta africana*) assemble in relation to a patchy and localised
food distribution.

Elephant seals (*Mirounga angustirostris*) provide a spectacular illustration of
female defence polygyny brought about by the social grouping of females. Each
year, females haul themselves onto remote beach locations (for example, Ano
Nuevo Island off California) to give birth to their pups. The shortage of beach
locations, the tendency to return to the same site and the fact that females
become sexually receptive only 1 month after giving birth leads to a large concentration of fertile females. Not surprisingly, the male seals fight viciously for access
to these ready-made harems. Intrasexual competition (male versus male) has led
in this species to pronounced sexual dimorphism: the males are several times

heavier than the females and have enormous probosci for calling and for fighting other males (see Chapter 5).

Female defence polygyny can also occur if the females are solitary, so long as the male is able to defend a territory containing several females. In fact, in over 60 per cent of mammalian species, males defend a territory that overlaps one or more female ranges.

As the size of a female group increases, it becomes less defensible from the intrusions of other males, so it becomes impossible for a single male to have exclusive sexual access to all the females, and he will be forced to tolerate the presence of other males. We then end up with multimale and multifemale groups characteristic of baboons and macaque monkeys. In such groups, mating behaviour is complicated, and access to females is determined largely but not entirely by the position of the male in a dominance hierarchy.

Comparative socioecology

The way in which the size of the female group influences the mating strategy of the male is also illustrated by a comparative socioecological approach. Peter Jarman's (1974) study of the habits of the numerous species of African antelope of the family Bovidae (cattle and antelope) is a classic in this respect. Jarman showed that a correlation existed between size, feeding size, social grouping and mating behaviour.

It is a well-established principle in biology that the metabolic rate of an mammal increases as its size decreases. This arises largely from the fact that the ratio of surface area to volume gets larger as a creature gets smaller. Since heat loss is a product of surface area, small animals lose heat relatively faster than large ones. Hence to maintain a given body temperature, small animals need to metabolise faster than large, and small antelope such as steinbock need to eat high-calorie food such as fruit and buds. This selective grazing does not favour the formation of herds since the best pickings in a new area would all be taken by the first to arrive. The best strategy for these feeders would be to feed in small groups in a territory with fairly dense vegetation. At the other end of the size spectrum, large species with lower metabolic rates can subsist on low-calorie food obtained by unselective grazing. Groups of these large creatures such as wildebeest and buffalo can wander together in a herd. The herd must move about to find fresh grass, but when it is found, it is likely to be in patches large enough to sustain a large herd and too large for a few individuals to monopolise anyway.

In both these contexts, the mating strategies employed by males can be interpreted as an attempt to maximise their reproduction in the context of the resource-dependent dispersal of females. Males of small species in which the females are not strongly grouped do best by attaching themselves monogamously to a female. Males of intermediate-sized species such as reedbuck, impala and gazelle may be able to command a harem of localised females. Males in a large herd cannot possibly command exclusive mating rights to all the females but vie with each other for mating opportunities.

Polyandry

Although there are plenty of examples of females copulating with more than one male, either in multimale and multifemale groups or through extrapair copulations, true polyandry, in the form of a stable relationship between one female and several males in which the sex roles are reversed and males assume parental responsibilities, is very rare. At first sight, both males and females linked polyandrously would appear to gain little. From the female's perspective, sperm from one male is sufficient to fertilise all her eggs, so why bother to mate with more than one male? From the male perspective, it is even worse: if a male is forced to supply some parental care, the last thing he should want is to share his mate with another male and face the prospect of rearing offspring that are not his own.

To understand the emergence of polyandry, we need to consider both food supply and predatory pressure. In the case of the spotted sandpiper (*Actitis macularia*) studied by Oring and Lank (1986), the productivity of the breeding grounds on Leech Lake, Minnesota, is so high that the female can lay up to five clutches of four eggs in 40 days. Her reproductive potential is limited not by food resources but by males to incubate and defend the eggs. In this situation, we observe **sex role reversal**: the females are larger than the males and compete with each other to secure males to incubate their clutches.

Many human societies are mildly polygynous or monogamous (see Chapter 8), there being very few polyandrous human societies. One of the best documented is the Tre-ba people of Tibet, where two brothers may share a wife. One reason for this arrangement seems to be as a means of avoiding the split of a family land-holding in a harsh environment where a family unit must be of a certain minimum size and where the tax system weighs against the division of property. This is not, however, simply a reversal of polygyny. Men are socially dominant over women, the younger brother's ambition is to obtain his own wife and, as in most societies, Tre-ba men acquire wives rather than the other way round (Crook and Crook, 1988). Moreover, when a Tre-ba family has daughters but not sons, polygyny is practised whereby the daughters share a husband and the family holding is passed on through them.

Polyandry exists briefly as a transition phase among the Pahari people of north India. Wives have to be purchased at a substantial price. Brothers may typically pool their assets to buy a wife, which they share. When they can afford it, another wife is taken. The eventual result is group marriage or polygynandry in which two or more husbands are married to two or more wives, and all the men are married to all the women. The Pahari are the only human society in which such polygynandry is the norm (Berreman, 1962).

Leks

When mating takes place on **leks** – a lek is the name given to the area where males guard a small patch of territory that they use for display – males neither protect nor provide resources for females nor supply any parental care. For males, this is sexual reproduction in its least committed form: they supply only genes. The word 'lek' comes from the Swedish word for play. A number of species practise leking, including peacocks, sage grouse, several birds of paradise and a number of antelopes, deer and bats.

Females visit the lek and, once they have carefully observed the display antics of the males, appear to make a choice and agree to copulate with one of them. The display can be visual, as in the case of the peacock that fans its train and the American sage grouse that struts and dances, or it can be aural, as in the case of hammer-headed bats (*Hypsignathus monstrosus*), which flap, call and buzz passing females.

Given that the males are offering nothing in the way of resources or care, it is difficult at first sight to see what induces females to comply with this strategy. It could be that mating in such circumstances also reduces female vulnerability to predation during mating. The zero investment of males in the protection of the females and young does not deter females from exercising a very careful choice, and a relatively few males usually receive nearly all the copulations (Figure 4.8)

It is likely that in these conditions females are making a choice for good genes (see Chapter 5). The ability of males to compete for display sites and then perform often complicated display rituals is a reflection of genotype. A female is attracted to males successful at leking since her offspring will inherit desirable features including, if they are sons, success on the lek site. The problem with this plausible answer is the so-called 'lek paradox'. If females mate with, for example, only 10 per cent of the males, after a few generations all the females and males will be nearly identical, and there will be little point in looking for the best male since they will all look the same. At this point, the females will have exhausted the possible range of genetic variation. It transpires that there is a way out of this paradox, to which we will return in Chapter 5.

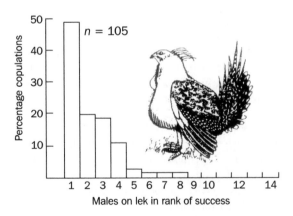

Figure 4.8   Variance of male reproductive success in the sage grouse (*Centrocercus urophasianus*) (adapted from Krebs and Davies, 1987)

Could there be an equivalent to leking for human courtship? At first sight, it would seem unlikely since human males are not as polygynous as sage grouse or peacocks, however much they would like to be. Without doubt, however, human males and females display to each other. Dunbar has made the amusing but plausible suggestion that mixed-sex human conversations may be functioning as a lek. In single-sex groups, both sexes spend much of the time talking about personal relationships and experiences. In all male groups, for example, the time spent on 'serious matters' such as religion, ethics or academic questions was often found to be as low as 0.5 per cent. In mixed-group settings, however, this percentage for males rose to 15–20 per cent, but with a much smaller rise for females. In addition, men spend about twice as much time talking about their own experiences than those of others, but this ratio is reversed for women. Dunbar concluded that such conversations serve as a type of social lek in which women practise networking for mutual support whereas men act to display their status to the opposite and the same sex (Dunbar, 1996a).

4.4.3 *Game theory: conflicts between rival strategies*

The previous sections have shown how animals seek to optimise their reproductive fitness in relation to ecological parameters. This can sometimes be best appreciated by considering the male and sometimes the female perspective, the approach illuminating how different mating patterns are favoured by different environments. The approach needs, however, to be supplemented by further ideas, for two reasons:

1. The strategies employed by either sex are not frequency independent, a frequency-independent strategy being one that can be pursued regardless of what other individuals in the population are doing. This is not the case in mating behaviour, in which success at finding a mate is strongly dependent on the strategies employed by others

2. The comparative socioecology approach does not sufficiently address the fact that the interests of the two sexes may be in conflict. In cases in which parental care is necessary, males may favour polygyny and females polyandry. So which system results or which strategy prevails?

One way of tackling situations in which two sets of strategies are in conflict is the use of **game theory**, which was pioneered by the British biologist John Maynard Smith (see also Chapter 11). The simplest way to appreciate the theory is to consider two players who each can play one of two or more strategies. The rewards they reap from each strategy depends on what the other player does.

As an illustration, consider the game of rock, paper and scissors. Children wave a fist at each other and then open their hands to reveal one of three possibilities. The scoring matrix is shown in Table 4.3. In this game, what is the best strategy to play? If one player consistently plays only one hand, it can eventually be beaten by some other as every move has another that can defeat it. In the language of game theory, we would say that no pure strategy is evolutionary stable. A population of, for example, rock players could be invaded and wiped out by paper players and so on. In fact, the best strategy is to play each move one third of the time on an unpredictable

BOX 4.1

Pay-off matrix for parental care game (after Maynard Smith, 1977)

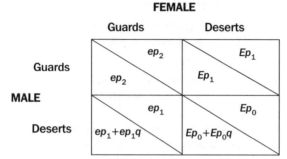

p_1 = survival probability of an egg guarded by one parent

p_0 = survival probability of a deserted egg

E = number of eggs produced by a deserting female

e = number of eggs produced by a guarding female

$E > e$

q = probability of male finding another mate after desertion

p_2 = survival probability of an egg guarded by both parents

Hence $ep_1 q + ep_1$ = number of eggs fathered by a male that deserts the first female and then finds another one

basis. This would be an evolutionary stable strategy (ESS), one that is resistant to being displaced by a rival.

Table 4.3   Scoring matrix for the game of rock, scissors and paper (after Barash, 1982)

	Rock	**Scissors**	**Paper**
Rock	0	+1	−1
Scissors	−1	0	+1
Paper	+1	−1	0

In the case of mating behaviour, two strategies can be considered. A male or a female could mate and then stay to help rear young, or desert and find another mate. Box 4.1 shows hypothetical conditions for a situation in which both sexes have to adopt one of two strategies: either to guard or to desert the offspring.

There are four ESSs depending on the values of the variables:

ESS 1:   male deserts and female deserts

            requires that $Ep_0 > ep_1$ or female guards
            and $p_0 (1+q) > p_1$ or male guards

ESS 2:   male guards and female deserts

requires $Ep_1 > ep_2$ or female guards
and $p_1 > p_0 (1+q)$ or male deserts

ESS 3:   male deserts and female guards

requires $ep_1 > Ep_0$ or female deserts
and $p_1(1+q) > p_2$ or male guards

ESS 4:   male guards and female guards

requires $ep_2 > Ep_1$ or female deserts
and $p_2 > p_1(1+q)$ or male deserts

We can see now that the ecological conditions discussed above can be viewed as setting the values of the various parameters in the models shown in Box 4.1. In the case of Arctic waders, polyandry may result from $Ep_1 > ep_2$ perhaps because the guarding effect of both parents (p_2) is only slightly greater than p_1 but the number of eggs that a female can produce if she leaves the nest to the male (E) is much greater than the number that can be incubated in the nest if she guards (e).

In many cases, if p_2 is not much larger than p_1, it pays males to desert if there is a good chance of finding another female. Females that have already invested heavily in offspring by, for example, internal fertilisation and gestation will be less inclined to desert since the cost of going through the whole process again may be too much to bear, so $ep_1 > Ep_0$.

A classic illustration of the operation of sexual conflict underlying mating behaviour is to be found in the work of Davies (1992) on the common hedge-sparrow or dunnock (*Prunella modularis*). The main ecological determinant of the dunnock's mating pattern is the size of a female's territory, which is in itself a function of the quality of foraging patches in the environment. A variety of mating patterns has been observed for this species: monogamy, polygyny, polyandry and polygynandry. Polyandry is observed when two males share a territory that was occupied by a single female. A dominance hierarchy is found such that the alpha male drives away the beta male from food and the female. In one study, the paternity ratios of these two males, as estimated by observations coupled with DNA fingerprinting, were 0.6 and 0.4 respectively. When two males share a territory occupied by two females, polygynandry results. Conflicts arise because each sex may do best by different mating systems.

Table 4.4 shows a calculation of the reproductive success of males and females under different mating regimes. The crucial point is that, in rich environments, the preferred male pattern is polygyny > monogamy > polyandry, whereas the preferred female pattern is exactly the opposite: polyandry > monogamy > polygyny. During the mating period, males in polyandrous situations are constantly fighting, the dominant male attempting to drive away his rival. Similarly, each polygynous female fights with the other and attempts to claim the male for herself. Polygynandry, involving two males and two females, results when the dominant male is unable to drive the other away and likewise the dominant female is unable to drive away the other female to achieve polyandry.

Table 4.4 Reproductive success of male and female dunnocks
in variable mating conditions (after Davies and Houston, 1986)

Mating pattern	Number of adults caring for young	Nestlings fledged (per female)	Nestlings fledged (per male)
Monogamy	One male, one female	5.04	5.04
Polyandry: both males feeding	One female, two males	6.75	alpha: 4.05 beta: 2.70
Polygyny	One female and part-time help of one male	3.82	7.64

The lessons here are important. The pattern of mating in the dunnock cannot be seen as simply the expression of the preferences of one sex but must instead be understood as the outcome of a conflict of interests between both sexes. The outcome of such contests is determined by the competitive abilities of the individual birds and also the food distribution since this determines range size and hence the ability of male or female dunnocks to monopolise mates (Krebs and Davies, 1991).

Conflicts in human mating strategies

Game theory may be of use in modelling human mating behaviour. If we consider the hunter-gathering days of human evolution, it would probably not pay a female to desert a newborn baby since it is so dependent on a mother's milk. There is, however, an asymmetry that lies at the heart of human reproductive behaviour. A human male could, by mating with for example 50 partners (if he could time his copulations accurately to coincide with ovulation by the female), increase his reproductive output by a factor of 50 compared with mating with one partner. The same argument does not apply to females: mating with 50 males would not increase her reproductive success 50-fold. If the male deserts, however, he faces the prospect that his child may not survive.

Desertion also brings other problems since a population of deserting males would constantly be moving from female to female. Any one male now faces the prospect that he is unsure whether by mating with a woman he becomes the father of the child. If women announced oestrus in the manner of chimps by swellings and pheromones, it would be easy to calculate the best time to impregnate a new female. This would of course lead to male rivalry, and this would also not serve the interests of females who are looking for a little more than sperm from their menfolk. One way in which to thwart a male's philandering intentions is to conceal ovulation. Any one male then does not know how best to time his sexual advances. A great deal of the evolution of human sexuality can be viewed in this light of the dynamics of the different but interacting strategies of males and females. This is explored more fully in Chapter 8.

4.5 The sex ratio: Fisher and after

4.5.1 *Why so many males?*

Let us recall some of the facts of human anisogamy, which are in many respects typical of mammals as a whole. Each ejaculate of the human male contains about 280×10^6 sperm, enough, if they were all viable and suitably distributed, to fertilise the entire female population of the United States. Moreover, they are produced at the phenomenal rate of about 3000 each second (Baker and Bellis, 1995). In contrast, the human female only produces about 400 eggs over her entire reproductive lifetime of 30–40 years. Now, the ejaculate of the male is of course not evenly distributed, and a male must impregnate the same woman many times to have a good chance of fathering a child. Even so, the longer period of fertility experienced by the male, the fact that females are incapable of ovulating when bearing a child or breast-feeding, and the heavy demands of childbearing that fall unevenly on females, all imply that a single male could, in principle and in practice, fertilise many women.

The obvious question that follows from this is why nature has bothered to produce so many men. It would seem that a species would do better in terms of increasing its number by skewing the sex ratio in favour of women, thereby producing fewer men. Men who remained would then be destined to mate polygynously with more women. Yet unfailingly, the ratio of males to females at birth for all mammals is remarkably close to 1:1.

The statistics of polygynous mating seem ever more wasteful. In cases where a few males fertilise the majority of females, such as in leking species, given a 1:1 sex ratio at birth, it follows that some males are not successful at all. In evolutionary terms, it seems as if their lives have been pointless and, for the parents that produced them, a wasted expenditure of paternal effort. It was Fisher who pointed a way out of this conundrum.

4.5.2 *Fisher's argument*

A superficial answer to the question of why roughly even numbers of human males and females are born is that every gamete (oocyte) produced by the female contains an X chromosome but that gametes produced by the male contains either a Y or X chromosome, these two types being produced in equal numbers. Consequently, there is an equal probability of a XX and a XY fusion, and it follows that boys (XY) are just as numerous as girls (XX). This is, in fact, the mechanism used for all mammals and birds (except that in birds the females are XY and the males XX).

This is of course only part of the answer. The X/Y chromosome system provides a proximate mechanism for sex determination, but we know that this is subject to some variation. In humans, it is estimated that, 3 months after conception, the ratio of males to females is about 1.2:1 and that because of the higher in utero mortality of male embryos, the ratio falls to 1.06:1 at birth. It evens out at 1:1 at age 15–20. What we are looking for of course is an ultimate evolutionary argument that explains the adaptive significance of the proximate mechanism. The argument that is now widely accepted was first provided by Fisher in his *The Genetical Theory of Natural Selection* (1930). Fisher's reasoning can be expressed verbally in terms of negative feedback. First we must rid ourselves of the species-level thinking that lies behind the view that species would better off with fewer males. Species might be better off, but selection cannot operate on species. Selection acts on genes carried by individuals, and what might seem wasteful at a group level might be eminently sensible at an individual level.

Consider the fate of a mutant gene that appeared and caused an imbalance of the sex ratio in favour of females. This could take the form of a gene influencing the probability of fertilisation or survival of the XY zygote in a positive way. Or a gene that influenced the number of X and Y gametes produced by the male. Let us further suppose that, for some reason, this gene gained a foothold and shifted the ratio of males to females to 1:2. Consider now the position of parents making a 'decision' (in the sense of the selection of possibilities over evolutionary time) of what sex of offspring to produce. In terms of the number of grandchildren, sons are more profitable than daughters since, in relative terms, a son will on average fertilise two females every time a daughter is fertilised once. More grandchildren will be produced down the male line than down the female line. It therefore pays to produce sons rather than daughters. In genetic terms, the arrival of a gene that now shifts the sex ratio of offspring in favour of males will flourish.

The argument of course also works the other way round. In a population dominated by a larger number of males, it is more productive of grandchildren to produce a female since she will almost certainly bear offspring whereas a male (given that there is already a surplus) may not. We can picture all this in terms of two negative feedback pressures tending to stabilise the ratio at about 1:1 (Figure 4.9).

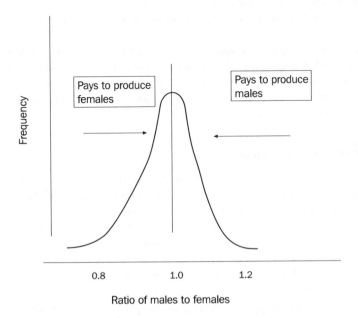

Figure 4.9   Pressures serving to stabilise the sex ratio

The logic of the argument also works for polygamous mating. Suppose only one male in ten is successful and fertilises ten females. It still pays would-be parents to produce an equal number of males and females even though nine out of ten males may never produce offspring because the one male in ten that is successful will leave many offspring and the gamble is worth it. This one successful male will have ten times the fertility of each female.

Fisher's argument has logic in its favour, but what of empirical confirmation? Interesting support for Fisher's theory has come from the work of Conover *et al.* (1990) on a species of fish called the silverside (*Menidia menidia*). The silverside is a small fish common in the Atlantic that has its sex partly determined by the temperature of the water at birth. A low temperature yields females and a high temperature males. This mechanism in itself probably has adaptive significance. A low temperature indicates that it is early in the season, and it is known that an increase in size boosts the reproductive performance of females more than males. Parents should prefer females to males early in the season since this gives the opportunity for growth to increase fertility. Conover *et al.* kept batches of fish in tanks at various constant temperatures. At first, the sex ratio drifted away from 1:1 in the way expected from the temperature effects, but after a few generations it returned to 1:1 in a way expected from the negative feedback effects of Fisher's argument.

4.5.3 *Testing and extending Fisher's argument*

Closer inspection shows that Fisher's reasoning predicts a 1:1 ratio in the following conditions:

1. The two sexes are equally costly to produce
2. Both sexes benefit or suffer equally from environmental variations.

There are other conditions (such as the requirement that mating is between non-relatives and the assumption that a gene cannot alter its own probability of transmission to offspring), which are of less relevance to evolutionary psychology, that we will not consider here (see Maynard Smith, 1989). Instead we will explore the two conditions listed above.

Costs of sons and daughters

If sons and daughters are unequal in their production cost, Fisher's line of reasoning predicts that the sex ratio will be biased in favour of the cheaper sex. The argument has been placed in a rigorous mathematical form but can also be expressed verbally (Maynard Smith, 1989). By 'cost' we mean the extent to which the production and care of an offspring reduces the parents' 'residual reproductive value' (Clutton-Brock *et al.*, 1991). Imagine a female about to produce sons and daughters in a population in which there are an equal number of males and females. Suppose sons are twice as costly as daughters. Within the 1:1 sex ratio prevailing, both sons and daughters face the prospect of equal reproductive success, but it can be seen that a mother could have produced more

BOX 4.2

Illustration of a hypothetical distribution of sex ratio when sons and daughters carry unequal costs

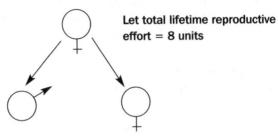

Let total lifetime reproductive effort = 8 units

Cost per offspring	Sons	Daughters
	2	1

Possible balance of offspring

Sons	Daughters	Total effort
0	8	8
1	6	8
2	4	8 *
3	2	8
4	0	8

The evolutionary stable strategy occurs when $m = kf$
* In this case; $m = 0.5 f$; that is, 2 sons and 4 daughters

total offspring and hence more grandchildren if she reduced the number of sons. Natural selection will thus favour a reduction in the output of sons until the extra cost of producing a son is balanced by the benefits arising from the fact that, as males become rare, they bring higher rewards from mating (Box 4.2). Eventually $m = kf$ where:

$$m = \text{number of males}$$
$$f = \text{number of females}$$
$$k = \text{ratio of costs of females to costs of males}$$

This 'extended Fisherian principle' could explain why the human sex ratio of males to females at birth is about 1.06:1 with negligible cross-cultural variation. Statistics show that male mortality is higher in the first few years after birth and before the end of the period of parental care, which implies that boys are cheaper to produce than girls (since, if they die earlier, boys need overall less care). This is compensated for by biasing the sex ratio in favour of males so that the overall total investment in boys and girls becomes the same.

Effect of environmental variations on the sexes: conditional sex expression and the Trivers–Willard hypothesis

Fisher's theory assumed that environmental effects (biotic and abiotic) acted equally on the phenotypes of males and females. In 1973, Trivers and Willard explored the consequences of questioning this assumption. Let us suppose that, in a population of mammals, females vary markedly in their condition and that a female in good condition will produce large, healthy young and females in poor condition, weaker sons. In most mammalian groups, since males compete for females, largeness of size carries more of an advantage for males than for females. In terms of increasing her number of grandchildren, our well-fed female should bias her offspring in favour of sons. This in essence is the **Trivers–Willard hypothesis**: females in good condition should favour sons, females in poor condition should (since the effect of poor condition has a smaller effect on the reproductive prospects of females) favour daughters.

4.5.4 *Empirical tests of the Trivers–Willard hypothesis*

Many biologists and anthropologists have looked assiduously for evidence of conditional sex expression. One problem to be faced is that other factors affect the sex ratio at birth, for example the unequal cost of sons and daughters already considered, these factors often being difficult to disentangle.

The Venezuelan opossum (*Didelphis marsupialis*)

This creature is a large rodent-like marsupial that lives in burrows in central Venezuela. Austad and Sunquist (1986) trapped and marked 40 virgin female opossums. Twenty were given extra nourishment (in the form of sardines left outside their burrows) while the other 20 were left to forage as normal. Every month, the animals were trapped and the sex of the babies in the pouches determined. *Didelphis marsupialis* is a polygynous species, and size in males

Table 4.5   Sex ratio manipulation in the common opossum (after Austad and Sunquist, 1986)

	Control group	Well-nourished group
Number of mothers	20	20
Number of babies	256	270
Ratio of males to females	1:1	1.4:1

carries an advantage in securing a harem. A summary of the results in shown in Table 4.5. The results are in keeping with the Trivers–Willard hypothesis.

Red deer

In a long-term study on the population of red deer on the Isle of Rhum, Clutton-Brock *et al.* (1986) observed the effect of the dominance rank of female red deer on the reproductive success of their offspring. Dominant mothers in the group tend to have better body condition, live longer and breed more successfully than their subordinates.

Because of polygyny, male offspring stand to benefit more from sturdy and well-adapted mothers than do females. It was also found that subordinate hinds were less likely to survive and breed the next season if they bore sons; in other words, sons cost more to produce than daughters. Both these facts, coupled with the Trivers–Willard hypothesis, lead to a prediction that subordinate females should prefer daughters and dominant females sons. Such predictions were supported.

Sex ratio differences within humans and other primates

Some primate species, such as spider monkeys (*Ateles paniscus*), show a sex ratio effect similar to that of red deer but for different ecological reasons. Unlike most monkeys, female spider monkeys leave their natal groups on reaching sexual maturity and males remain at home to breed. High-ranking mothers of *A. paniscus* are then in a position to assist the status of their sons and hence to improve their reproductive prospects. As expected from the predictions of the Trivers–Willard hypothesis, there is found to be a bias towards male offspring in high-status mothers and towards female offspring in low-status mothers (Symington, 1987).

Numerous studies have shown that the human sex ratio is open to various influences, but most of these studies raise more questions than they provide answers. One such is the 'returning soldier effect': men returning from wars are more likely to produce sons than daughters. One easily refuted explanation of this is that it is to replace the missing men. Such a strategy would not make much genetic sense since the new children will mate with their contemporaries rather than with war widows or women of the fighting generation.

As a broad and perhaps tentative generalisation, we could classify the spectrum of human societies as 'female exogamous patriarchies' in that females more often than males leave the home to marry, and sons inherit their father's (or sometimes mother's) status more so than daughters. On this basis, we would predict that high-status men and women should bias their offspring in favour of males. The question, which remains largely unresolved, is whether or not they do. Mueller (1993) has provided evidence suggesting that high-status males tend to father more sons than daughters (Table 4.6). The mechanism by which this effect is achieved is unclear. One interesting possibility is that a gonadotrophic hormone in the mother can increase the proportion of girls, and testosterone in fathers the proportion of boys. If social rank has a phenotypic effect on hormone production, here at least is a mechanism for sex determination. Other proximate

Table 4.6 Sex ratios of the children of high-status males (data from Mueller, 1993)

Sample population	Source	Sons	Daughters	Ratio of sons to daughters	Expected average	Significance
US elite 1860–1930 1014 males	American Who's Who	1180	1064	1.109	1.06	P< 0.005
German elite 1830–1939 1757 males	German Who's Who	1473	1294	1.138	1.0512	P< 0.001
British industrialists 1789–1925 1179 males	Jeremy, D. J. (1984–1986) Dictionary of Business Biography	1789	1522	1.1754	1.06	P<0.001

mechanisms could include a differential mobility or survival of the X and Y sperm, or a differential mortality of male and female embryos. Daniela Sieff (1990) provides an excellent review of biased sex ratios in human populations.

SUMMARY

■ Sexual reproduction carries costs and benefits for individual organisms. The formidable cost of sex is probably offset by the genetic variation conferred on offspring: genes for sex may find themselves in new winning combinations. Such variation is invaluable in enabling organisms to compete with others.

■ At a superficial level, the mating behaviour of animals can be described in terms of species-characteristic mating systems. A deeper understanding is gained, however, by looking at the strategies pursued by individuals as they strive to maximise their reproductive success.

■ Such strategies are influenced by a range of factors, for example dispositions inherited from distant ancestors (phylogenetic heritage), ecological conditions and the behaviour and distribution of the opposite sex. In many cases, mating should be seen as the outcome of competition between the two sexes as they pursue different fitness-maximising strategies.

■ Even where there is to be found considerable variance in the reproductive success between males and females, the sex ratio remains remarkably close to 1:1. The best ultimate explanation of this so far is that of Fisher, who suggested that natural selection gives rise to stabilising feedback pressures tending to maintain unity. Departures from a 1:1 ratio are interesting in themselves and may have adaptive significance.

KEY WORDS

Anisogamy ■ Asexual reproduction ■ Clone ■ Conspecifics
Evolutionary stable strategy ■ Game theory ■ Gamete
Isogamy ■ Lek ■ Lineage ■ Major histocompatability complex
Parthenogenesis ■ Polygamy ■ Sex ratio ■ Sex role reversal
Trivers–Willard hypothesis ■ Zygote

FURTHER READING

Alcock, J. (1998) *Animal Behaviour: An Evolutionary Approach*. Sunderland, MA, Sinauer Associates.
A good general book on evolution and animal behaviour. Contains only one short chapter on humans, but see Chapters 10, 12 and 13 for mating theories.

Rasa, A. E., Vogel, C. and Voland, E. (1989) *The Sociobiology of Sexual and Reproductive Strategies*. London, Chapman & Hall.
A useful series of case studies on humans and other animals.

Short, R. V. and Balaban, E. (1994) *The Differences Between the Sexes*. Cambridge, Cambridge University Press.
A valuable series of specialist chapters by experts. Covers humans and non-humans.

5

Sexual Selection

The senses of man and of the lower animals seem to
be so constituted that brilliant colours and certain
forms, as well as harmonious and rythmical sounds,
give pleasure and are called beautiful; but why this
should be so we know not

(Darwin, 1871)

This chapter examines the selective force that operates on males and females as a result of the phenomenon of sexual reproduction and its outcome in shaping the behaviour and morphology of animals. The selective force is a consequence of the fact that a sexually reproducing animal has to surmount a number of hurdles before it can be confident that its gamete has fused with another. It must find a mate, make a judgement on its suitability as a prospective partner and be judged in turn. Once copulation has taken place, competition between males is not necessarily over: the reproductive tract of the female may carry sperm from other males who have also reached this far. Competition now shifts to the level of sperm itself. Sperm competes against sperm in the struggle to fertilise the egg. Even here, the female is no passive recipient and may herself exert some choice over the sperm she wants to retain.

In this chapter, we will show how all these hurdles have left their mark on the physical and behavioural characteristics of animals. Features such as size, behavioural tactics, colouration, the possession of appendages for fighting and the number and type of sperm produced by males, have all been moulded by the force of sexual selection. The theoretical principles established in this and the preceding chapter are then applied to humans in Chapters 8 and 9.

5.1 Finding a mate

5.1.1 Natural and sexual selection

Natural and sexual selection form the twin pillars of Darwin's adaptationist paradigm. We should be wary, however, of overstating the distinction between these two forms of selection: whether you survive to reproduce because you can run fast to avoid predators, or because you are successful in attracting mates, the same principle of the differential survival of genes is in operation. Indeed, some features that help in avoiding predators, such as body size, may also be of assistance in securing a mate. One way to view the distinction is shown in Box 5.1.

BOX 5.1

Natural and sexual selection as components of total selection

Selection: The differential survival of genes

Natural selection: Traits favoured by non-sexual aspects of survival, for example the avoidance of predators and metabolic efficiency

Natural and sexual selection: Traits favoured by natural and sexual selection, for example size, pathogen resistance, symmetry and motor co-ordination

Sexual selection: Traits favoured by sexual selection (competing for mates) but disfavoured by natural selection, for example bright colours and courting displays.

5.1.2 *Inter- and intrasexual selection*

As a rough guide, the degree of choosiness that an individual displays in selecting a partner is related to the degree of commitment and investment that is made by either party. Male black grouse that provide no paternal care will mate with anything that resembles a female black grouse, but females, mindful of their onerous parental duties, are more discriminating. Likewise, male chimpanzees provide little care for their young and are consequently not particularly discrim-inating in their choice of mate – as long as the female has that irresistible pink swelling announcing oestrus. A male albatross, in contrast, will mate for life and is consequently very choosy about his choice of partner. Among humans, both males and females have a highly developed sense of male and female beauty, and this aesthetic sensibility is similarly consistent with a high degree of maternal and paternal investment. The more investment that an individual makes, the more important it becomes to choose its mate carefully. All this decision-making results in a selective force, complementary to natural selection, that is known as sexual selection.

We should really distinguish between two types of sexual selection. For reasons already given (see Chapter 4), the sex ratio usually remains close to 1:1, so where conditions favour polygyny, males must compete with other males. This leads to **intrasexual selection** (intra = within). Intrasexual competition can take place prior to mating or after copulation has taken place. On the other hand, a female investing heavily in her offspring or capable of raising only a few offspring in a season or a lifetime needs to make sure that she has made the right choice. There will probably be no shortage of males, but the implications of a wrong choice for the female are graver than for the male, who will be seeking other partners anyway. Females in these conditions can afford to be choosy. This leads to **intersexual selection** (inter = between) (Figure 5.1). The next section examines the intrasexual selection that results from competition before copulation.

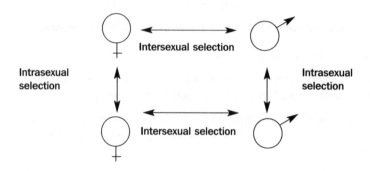

Figure 5.1   Inter- and intrasexual selection

5.2 Consequences of intrasexual selection

5.2.1 *Sexual dimorphism: size and weaponry*

Fighting between males over access to females is a sight so common that Darwin called it the 'law of battle'. Such contests are often spectacular affairs and provide good footage for the makers of natural history films. Darwin argued that intrasexual selection was bound to favour the evolution of a variety of special adaptations, such as weapons, defensive organs, sexual differences in size and shape and a whole range of subtle devices to threaten or deter rivals. Some of these structures will be expensive for their possessors and may reduce their ecological (as opposed to sexual) viability.

Such intrasexual contests will lead to arms races since mating success is a product of relative rather than absolute size. It could be imagined that as the escalation of body size or weaponry proceeded through each generation, so natural selection would increase the mortality of males and as a result adult males could begin to become rare. It has even been suggested that the trend towards larger size among some ancient mammals may have led to their extinction (Maynard Smith and Brown, 1986). The whole process of size escalation dampened by natural selection has been modelled on a number of occasions (see, for example, Parker, 1983), the most common result being that, at stable equilibrium, male traits are distributed polymorphically about a mean that is shifted from the ecological optimum. In short, sex has led males to grow too large, and burdened them with appendages that are too demanding for their own ecological good.

The importance of size is illustrated by a number of seal species. During the breeding season, bull elephant seals (*Mirounga angustirostris*) rush towards each other and engage in a contest of head-butting. Such fighting has led to a strong selection pressure in favour of size, and male seals are consequently several times larger than females. Elephant seals are in fact among the most sexually dimorphic of all animals. In the northern elephant seal (*M. angustirostus*), a typical male is about three times heavier than a typical female. The mating system is described as female defence polygyny and, to defend a sizeable group of females, a male needs to be large. Competition between males is intense and many males die before reaching adulthood without ever having mated. The variance in the reproductive success of males is correspondingly large (Figure 5.2).

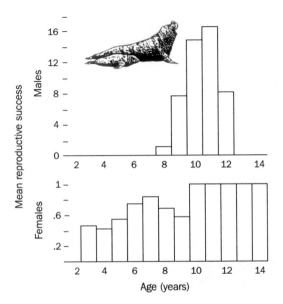

Figure 5.2   Mean reproductive success (measured by number of pups weaned) of male and female elephant seals (*Mirounga angustirostris*) (after Le Boeuf and Reiter, 1988, in Andersson, 1994)

In male–male contests, it may also benefit males to possess fighting weapons. The walrus, elephant and hippopotamus all carry conspicuous tusks. Of the 40 species of deer left today, 36 develop antlers, and in 35 of these species antlers are an exclusively male characteristic. In the case of the European red deer (*Cervus elaphus*), the body weight of males is about one and a half times that of the females, and the males carry large antlers. Males battle with their antlers during the rutting season, but during the rest of the year they are tolerant of each other and often move about in groups. It seems probable that antlers also serve as symbols of dominance (Lincoln, 1972).

Some of the most spectacular examples of such weapons are found in beetles such as the stag beetle. The males have large horn-like jaws, absent in the females, which are used only to fight other males. Such differences between males and females are referred to as **sexual dimorphism**. As one would expect, the greater the prize in intramale contests, the greater the degree of sexual dimorphism. Figure 5.3 shows how the degree of dimorphism in body size relates to the size of a harem for seals and ungulates.

(a)

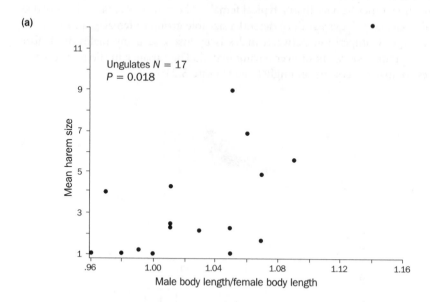

(b)

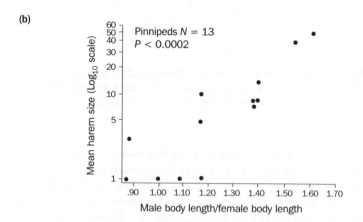

Each data point represents a species. Harem size estimated from females per breeding male or number of copulations per male

Figure 5.3   Sexual dimorphism in relation to harem size for (a) ungulates (deer and antelope) and (b) pinnipeds (seals) (Alexander et al., 1979)

Figure 5.4   Intrasexual selection in action:
male southern elephant seals (*Mirounga leonina*) fighting

The victor in this contest will reap the genetic benefits of fertilising a harem of
several dozen females. The value of the prize ensures that such contests are fierce
and that males are about seven times heavier than females

5.2.2 *Variance in reproductive success – Bateman's principle*

Bateman (1948) first quantitatively documented the differential variance in reproductive success between male and female animals (as illustrated in Figure 5.2 above for elephant seals) in his classic work on *Drosophila*. The greater variation in the reproductive success of males compared with females has become known as Bateman's principle.

Bateman's work was neglected until it was revived in 1971 by Trivers, then a graduate student, in a paper presented at a symposium commemorating the centenary of Darwin's 1871 work on sexual selection. Trivers illustrated Bateman's principle by his own work on the Jamaican lizard (*Anolis garmani*). Trivers found that the variance in reproductive success was larger for males than females and that large males tended to have more reproductive success (Trivers, 1972). The key concept to note here is that a difference in the variance in reproductive success between males and females indicates the operation of **intrasexual competition**. If male variance is greater, this tends to suggest that the mating system is averaging out at polygyny. These signs and principles are important when we come to examine human mating behaviour in later chapters.

It would be wrong to conclude that males are always larger than females. A useful test of the principle that large body size is favoured in the sex that competes for the other is to look at cases in which the usual sex roles are reversed. If we find species where males invest more than females in the production of offspring, we would predict that female reproductive success should vary more than that of males, that females should be larger than males and that males should be careful in their choice of mating partner. A number of species are known that bear out these predictions fairly well. In the common British moorhen (*Gallinula chloropus*), for example, males perform about 72 per cent of the incubation and lose about 10 per cent of their body weight as a consequence. Petrie (1983) has observed that competition for mates is more intense among females than males and that heavier females win fights more often. Females are larger than males and occasionally mate polyandrously.

Intrasexual competition is not the only cause of sexual dimorphism. Darwin suggested in 1871 that the larger size of some female animals could be a consequence of the fact that large size favours increased egg production. In most

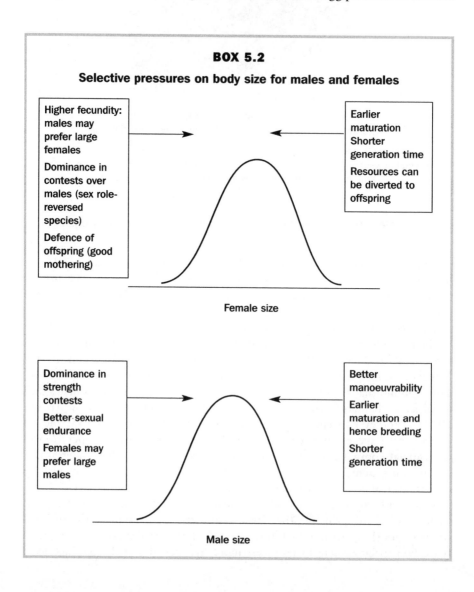

BOX 5.2

Selective pressures on body size for males and females

Higher fecundity: males may prefer large females

Dominance in contests over males (sex role-reversed species)

Defence of offspring (good mothering)

Earlier maturation
Shorter generation time

Resources can be diverted to offspring

Female size

Dominance in strength contests

Better sexual endurance

Females may prefer large males

Better manoeuvrability

Earlier maturation and hence breeding

Shorter generation time

Male size

mammalian species, males are larger than females, and intrasexual competition seems able to account for this, but there are some species, for example rabbits, hares and two tribes of small antelope (*Cephalophini* and *Neotragini*), in which females are the larger sex. In these cases, natural selection seems to override sexual selection. Ralls' explanation is the 'big mother hypothesis' that large mothers can produce larger babies with a better chance of survival. A larger mother can also provide better maternal care, such as defending and carrying the young (Ralls, 1976). A summary of the selective factors that influence body size in males and females is shown in Box 5.2.

Another physiological characteristic that can assist males is 'sexual enthusiasm' or the capacity to be easily aroused. In many polygynous species, males have a low threshold for sexual arousal. Some species of frog will, in the mating season, cling to anything that resembles a female frog, and males will often attempt to mate with the wrong species and even the wrong sex. Another feature of the sex drive of the male is the 'Coolidge effect', so named after the United States' President Coolidge. The story goes that, while visiting a farm, President and Mrs Coolidge were shown a yard containing many hens and only one cockerel. When Mrs Coolidge asked why only one cockerel was necessary, she was told that he could copulate many times each day. 'Please tell that to the President', she said. When the President was informed, he asked whether the cockerel copulated with the same hen and was told no; his reply was 'Tell that to Mrs Coolidge' (Goodenough *et al.*, 1993). The Coolidge effect has been observed in many species. In a study on Norway rats (*Rattus norvegicus*), Fisher (1962) found that whereas a male rat with a single female reached sexual satiation after about 1.5 hours, some males could be kept sexually active for up to 8 hours by the introduction of novel females at appropriate intervals.

5.3 Parental investment, reproductive rates and operational sex ratios

5.3.1 Problems with the concept of parental investment

When Trivers advanced his concept of **parental investment** in 1972, it seemed to promise, and indeed to a degree did deliver, a coherent and plausible way of examining the relationship between parental investment, sexual selection and mating behaviour. The sex that invests least will compete over the sex that invests most, while the sex that invests most will have more to lose by a poor match and will thus be more exacting in its choice of partner.

Trivers defined parental investment as:

> any investment by the parent in an individual offspring that increases the offspring's chance of surviving (and hence reproductive success) at the cost of the parent's ability to invest in other offspring. (Trivers, 1972, p. 139)

Using this definition, Trivers concluded that the optimum number of offspring for each parent would be different. In the case of many mammals, a low-investing male will have the potential to sire more offspring than a single female could produce, and a male will therefore increase his reproductive success by increasing the number of his copulations. It has proved very difficult, however, to measure such terms as increase in 'offspring's chance of surviving' and 'cost to parents', so deciding which sex invests the most is not always easy.

In considering the expenditure of time and energy that an organism makes in mating, it is helpful to divide reproductive effort into mating effort and parenting effort such that:

$$\text{Total reproductive effort = Mating effort + Parenting effort}$$

We have already noted that, as a broad generalisation for mammals, we could say that mating and parenting efforts are unequally distributed across each sex. Although this distinction is useful, difficulties still arise in deciding how to allocate particular activities to the two categories. A **nuptial gift** from a male to a female or the effort a male makes in guarding a valuable resource could be thought of as mating effort since it enables him to attract females, or as parental effort in that his offspring may benefit from the resources provided. One concept that may help to circumvent these difficulties is that of potential reproductive rates.

5.3.2 *Potential reproductive rates: humans and other animals*

It can be misleading to focus on anisogamy as a clear sign of an unequal invest-ment by males and females. It is true that males produce sperm in vaster numbers than females produce ova, and that sperm are minute compared with eggs, but we must remember that males deliver millions of sperm, together with seminal fluid, in the hope of reaching one egg. In terms of energy investment, that of the ejaculate of the male probably exceeds that needed to produce one egg by the female. In mammals, it is not the size of the egg but the involvement in gestation and nurturing that places limitations on the reproductive opportunities of females.

On this theme, Clutton-Brock and Vincent (1991) have suggested that a fruitful way of understanding mating behaviour is to focus on the potential offspring production rate of males and females rather than trying to measure investment *per se*. As a guide to understand sexual selection, these authors suggest that it is important to identify the sex that is acting as a 'reproductive bottleneck' for the other. This approach works particularly well for some species of frog, bird and fish in which males are responsible for parental care. In some of these species, brightly coloured large females compete for smaller, duller males while in others, large, brighter males compete for choosy females – even though the paternal care of offspring (and thus high male parental investment) is common to both sets. Table 5.1 shows two cases in which, even though males provide parental care, it is the reproductive rate rather than the amount of parental investment that predicts selection.

In the same study, data on the potential reproductive rates of males and females from 29 species in which high levels of paternal care are found were extracted. The results are summarised in Table 5.2. The general conclusion is that the sex with the highest potential reproductive rate competes for that with the least and that this is therefore a better predictor of competition than is investment as such.

Table 5.1   Two examples of sexual selection in relation to the potential
reproductive rate of the two sexes (data from Clutton-Brock and Vincent, 1991)

Species	Behaviour	Ratio of female to male reproductive rate	Dimorphism/ sex-specific behaviour
Three-spined stickleback (*Gasterosteus aculeatus*)	Males can guard up to ten clutches of eggs at any one time on their own territory Females can lay only one clutch every 3–5 days	<1	Males brightly coloured
Pipe-fish (*Nerophis ophidion*)	Males carry fertilised eggs Females can lay more eggs in a season than males can carry	>1	Females compete for males

Humans are a special case in point in that the range of parental investment
possible from a male ranges from near zero, if the male deserts, to equal or more
than that of the female. Given this wide range, it is difficult to measure the investment that human males make. One could turn to hunter-gatherer tribes, but
again there is cross-cultural diversity as well as variation within a culture. Another
approach might be to look at the potential reproductive rate. The record often
claimed for the largest number of children from one parent is 888 for a man and
69 for a woman. The father was Ismail the Bloodthirsty (1672–1727), an
Emperor of Morocco, the mother a Russian lady who experienced 27 pregnancies
with a high number of twins and triplets. It is a safe bet that you are more
astonished by the female record than the male.

Table 5.2   Intersexual competition for mates in cases of high paternal
investment in relation to reproductive rates out of total of 29 species
examined (after Clutton-Brock and Vincent, 1991)

Ratio of female to male reproductive rate	Competition for mates more intense in males	Competition for mates more intense in females
<1	Fish 10 species Frogs 3 species	0 species
>1	Fish 1 species Birds 1 species	Fish 3 species Birds 11 species

The figure of 888 looks extreme compared with most cases of fatherhood but
would *prima facie* seem to be a practical possibility. Ismail died at the age of 55

and could have enjoyed a period of fertility of 40 years. Over this time, he could have had sex with his concubines once or twice daily. The record claimed for Ismail has, however, recently been questioned by Dorothy Einon of University College London (Einon, 1998). She analyses the mathematical probability of conception by members of his harem. The problem for a breeding male with access to a large number of females is, first, that he is uncertain when they are ovulating. The fact that ovulation takes place 14–18 days before the next menstruation was not known until 1920. Copulating with a woman once every day over her ovarian cycle would only give a probability of hitting the right day of about 10 per cent, which could be raised to 15 per cent if days of menstruation were avoided. Second, only half of all menstrual cycles are fertile. Further reductions then have to be made for probabilities of conception, implantation and miscarriage. The end result becomes that if Ismail had coitus three times per week, without interruption caused by illness or exhaustion, he would have produced a lifetime total of 79 children, and with coitus 14 times per week a total of 368 children.

It is of course possible that men subconsciously know when ovulation is taking place and are thus able to target their reproductive efforts better. Even so, Einon's calculations give us pause to reflect that the male reproductive rate is not as high at it may seem at first sight. It is reduced of course by the concealment of ovulation. If Ismail knew exactly when women were ovulating, he could direct his efforts accordingly. The concealment of ovulation may have evolved as a tactic by females to elicit more care and attention from males (see Chapter 8).

It is probably true to say that, in modern *Homo sapiens*, the limiting factor in reproduction resides marginally with the female. This would by itself predict some male versus male competition and sexual selection, and certainly an increased intensity of these factors in the evolutionary lineage of the hominoids before ovulation was concealed. We should also note that most men in history have not been emperors, and the harem that Ismail enjoyed would not have been a regular feature of our evolutionary past. These issues are examined again in Chapter 8.

5.3.3 *The operational sex ratio*

The potential reproductive rate and the **operational sex ratio** are closely related concepts. In Chapter 4, it was suggested that the spatial distribution of females determines the environmental potential for polygyny. Spatially clumped females could in principle be monopolised by a male who could thereby achieve polygyny, but such spatial concentration is only useful to the male if the females are fertile. In this respect, we can think of females as also being temporally distributed in the sense of their sexual receptivity at any one time. This idea is contained in the concept of the operational sex ratio:

$$\text{Operational sex ratio} = \frac{\text{Fertilisable females}}{\text{Sexually active males}}$$

When this ratio is high, females could be more willing to mate than males and might engage in competition. If the male pipe-fish, as shown in Table 5.1, can hold fewer eggs in a season than a female can lay, the operational sex ratio exceeds 1 and is said to be female biased. In these circumstances, females will compete for males. When the operational sex ratio is low, this situation is reversed and males will vie with other males for the sexual favours of fewer females.

Of the many factors that can affect the operational sex ratio, food availability is one of the most important. Studies on the orthopteran katydids or bush crickets (*Tettigonidae*) have shown this effect dramatically. Male katydids transfer sperm in a large nutritious spermatophore during copulation. This nuptial gift provides an important source of food for the female and affects her fecundity (Gwynne, 1988). A number of studies have been carried out on these creatures. The overall effect of food availability on male and female behaviour across several species of bush cricket is shown in Table 5.3.

Table 5.3   Food supply, operational sex ratios and intraspecific competition among various species of bush cricket (data from various sources, reviewed by Andersson, 1994)

	Food plentiful	**Food scarce**
Males	Rapid production of spermatophores	Production of spermatophores slow and difficult
Operational sex ratio	Reduces, that is, male biased	Increases, that is, female biased
Competition	Reduction in female competition for males	Females compete for males
Male investment	Low*	High*
Reproductive bottleneck	Females	Males

*The fact that we can say that male investment is high when food is scarce and the production of spermatophores is slow stresses the need to define investment carefully. When food is limited, a small spermatophore produced by a male may represent a high investment of time and energy.

The bush crickets illustrate how food supply, operational sex ratios and sexual competition are intertwined (Table 5.3). A general model of the way in which these and other factors may influence the operational sex ratio, and the way this in turn affects mating competition, is summarised in Figure 5.5. The manner in which intrasexual competition varies with the operational sex ratio can also be conjectured along the lines shown in Figure 5.6.

5.3.4 *The operational sex ratio and humans*

Using the concept of the reproductive bottleneck, it is tempting to say that females are the limiting resource for male fecundity. After all, a man could impregnate a different woman every day for a year whereas over this same period a woman can become pregnant only once. But we need to proceed carefully. Imagine a male that mates with 56 different women over 56 days and a female that mates with 56 different men over the same period. The woman is likely to

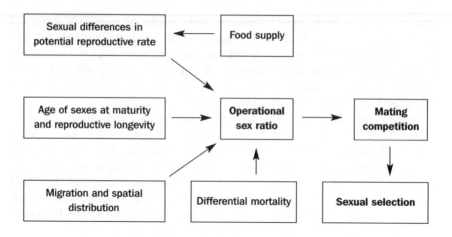

Figure 5.5 Influences on the operational sex ratio and
the relationship between this and sexual selection

become pregnant and bear one offspring in the same year. Using the reasoning
advanced by Einon earlier, if the male avoids the time of menstruation, he has
about a 15 per cent chance of impregnating a woman during her fertile period.
Only half of the female ovarian cycles will be fertile, some women will be infertile
themselves anyway, and implantation will only take place about 40 per cent of the
time. The number of women a man could expect to make pregnant is one. Over
a year, this could be raised to about 6 $(365/56 = 6.5)$. Women are in one sense a
limiting resource but not to the extreme sometimes claimed. It is, however,
significant that men engage in competitive display tactics and are more likely to

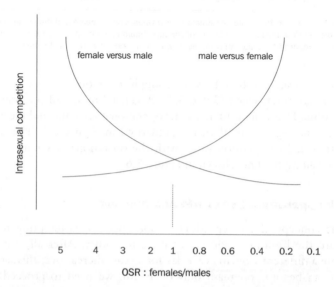

Figure 5.6 Intrasexual competition and the operational sex
ratio (OSR) (adapted from Kvarnemo and Ahnesjo, 1996)

take risks than are women. It is also men who tend to pay for sex – one way of increasing the supply of the limiting resource.

The operational sex ratio (females to males) for a group of humans with a 1:1 sex ratio will be less than one if we measure it in terms of males or females that are fertile. There will be more sexually fertile males than females; this arises from the fact that men experience a longer period of fertility compared with women. It is counterbalanced to some degree, but not entirely, by the higher mortality rate for men than women. The picture is, however, complicated if the population is growing. In these circumstances, the fact that women tend to prefer to marry slightly older men will mean more younger women than there are slightly older men available, since the cohort of marriageable men will be smaller than the number of marriageable women in the expanding cohort below. Guttentag and Secord (1983) have argued that this can in itself be a contributory factor to the development of social mores. In the United States from 1965 to the 1970s, because of the post-war baby boom, there was an oversupply of women, with the effect of decreasing male–male competition and increasing female–female competition. This allowed men to pursue their own reproductive preferences, especially in terms of an increased number of partners, to a greater extent than women could pursue theirs. These authors suggest that this could be a factor contributing to the liberal sexual mores of those decades, characterised by a high divorce rate, a lower level of paternal investment and a relaxed attitude to sex. They stress that sex ratios by themselves are not a sufficient cause for such social changes but that they may be part of the equation (Guttentag and Secord, 1983).

A more realistic application of sex ratio thinking may be found in the analysis of traditional cultures in which social values shift less rapidly. In South America, there are two indigenous Indian groups with a different sex ratio. The Hiwi tribe shows a surplus of men while the Ache people have a sex ratio of females to males of about 1.5 (Hill and Hurtado, 1996). The ecology of the two groups is otherwise similar, but whereas among the Ache people extramarital affairs are common and marriages are unstable, marital life is more stable among the Hiwi. This pattern is what one would expect from the anticipated effect of sex ratio on mating strategies.

A man who attempts to impregnate several women in the course of a year faces another problem apart from the concealment of ovulation: other men will probably want to do the same thing. The reproductive tract of a female could therefore contain the sperm of more than one male at any one time. This is where sperm competition begins.

5.4 Post-copulatory intrasexual competition

5.4.1 Sperm competition

It may seem that once copulation has taken place, intrasexual competition is over; one male must surely have won. The natural world, however, has more surprises in store. Some females mate with many males and retain sperm in their reproductive tracts; such sperm compete inside them to fertilise the egg. The concept of **sperm competition** illuminates many features of male and female anatomy in non-human animals. Male insects are particularly adept at neutralising or displacing sperm already present in the female. The male damselfly (*Calopteryx maculata*) has evolved a penis designed both to transfer sperm and, by means of backward-pointing hairs on the horn of the penis, to remove any sperm already in the female from a rival male.

We should not, however, think of females as passive in this process of sperm competition. A female may choose to mate with many males (as is the case with female chimps) to ensure that the sperm reaching her egg is competitive. The female may also exercise choice over the sperm once it is inside her (Wirtz, 1997). Female sand lizards (*Lacerta agilis*) have been reported to accept sperm from nearly every male that courts them, including close relatives. Given that mating between close relatives is likely to reduce genetic fitness, it is a plausible hypothesis that females exert some control over the sperm that they eventually allow to fertilise their eggs. Olsson *et al.* (1996) have found support for this. When males are genetically similar to females, the probability of producing an offspring from this union is reduced.

Many female insects store sperm that they use to fertilise their eggs at oviposition (egg-laying) as the eggs pass down the female's reproductive tract. It has been suggested that the function of the female orgasm in humans is to assist the uptake of sperm towards the cervix (Baker and Bellis, 1995). Thornhill *et al.* (1994) carried out a study to show that the bodily **symmetry** of the male is a strong predictor of whether or not a female will experience a copulatory orgasm. Symmetry is thought to be an indicator of genetic fitness and the possession of a good immune system (Thornhill *et al.*, 1994). The orgasm therefore ensures that sperm from exciting and desirable males, who are presumably genetically fit and unlikely to be transmitting a disease, stand a good chance of meeting with the female's egg. In a sense, the human female is extending her choice beyond courtship (Baker and Bellis, 1995).

In some ways, all intrasexual competition is a form of sperm competition in that the winning male is the one which produced the sperm that fertilises the egg. The term 'sperm competition' is, however, usually reserved for interejaculate competition, that is, competition between sperm from different males after ejaculation. Sperm competition also takes place in species in which the eggs are fertilised outside the female's body, as in many invertebrates, fish and amphibians, since several males can release their sperm in the vicinity of eggs from the female.

The theory of sperm competition was first developed by Geoffrey Parker, a biologist at the University of Liverpool, in a classic series of papers (Parker, 1970). Its importance was first recognised in insects, but the advent of sophisticated paternity assignment techniques such as DNA fingerprinting has revolutionised the detection of sperm competition. There is now abundant evidence for the existence of sperm competition in virtually every animal taxon (Birkhead and Parker, 1997).

5.4.2 *Anisogamy and sperm warfare*

The theoretical rationale for the production of a large number of small mobile sperm by males has already been discussed in the context of anisogamy (see Chapter 4). In this respect, one could argue that the very existence of two sexes is a result of sperm competition. Anisogamy probably began in conditions of external fertilisation, and the advantages of small mobile sperm are clear to see. However, in situations where a single male fertilises an egg internally, we may question the point of retaining anisogamy. If there are no competitors, the millions of sperm produced by monogamous mammals are simply wasted. So why not return to isogamy and have males producing larger gametes, say 50 per cent of the optimum size for a zygote? The answer is, as pointed out by Parker (1982), that only a mild degree of sperm competition is required to maintain anisogamy.

Consider cattle. A cow's ovum is typically about 20 000 times larger than a bull's sperm. If a mutant bull were to appear that reduced the number of sperm by half but increased their size by a factor of two, the biomass of the fertilised egg would be increased by 1 in 20 000, or 0.005 per cent – a trivial amount. If this bull and another now mate with the same cow, and sperm competition takes place, the chances of the mutant bull (assuming that the larger sperm behave the same as the smaller ones) are reduced from 50 per cent to 33.3 per cent – a significant reduction. It follows from this argument that even a mild degree of sperm competition can maintain anisogamy. It also follows that animals that are anisogamous and classified as monogamous have developed monogamy fairly recently in their evolutionary lineage or else are not 100 per cent monogamous.

The more sperm produced, the greater the chance of one finding an egg: 50 million sperm are twice as effective as 25 million and so on. In situations in which sperm competition is rife, we would expect males to produce more sperm than are typically produced when sperm competition is less intense. This could be expected to apply both within and across species. This prediction has been supported indirectly between species by measurements on the level of sperm expenditure as measured by testis size. Species facing intense sperm competition have larger testes than those in which sperm competition is less pronounced (see Chapter 8). Measurements of the size of sperm do not, however, fit so neatly and it may be that the ability of larger sperm to swim faster is a confounding variable.

Within a given species, it is at least a theoretical possibility that males could adjust the number of sperm they ejaculate according to the risk of sperm competition. Given that it requires energy to produce sperm, it would be in the interests of any male to reduce the number of sperm introduced into a female when he suspects the level of competition to be low – as with a socially monogamous partner that he has guarded – and conversely increase the number in the case of **extrapair copulation** with an already mated female.

Baker and Bellis at Manchester University provide evidence to support the idea that the number of sperm in the ejaculate of men is adjusted according to the probability of sperm competition taking place. In one study, when couples spent all their time together over a given period, the male was found to ejaculate about 389×10^6 sperm during a subsequent sexual act. When the couple only spent 5 per cent of their time together, men typically ejaculated 712×10^6 sperm. Baker and Bellis interpret this as being consistent with the idea that the male increases the number of sperm in the latter case to compete better against rival sperm in the case of any infidelity on behalf of the female. Baker and Bellis have been successful in generating new ideas in an area of research that faces innumerable experimental and ethical difficulties (Baker and Bellis, 1995). They have also been successful in disseminating their ideas, helped partly by a prurient media and partly by the popularisation of their work in such books as *Sperm Wars* (Baker, 1996). Some aspects of their work, especially the sensational and lurid presentation in the latter work, have caused some concern in academic circles (Birkhead *et al.*, 1997).

In the 'sperm wars', males can adopt various tactics: they can produce sperm in large number, attempt to displace rival sperm, insert copulatory plugs or produce sperm that actively seek out to destroy rivals. In moths and butterflies (the Lepidoptera) males produce two types of sperm. One type, 'eupyrene' sperm, carries genetic material and can fertilise eggs. The other type, 'apyrene' sperm, typically furnishes half the number in any ejaculate but is lacking in genetic material and thus cannot fertilise the female egg. The function of apyrene sperm is something of a mystery. One intriguing hypothesis proposed by Silberglied *et al.* (1984) is that apyrene sperm play a role in sperm competition. Either they 'seek and destroy' active sperm from other males, or they serve as a cheap 'filler' that reduces the receptivity of females to further matings. Baker and Bellis (1995) have developed this into a 'kamikaze sperm hypothesis', claiming that a wide variety of animals, including humans, produce sperm whose function is to block or destroy rival sperm.

Copulatory plugs are plugs of a thick, viscous material left by the male in the reproductive tract of the female. They could be a functionless artefact of insemination, but it seems much more probable that they either help to seal in the sperm from the last male to deposit sperm, or serve as 'chastity enforcers' to reduce the likelihood of successful insemination by a rival male (Voss, 1979). In the common honeybee (*Apis mellifera*), males produce mating plugs that attempt to seal off the reproductive tract of the queen to prevent further injections of sperm from competitors. Interestingly, the female is not as concerned to secure the sperm from only one male, and in this case the queen still manages to carry multiple-origin sperm in her body. In the case of deer mice (*Peromyscus maniculatus*), experiments by Dewsbury (1988) suggest that the plug serves to retain the sperm from a male. Perhaps the most determined example of copulatory plugging occurs when males of the fly *Johannseniella nitidia* leave behind their genitalia while the rest of their body is eaten by the female. Baker and Bellis (1995) claim that there is evidence for somewhat less extreme copulatory plugging in human mating.

Long before sperm competition takes place, however, a male has to be accepted by a female or vice versa. Passing this test of approval has also left its mark on anatomy and behaviour. It has given rise, of course, to the force of intersexual selection, and it is to this process that we now turn.

5.5 Intersexual selection

5.5.1 *Mechanisms of intersexual selection*

In sexually reproducing species, the outcrossing of genes is the gateway through which all genes must pass. The fusion of gametes to yield a fertilised zygote represents a type of genetic rite of passage. Darwin realised that, during all the preliminaries to this process, when mate choice takes place the preferences of one sex can exert a selective pressure on the behaviour and physical features of the other. This is the essence of intersexual selection.

Darwin's suggestion that female choice could over time bring about an extreme change in the appearance of males was poorly received in the patriarchal climate of Victorian Britain (Cronin, 1991). Darwin's insight has, however, emerged triumphantly since the 1970s and is now the basis of a flourishing school of research. As noted in Chapter 2, Darwin had difficulty in explaining in adaptionist language why females find certain features attractive. Numerous ancillary theories have emerged recently to address this problem, tending to fall into two schools: the 'good sense' school and the 'good taste' school.

The good taste school of thought stems largely from the ideas of Fisher, who tackled the problem in the 1930s. Consider a male character such as tail length that females may find attractive. Fisher argued that a runaway effect would result, leading to long tails, if some time in the past an arbitrary drift of fashion led a large number of females in a population to prefer long tails. Once this fashion took hold, it would become despotic and self-reinforcing. Any female that bucked the trend, and mated with a male with a shorter tail, would leave sons with short tails that were unattractive. Females that succumbed to the fashion would leave 'sexy sons' with long tails and daughters with the same preference for long tails. The overall effect is to saddle males with increasingly longer tails, until the sheer expense of producing them outweighs any benefit in attracting females. But since attracting females is fundamental, very long tails indeed could be produced by this process. In this Fisherian view, tail length need serve no other purpose than a simple fashion accessory to delight the senses of the opposite sex.

The good sense view suggests that an animal is responding to, and estimating the quality of the genotype of, a prospective mate through the signals that he or she sends out prior to mating, or alternatively, that a judgement is made on the level of resources that a mate is likely to be able to provide. The idea that females are choosing good genes when selecting a mate was also suggested by Fisher, who in 1915 spoke of the 'profitable instincts of the female bird' in choosing features such as 'a clearly-marked pattern of bright feathers', which afforded 'a fairly good index of natural superiority' (quoted in Andersson, 1994, p. 27). The idea was raised again and developed by Williams (1966) and others, now being one of the most promising lines of inquiry in sexual selection theory.

Table 5.4 shows a breakdown of the different possibilities of intersexual selection. This division of sexual selection into the neat categories shown should not, however, deceive us into thinking that the natural world is so simple. Darwin himself despaired of the possibility of being able to distinguish between natural and sexual selection in all cases. Pheasants, for example, have long tails, which would seem to be a classic case of intersexual selection. The tarsal spurs on their legs used in fighting would appear to be products of intrasexual selection. Yet, in the case of the golden pheasant (*Chysolophus pictus*), the tail is used for support while fighting, and among ring-tailed pheasants (*C. colchinus*), the females judge males according to the quality of their spurs (Krebs and Davies, 1991). The picture is also clouded by the fact that, in some species, the females seem to encourage the males to fight and then choose the winners. This is found with the spider *Linyphia litigosa*: females attract into their nests a succession of males that fight one another to stay there until she is ready to mate (Watson, 1990).

At the gene level, the distinction is of course even more blurred. Both intra- and intersexual selection are forms of competition between male genotypes, whether they are instrumental in displaying to a female or fighting another male. With these qualifications in mind, we will examine the good taste and good sense views in turn.

Table 5.4   Mechanisms of intersexual competition

Category	Mechanism
Good taste (Fisherian runaway process)	Initial female preference becomes self-reinforcing. A runaway effect results in elaborate and often dysfunctional (in terms of natural selection) appendages, for example the peacock's train
Good sense (genes)	The female may use signals from the male to indicate the resistance of the male to parasites. Symmetry could be inspected as a clue to general metabolic efficiency
Good sense (resources)	The female may inspect resources held by the male and his willingness to invest resources (potential for good behaviour). This could also serve as indication of genetic quality

5.5.2 *Good taste: Fisher and runaway sexual selection*

For many of Darwin's followers, the problem with his theory of sexual selection was that it did not adequately explain the origin and adaptive purpose of female choice. This perceived weakness was attacked in ironic tones by the Nobel Prize winning geneticist Thomas Hunt Morgan:

> Shall we assume that still another process of selection is going on... that those females whose taste has soared a little higher than that of the average (a variation of this sort having appeared) select males to correspond, and thus the two continue heaping up the ornaments on one side and the appreciation of these ornaments on the other? No doubt an interesting fiction could be built up along these lines, but would anyone believe it, and if he did could he prove it? (Morgan, quoted in Andersson, 1994, p. 24)

The irony backfires since the answer to Morgan's questions is probably yes. The idea of female taste that could not be entertained seriously by Morgan was taken up and developed by Fisher into one of the classic theories for the existence of conspicuous male traits.

Fisher's original account of his model was rather brief, but later refinements have established that a number of conditions need to hold if it is to function properly. In the simplest accounts, these conditions are:

- Variation in a male trait that is heritable
- Variation in female preference that is heritable

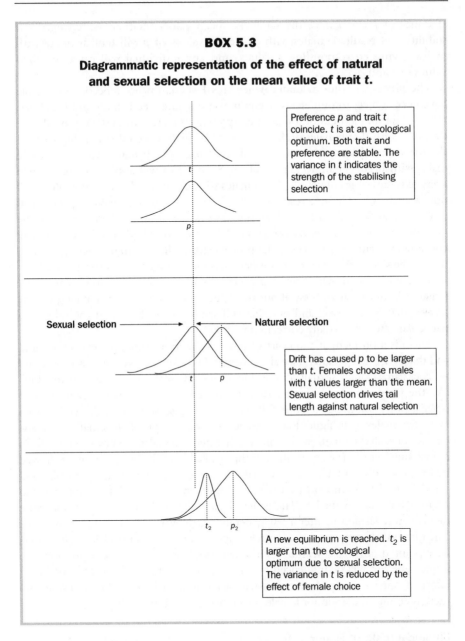

BOX 5.3

Diagrammatic representation of the effect of natural and sexual selection on the mean value of trait *t*.

Preference *p* and trait *t* coincide. *t* is at an ecological optimum. Both trait and preference are stable. The variance in *t* indicates the strength of the stabilising selection

Sexual selection —————→ ←————— Natural selection

Drift has caused *p* to be larger than *t*. Females choose males with *t* values larger than the mean. Sexual selection drives tail length against natural selection

A new equilibrium is reached. t_2 is larger than the ecological optimum due to sexual selection. The variance in *t* is reduced by the effect of female choice

• Individuals having genes for both trait and preference (linkage disequilibrium) but only expressing the character appropriate to their sex. Thus males carry preference genes for a particular trait as well as genes prescribing that trait, but only the genes for the trait are expressed. It is also assumed that genes for high trait values are linked with genes for preferring high trait values.

Suppose that females have a preference (*p*) for a particular tail length, and that this and the value of tail length in males (*t*) are both normally distributed.

If values of p and t have the same the mean values, neither will change, and stability will result. Females with particular values of p will tend to mate with males having corresponding t values, and any individual will carry matched values of t and p.

The precise conditions under which this Fisherian process becomes unstable have been subject to numerous attempts at modelling (see Harvey and Bradbury, 1991). The simplest condition is when (perhaps through drift) the mean value of the tail length that females prefer (expressed by preference value p) is higher than the mean population value of the male trait itself (t). Females will thus choose males with higher than average t values. Such males will also carry higher than average values of p; conversely, low-t males with correspondingly low p values will not fare as well. The net result will be that, in the next generation, the mean values of p and t will both increase, but since p is now larger than t (because the next generation contains descendants of females with high p values), this will continue in a runaway process. The process stops when natural selection forces a halt to the size of the male trait as it becomes increasingly burdensome (Box 5.3).

We could still ask why a set of p genes should exist at all, in other words why females should be at all fussy about the value of any trait such as tail length. The answer that Fisher gave, and one that still seems probable, is that the value of a particular trait was initially correlated with fitness.

There is a problem at the heart of Fisher's theory of runaway sexual selection, and that is the 'lek paradox' (see also Chapter 4). Leking species such as grouse and peacocks are strongly sexually dimorphic: males have elaborate ornaments that look like the products of a Fisherian runaway process. Since males in leking species practise polygyny, the generation following a mating season will be descended from just a few males with higher than average t values. The problem is that, after a few generations, all the variation in t will be exhausted. All males will be descended from a small number of male ancestors, and they will all have the highest t values allowed by the gene pool. Put another way, males will tend to converge on a particular trait value. If all males eventually have the same value, there is no point to choosing, and t cannot increase any further. This is the lek paradox: how can a trait value t increase in a runaway fashion if, after a few generations, the variation in t reduces to zero? The most satisfactory answer so far relies upon mutation. Lande (1981) and others have pointed out that a feature such as tail length will not be the expression of a single gene but will be influenced by a large number of polygenes. The combined effect of outcrossing and mutation will, according to Lande, always ensure that there is enough variation for female preference to exert its effect.

Empirical tests of Fisher

It has in fact proved difficult to test Fisher's ideas. One approach has been to examine the geographical variation of both trait and preference. They should be found in close proximity if Fisher's ideas about the co-evolution of male traits and female preferences are correct. A number of studies have shown this to be the case for the guppy (*Poecilia reticulata*). The distribution of bright orange colouration in males was found to be strongly correlated with the strength of female preference for orange (Endler and Houde, 1995).

Further evidence that is at least consistent with Fisher's ideas comes from breeding experiments on sticklebacks. It has been known for many years that some females find the bright red breeding patch on a stickleback highly attractive. The redness of the patch on males is geographically variable. Bakker (1993) bred from male sticklebacks with either dull or bright red nuptial colouration. It was found that the sons of bright red males also tended to have bright red patches and the daughters of red males preferred red to dull males. The daughters of dull males showed no particular preference for male colour. Such experiments offer support for the linkage that Fisher posited between male traits and female preferences. The general problem with much of the empirical evidence so far is that it does not give unique support for Fisher, other processes still being possible. Andersson refers to the problem of deciding between Fisherian and other mechanisms as 'one of the greatest present challenges in the field of sexual selection' (Andersson, 1994, p. 51).

5.5.3 *Good genes and indicator mechanisms*

The **good genes** dimension of good sense would explain why, in polygynous mating systems, females share a mate with many other females, even though there may be plenty of males without partners and despite the fact that males contribute nothing in the way of resources or parental care. Females are in effect looking for good genes. The fact that the male is donating them to all and sundry is of no concern to her. It is suggested that the female is able to judge the quality of the male's genotype from the **'honest signals'** he is forced to send. Thus, for example, size, bodily condition, colour of plumage, symmetry and size of territory held are all signals providing the female with information about the potential of her mate. Human females find some men sexually attractive even though they know they may be unreliable, philandering and something of a cad.

There are several ways in which selected features may be a signal from a male of genetic prowess. Two models we will consider are handicap and parasite exclusion.

Handicap models

Imagine that two competitors enter a running race. One is dressed in appropriate sportswear, the other is similarly attired but with the addition of a rucksack full of pebbles. The result of the race is a dead heat. Which competitor is most impressive? Most would agree the contestant with the extra load is probably the most physically fit and able. The analogy is not perfect, but this is roughly how the **handicap principle** proposed by Zahavi (1975) is supposed to work. Females choose males that sport a costly handicap since the very fact that they have survived with such a handicap is itself an indication that they are genetically fit. As Zahavi (1975, p. 213) says:

> The handicap principle as understood here suggests that the marker of quality should evolve to handicap the selected sex in a character which is important to the selecting sex, since the selecting sex tests, through the handicap, the quality of its potential mate in characters which are of importance.

Since its publication, this idea has met with a very mixed response, and the literature has become technically complicated. Intuitively, it might seem as if any handicap carried by a male simply cancels out any superior fitness that a male may possess in developing one. Although population geneticists originally rejected the idea, support has been rallying over the last few years. Zahavi's original idea has been refined into a number of versions. In one, the 'qualifying handicap', only males with a high viability can survive to display the handicap. The handicap serves as a sort of fitness filter. In another, the 'revealing handicap', males perform some onerous task such as growing a long tail or developing bright colours as an 'honest signal' of their otherwise hidden qualities.

Whatever the outcome of Zahavi's theory, it has been valuable in introducing the concept of 'honest signalling' into the study of animal behaviour. It is now generally accepted that many animals, including humans, carry 'badges' advertising their worth. It is probable that such badges generally evolve towards honest advertising, not because honesty is a virtue but because **dishonest signals** are eventually ignored. One such honest signal could be the possession of a good immune system to ward off parasites.

Coping with parasites – the model of Hamilton and Zuk

William Hamilton and Marlene Zuk (1982) tackled the question of how male ornamentation could be a reliable indication of the male's health and nutritional status. They suggested that secondary sexual characteristics such as elaborate ornamentation are indicators of the parasite burden of the host since a male infected by parasites produces a poorer display than males not so infected. The conditions in which this process could operate are as follows:

- Host fitness decreases with increasing parasite burden
- Ornament condition decreases with increasing parasite burden
- Resistance to parasites has a heritable component
- Female choice favours the most ornamented males since these are the least parasitised
- Host and parasite are locked into a genetic arms race, each striving to stay ahead of the other's resistance.

The last point also helps to explain the lek paradox – the problem that after a few generations of choice, males will tend to look alike and females will have no variation to choose from. In a world of parasite-infected males, the females will always be choosing a slightly different set of genes since males must constantly change genes to keep parasites at bay. If the genes 'stood still', males would quickly show signs of increased parasite load and appear less attractive to females. Since Hamilton and Zuk (1982) published their ideas, a mass of experimental and observational data have accumulated in relation to their hypothesis. On the whole, the data have been positive in its support. It has proved most difficult to establish whether or not parasite resistance is heritable, but the evidence is generally favourable. For three species (guppy, pheasant and swallow) all five conditions cited above are found to hold (see Andersson, 1994, for a review).

Parasite resistance may also apply to human mate choice. The American evolutionary psychologist David Buss has done much work on delineating the features that humans find attractive in the opposite sex. Buss and his co-workers found that males in all cultures rated female beauty very highly, but especially so where there was a risk of serious parasitic infection such as malaria or schistosomiasis. If you live in a culture where parasites are common, it is even more important to choose your mate carefully and use every cue available to estimate his or her physical health status (Gangstead and Buss, 1993).

Males and females can send signals about their health and reproductive status in a variety of ways. One time-honoured principle of fashion is that 'If you have it flaunt it; if you haven't, hide it.' This applies to cosmetics as much as clothes. One study has suggested a function of bodily markings such as those produced by the scarification and tattooing practised in some cultures. Devendra Singh and Matthew Bronstad of the University of Texas found a correlation between the degree of bodily marking on females and the strength of pathogen prevalence in that culture. They hypothesise that females draw attention to sexually dimorphic features such as waist and breast measurements by the use of body markings. In the cultures examined, they did not find any correlation for male markings (Singh and Bronstad, 1997).

Hamilton and Zuk's work has been of enormous heuristic value, but there remains the problem that much of the experimental work is still open to other interpretation. It could be, for example, that females are simply avoiding the transmission of parasites to themselves and not judging the condition of the males with a view to choosing good genes. In many relationships, however, females are looking for something more substantial than just parasite-resistant genes.

5.5.4 *Good resources and good behaviour*

In the leking species discussed earlier, the males provide nothing except a few drops of sperm. The females have come to expect nothing except genes, so if they are choosy, it will be for good genes. In many species, however, males are expected to bring something to mating in addition to their DNA. In effect, the female may only consent to mating once she has exacted some resources from the male. She may thus judge the ability of the male to provide resources before and after copulation. Resource provision could be in the form of a nesting site, food, territory, parental care or some combination of these.

The donation of resources and genes may be linked. A courtship period in which nuptial gifts are exchanged is a common sight in the animal world. A female may use this period to assess whether the male is able to gather resources and, just as importantly, be willing to devote them to the relationship. In many bird species (for example, European crossbills), males pass food to the females during courtship. A female could use this as a signal of foraging ability and a willingness to feed offspring. Such resources may be crucial in raising young and may, as a bonus, also indicate the genetic fitness of a male to gather such resources. In human hunter-gatherer societies, there is evidence that food is often exchanged for sex and that wealthy men are able to secure more partners. In modern societies, the phenomenon of the 'sugar daddy' is well known. Rich and powerful men seem to be able to attract younger and highly attractive females as their partners. This dimension to human mating will be examined in more thoroughly in Chapter 9.

Intra- and intersexual selection often overlap. We have already noted that where one sex provides a significant commitment in terms of parental investment, such as a large egg, nurturing a fetus or caring for the young, this sex is often the limiting factor for the reproductive success of the other sex. The least investing sex will compete for the sex that constrains its reproductive potential. The limiting sex can afford to be choosy and may look for good genes, good behaviour (indications of future investment) or more tangible resources. The insect world is full of examples of males transferring nutritious offerings before or during sex. In the scorpionfly *Harpobatticus nigriceps*, females lay more eggs with males who provide larger gifts of prey that they have captured. Females also lay more eggs fertilised by larger males, so females may in this species be choosing for good genes as indicated by large size of male, and for good resources (which may indirectly indicate good genes anyway) by preferring larger prey offerings (Trivers, 1972).

In previous discussions on resource defence polygyny, we saw how this system allows the males to achieve polygyny. From a female perspective, it may be that females are choosing males who are able to provide resources. Studies have shown that the number of females per male is strongly influenced by the resources in the territory of the male (see, for example, Kitchen, 1974). In polygyny, there will of course be male competition, and only vigorous males will be able to command the territory to attract females. In these cases once again, it is not always easy to establish whether females are choosing good genes, good resources or both.

Courtship has other functions as well as providing a forum for the inspection of genes and the passing of gifts. Monogamous species often engage in lengthy courtship rituals prior to actual mating. Courtship has many functions, such as the identification of species and the advertisement of readiness to mate, but it seems more than likely that courtship also enables each sex to 'weigh up' its prospective partners in terms of their commitment to a relationship. In the case of the behaviour of the common tern (*Sterna hirundo*), it looks highly probable that females choose males on the basis of their willingness to invest. When on the feeding grounds, females rarely provide their own food and instead rely upon males to bring food to them. The amount of food brought by the male strongly influences the size of the clutch laid. In the first phase of courtship, the male provides food as part of pair formation, and a female will only pair with a male who is carrying a fish (Trivers, 1972). The female is in effect acting to increase the investment made by the male. The more investment a male makes, the less likely he is to desert.

5.5.5 *Case studies: the peacock and the widow-bird*

The peacock's train

The peacock's train has become a paradigm case for the theory of sexual selection. The train looks precisely like the product of some crazy runaway Fisherian process (Figure 5.7). It is clearly a handicap and, in the Indian native home of the peafowl (*Pavo cristatus*), tigers often bring down male birds by their

Figure 5.7   The result of intersexual selection: a male of
the common peafowl (*Pavo cristatus*) displaying

train as they struggle to take flight. It could also be an honest signal that males
with highly elaborate and 'beautiful' trains are relatively free from parasites.
Deciding between these hypotheses has proven very difficult, but the work of
Petrie *et al.* (1991; Petrie, 1994) on the peafowl population of Whipsnade Park in
the United Kingdom has thrown considerable light on the factors involved.

The behaviour of peafowl is typical of leking species in that males attempt to
secure a display site within the lek and only those which secure a site will display.
Females never mate with the first male to court them and will reject several before
deciding. As in many polygynous mating systems, there is a large variance in
reproductive success. The key predictor of reproductive success in the first studies
turned out to be the number of eye spots in the train (Figure 5.8).

It could be argued of course that the number of spots correlates with some other
variable such as age of the male or his overall symmetry, or that the number of spots
is related to something that males use in intrasexual competition. The question also
arises of what peahens gain by choosing males with plenty of spots. In an effort to
address this last question, Petrie took eight free-ranging displaying males (whose
mating success varied) from Whipsnade Park and transferred them to pens, where
they were each mated individually with four randomly chosen young peahens. The
eggs and the young that resulted from these matings were carefully measured. It
was found that the weight of the young after 84 days and their chances of surviving
when introduced back into Whipsnade Park (both of which can be used as a indication of condition and fitness) varied strongly with the average area of each eye spot
on the male's train and the overall length of the train (Figure 5.9).

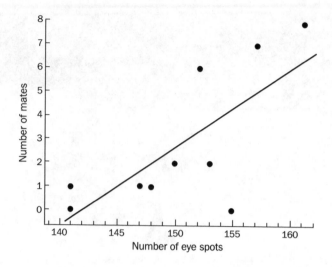

Figure 5.8   Relationship between the number of eye spots and mating success in a group of ten peacocks (data from Petrie *et al.*, 1991)

These results suggest that peahens may be choosing peacocks for good 'viability' genes for their offspring and that Fisherian runaway selection may be coupled with something that also indicates the genetic quality of the males.

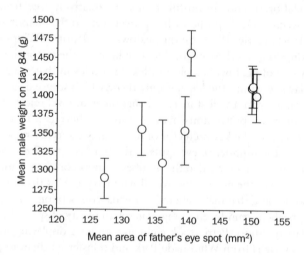

Figure 5.9   Relationship between the eye spot area of males and the fitness of offspring (from Petrie, 1994)

The widow-bird

Another way in which to approach the problem of what females look at in assessing males is to manipulate experimentally the male character that attracts females. Andersson (1982) did this for an African bird called the long-tailed widow-bird (*Euplectes progne*). Early in the breeding season, Andersson caught 36 widow-birds, having first recorded the number of nests in their territories, which in this species is an indication of the number of females that a male has attracted. Andersson then divided them into four groups and, by cutting off portions of their tails, manipulated their tails as follows:

- Group 1: males whose tails were lengthened by replacing a cut portion with a piece longer than the one removed
- Group 2: males whose tails were shortened by replacing the piece cut off with a shorter piece
- Group 3: males whose tails were kept the same length by gluing back the piece cut off
- Group 4: males whose tails were left untouched.

After this cosmetic surgery, the birds were released and the number of additional nests that each male secured was counted. As Figure 5.10 shows, the elongated-tail birds fared better than any of the others, even those who were returned to the wild in the same condition. Females prefer long tails, and the longer the better it seems. The results seem to indicate that the perceptual apparatus of the female long-tailed widow-bird is geared up to prefer a length even above the male population average. As Francis Bacon commented in the 17th century, 'there is no beauty without some strangeness of proportion'.

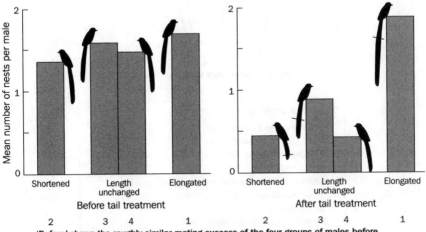

'Before' shows the roughly similar mating success of the four groups of males before any treatment applied. 'After' shows mating success after tail manipulation
1. Tails lengthened by adding a length of tail from another male
2. Tails shortened by replacing the cut portion with a shorter piece
3. Tails cut and replaced, thereby retaining the original length
4. Tails left untouched

Figure 5.10   Effects of manipulation of tail length on the breeding success of male widow-birds (data from Andersson, 1982; modified from Hall and Halliday, 1992)

5.5.6 *Sexual selection in humans – some questions*

Humans show sexual dimorphism in a range of traits (Figure 5.11), and it is probable that many of these are the results of sexual selection. The fact that human infants need prolonged care would ensure that females were alert to the abilities of males to provide resources. In addition, the fact that a female invests considerably in each offspring would make mistakes (in the form of weak or sickly offspring that are unlikely to reproduce) very expensive. It has been estimated that human females of the Old Stone Age would have raised successfully to adulthood only two or three children. Females would therefore be on the look-out for males who showed signs of being genetically fit and healthy, and who were able to provide resources. Both of these attributes, genetic and material, would ensure that her offspring receive a good start in life.

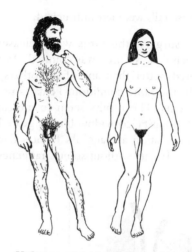

Males compared with females

On average, males have:
- Greater upper body strength
- More facial and bodily hair
- Greater height and mass
- Deeper voices
- Riskier life histories and higher juvenile mortality
- Later sexual maturity
- Earlier death
- Broader and more prominent chins
- Lower levels of fat deposited on buttocks and hips

Figure 5.11   Human sexual dimorphism

One problem to address is whether features that attract the opposite sex are the products of sexual selection for good genes (health, fertility, parasite resistance and so on) or the products of an arbitrary Fisherian runaway process. If features were the result of the latter, we would expect some or all of the following:

- The expression of the trait is not correlated with any other reliable indicator(s) of fitness
- Differences in traits between people are based on genetic differences
- Any cross-cultural differences are not related to ecological factors; hence trait is arbitrary with respect to natural selection
- Extreme expressions of the trait will be more attractive than average ones: size does matter.

On the other hand, the good genes argument would predict that:

- The trait is correlated with a variety of fitness indicators, for example immuno-competence, fertility and metabolic efficiency
- Symmetrical traits will be preferred to unsymmetrical ones. The logic here is that sexually selected traits are challenging to physiological mechanisms that develop and maintain symmetry, and hence revealing of overall genetic efficiency and fitness. Parasites or environmental stress reduce symmetry.

The theory is clear enough, but it turns out that when tackling human morphology, it is difficult to find crucial evidence that will falsify one approach and support another. Two particular enigmatic features of human anatomy that may have been shaped by sexual selection are the female breast and the human brain.

The female breast

In Western cultures, and probably many more, there is a fascination with the enlarged mammary glands of the human female. There are strong cultural mores about when they can be revealed, or should be concealed or only half revealed. The femininity of a woman is strongly associated with her breasts. Women sometimes pay large sums of money and experience much discomfort to have them reduced or enlarged. Bra manufacturers expend much time and effort researching how best to make a product to support them in the right shape. There is agreement between the sexes that they are essential objects of desire, but what are they for? Most people would take this to be a pointless question since it is obvious that they are there to provide infants with milk. A consideration of the facts below forces a rethink on this issue:

- Breasts are strongly sexually dimorphic and appear at puberty
- Permanently enlarged breasts are not found among any other primates: most primates have enlarged breasts only during pregnancy and lactation
- Large breasts, although attractive to males, interfere with locomotion, and women athletes engaged in running sports tend to have small breasts
- There is some cross-cultural variation in breast morphology but with no obvious ecological correlates
- The size of a woman's breasts bears very little relationship to her ability to lactate. Women could supply the necessary nutrition to a baby with much smaller breasts.

Figure 5.12   Young men and women at a disco. Sexual selection in action or just having a good time?

It looks then as if breasts have not been shaped by natural selection, women would move better without them and their permanently enlarged state is not essential in order to supply milk. Acting as a storage device for fat is a possibility, but storage around the waist would be mechanically more efficient. Breasts are thus prime candidates for good genes or runaway sexual selection. Some studies have shown that breast symmetry correlates with fertility, which suggests a role in the honest advertisement of good genes. The fact that breast size is not negatively correlated with asymmetry runs counter to this, however, since a sexual trait that increases in size should become more asymmetrical as the demands of size growth take their toll on symmetry (Thornhill and Gangestad, 1994).

Until more conclusive evidence is forthcoming, the consensus seems to be that they have been sexually selected, but the precise mechanism is not certain.

The brain

One human feature that some have claimed shows signs of runaway sexual selection is the brain. Between about 6 million and 3 million years ago, our ancestors roamed the African savannah with brains about the size of those of a modern chimpanzee (450 cm³). Then, 2 million years ago, there began an exponential rise in brain volume that gave rise to modern humans with brains of about 1300 cm³. A tripling of brain size in 2 million years is rapid by evolutionary standards. One force that can bring about such rapid change is sexual selection (Miller, 1998).

Miller suggests that humans would have examined potential partners to estimate heath, age, fertility, social status and cognitive skills. It is this latter criterion that might have set up a runaway growth in brain size. Miller sees this as beginning with females choosing males who were amusing and inventive, and had creative brains. Language accelerated the process since the exchange of information could then be used to judge the suitability of a potential partner. Although brain growth was driven by female choice, both sexes gradually acquired larger brains since brains were needed to decode and appreciate inventive male displays. Miller makes a remarkable, and what will prove to be a controversial, assertion:

> Males produce about an order of magnitude more art, music, literature... than women, and they produce it mostly in young adulthood. This suggests that... the production of art, music, and literature functions primarily as a courtship display. (Miller, 1998, p. 119)

It is an intriguing thought that much of what passes for culture may be a form of sexual display. The view that art and literature represent the outpourings of testosterone-fuelled males strutting their stuff is a wonderful image destined to infuriate at least half of the academic community and most female artists and writers, but it has a poetic plausibility that may carry more than a few grains of truth when we consider the strength of the libido and the sexual activities of successful artists and musicians.

There are of course other theories to account for the rapid brain growth (encephalisation) of hominids. The 'Machiavellian intelligence' hypothesis suggests an arms race between mind-reading and deception. In this view, success depended on anticipating and manipulating the actions of others. Large brains helped early humans to understand each other's minds; this allowed deception, which in turn stimulated brain growth to detect and avoid deception, and so the process ran on, causing an escalation of brain size. This is examined in more detail in the next chapter.

SUMMARY

▨ Sexual selection results when individuals compete for mates. Competition within one sex is termed intrasexual selection, and typically gives rise to selection pressures that favour large size, specialised fighting equipment and endurance in struggles.

▨ Individuals of one sex also compete with each other to satisfy the requirements laid down by the other sex. An individual may require, for example, some demonstration or signal of genetic fitness or the ability to gather and provide resources. Selective pressure resulting from the choosiness of one sex for the other is studied under the heading of intersexual selection. Such pressure often gives rise to elaborate courtship displays or conspicuous features that may indicate resistance to parasites, or may possibly be the result of a positive feedback runaway process.

▨ The precise form that mating competition takes (such as which sex competes for the other) is related to the relative investments made by each sex and the ratio of fertile males to females. If females, for example, by virtue of their heavy investments in offspring or scarcity, act as reproductive bottlenecks for males, males will compete with males for access to females, and females can be expected to be discriminatory in their choice of mate.

▓ In cases where a female engages in multiple matings and thus carries the sperm of more than one male in her reproductive tract, competition between sperm from different males may occur. The theory of sperm competition is successful in explaining various aspects of animal sexuality, such as the high number of sperm produced by a male, the frequency of copulation and the existence of copulatory plugs and infertile sperm.

▓ It is probable that many features of human physiognomy and physique have been sexually selected. In examining females, males can be expected to look for features that indicate youth and fertility (nubility), health and resistance to parasites. Females can be expected to look for strength, wealth, health and status as well as parasite resistance in prospective male partners. Symmetry is an attribute valued by both sexes and may correlate with physiological fitness. The rapid increase in brain size among hominids that started about 2 million years ago is a candidate for sexual selection.

KEY WORDS

Dishonest signals ■ Extrapair copulation ■ Good genes
Handicap principle ■ Honest signals ■ Intersexual selection
Intrasexual selection ■ Nuptial gift ■ Operational sex ratio
Parental investment ■ Sex role reversal ■ Sexual dimorphism
Sperm competition ■ Symmetry

FURTHER READING

Andersson, M. (1994) *Sexual Selection*. Princeton, NJ, Princeton University Press.
Extremely thorough book that reviews a wide range of research findings. Tends to concentrate on non-human animals.

Geary, D. C. (1998) *Male, Female: The Evolution of Human Sex Differences*.
Washington DC, American Psychological Association.
Geary explains the principles of sexual selection and how these can be used to understand differences between males and females. Good discussion of the evidence for real cognitive differences between males and females.

Gould, J. L, and Gould, C. G. (1989) *Sexual Selection*, New York, Scientific American.
Readable, well structured and well illustrated. Its main drawback is a lack of references in the text to support the evidence. Mostly covers non-human animals.

Ridley, M. (1993) *The Red Queen*. London, Viking.
An enjoyable and well-written account of sexual selection theory and its application to humans.

6

The Evolution of Brain Size

Plac'd on this isthmus of a middle state,
A being darkly wise, and rudely great:
With too much knowledge for the Sceptic side,
With too much weakness for the Stoic's pride,
He hangs between; in doubt to act or rest,
In doubt to deem himself a God or beast;
In doubt his Mind or body to prefer.
Born but to die, and reas'ning but to err.

(Pope, 'Essay on Man')

We share much of our basic anatomy and physiology with other animals: we eat, breathe, reproduce and care for our young in ways that can be found throughout the class Mammalia. Yet we also know that, in so many ways, we are fundamentally different from all other living things. Even the distance to our nearest relatives among the great apes seems to be an unbridgeable gulf. To a visitor from Mars, the obvious feature of this divide would probably appear to be our culture and our natural disposition to use language. Chimpanzee culture, if it exists, is minimal, and compared with chattering humans, they are remarkably silent in their affairs. It is a reasonable assumption that these differences must be a product of the size or structure, or both, of human brains. Why the human brain grew so large, a process called encephalisation, is one of the most significant puzzles in the study of human evolution, and, understandably, there has been no shortage of theories.

This chapter considers the process of encephalisation and some of the factors that may have contributed to it. It begins by examining our evolutionary relationship to the great apes – a knowledge of which is helpful in elucidating the distinctive features of human intelligence – and concludes with an examination of recent ideas suggesting that human brains grew large to help people to cope with the social complexity of large groups. This radical approach suggests a function for the evolution of brain size and the origin of language itself. The comparisons between humans and non-human primates established in this chapter are necessary precursors to Chapters 7 and 8, on language and sexuality respectively.

6.1 Humans and the great apes

Trying to reconstruct the social behaviour of early **hominids**, and hence adduce reasons why the human brain grew in size, may seem like a daunting, if not impossible, task, yet a careful sifting of the evidence from palaeontology and archaeology, coupled with information about the socioecology of contemporary apes, enables some progress can be made. To facilitate this, a brief discussion on classification and **taxonomy** is necessary.

6.1.1 *A note on cladistic taxonomy*

Although biologists had classified species long before the acceptance of any theory of evolution, it became clear in the years following Darwin's *On the Origin of Species* that classification should relate in some way to the reconstruction of evolutionary (that is, phylogenetic) pathways. At the onset of this enterprise, however, a tension arises between the need for a robust system of classification and the requirement that phylogenetic trees should be able to be modified as more evidence becomes available. This tension accounts for some of the confusion and debate over classification that still remains.

Humans, chimpanzees and gorillas share some features in common, but all unambiguously belong to different species. The concept of a species has a fairly clear biological definition: organisms are said to belong to the same species if they can interbreed to produce fertile offspring. Some different species, such as donkeys and horses, are closely related and produce viable offspring, but such offspring are inevitably infertile. On this basis, present-day humans all belong to the same species called *Homo sapiens* (literally 'wise humans'). Through time, a single species can give rise to others through a process that Darwin and his followers called transmutation, which today is usually known as **speciation**. Speciation occurs when populations are reproductively isolated by geographical barriers such as islands or mountain ranges, isolated temporally by gradually breeding at different times, or separated in some other way. By gradual mutation and genetic drift, the two populations of what was once a single species reach a state in which gene flow between them ceases and cannot be revived by the removal of the barrier (Figure 6.1).

Where speciation is not quite complete, we often observe the existence of subspecies. In the United States, there are distinct geographical populations of deer mice (*Peromyscus maniculatus*) in which gene flow between the groups is very limited. Hence we have the subspecies *Peromyscus maniculatus borealis* in the north and *Peromyscus maniculatus sonoriensis* in the south. In time, these may drift further apart to form distinct species. There is still debate in early hominid studies over whether Neanderthal man and early *Homo sapiens* were really distinct species or whether they did, in fact, interbreed. Neanderthals were the last hominid species to die out. They lived in Europe as recently as 30 000 years ago and coexisted, at least in time, with *Homo sapiens*. The consensus now seems to be that Neanderthals were a subspecies, hence *Homo sapiens neanderthalenis*.

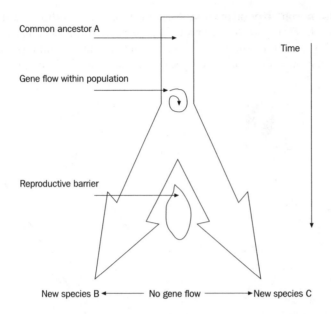

Figure 6.1   Diagrammatic illustration of speciation around a reproductive barrier

Groups of species may be lumped together into taxa, such as the class of organisms that are mammals called Mammalia. Systematics is the science dealing with the classification and organisation of organisms. A traditional classification of animals in the classes appropriate to humans is shown in Figure 6.2.

Kingdom: Animalia
    Phylum: Chordata
        Class: Mammalia
            Order: Primates
                Infraorder: Anthropoidea
                    Superfamily: Hominoidea
                        Family: Hominidae
                            Genus: *Homo*
                                Species: *Homo sapiens*

Figure 6.2   Traditional taxonomy of the human species

Classical taxonomy from the time of Darwin has been greatly concerned with the amount of evolutionary divergence that has occurred between different taxa. More recently, cladistic taxonomy has reinvigorated this old discipline by suggesting that more attention should be paid to the order of branching (Greek *klados*, branch) in phylogenetic lineages, as indicated by either macroscopic

similarities in morphology or microscopic similarities such as those occurring at
the molecular level. Figure 6.3 shows how a cladogram can be constructed for
five vertebrates using the notion of a branch point. A branch point represents the
most recent common ancestor for all species downstream of that point.

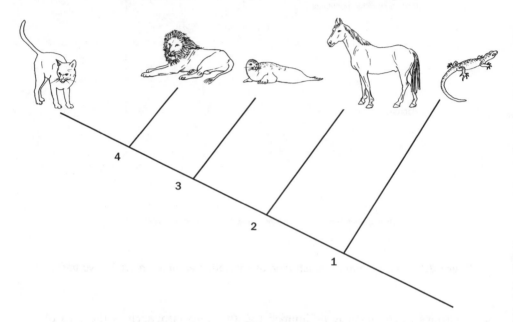

Figure 6.3   Cladogram for five vertebrates

All the vertebrates in Figure 6.3 are said to share the *primitive* feature of five
toes. This feature must, therefore, have existed in the common ancestor before
point 1. Beyond each of the branching points, we have derived *characters* that
enable us to separate out the branches. Hence the species at point 1 represents
the last common ancestor of lizards, horses, seals, lions and cats since beyond this
point we find hair and mammary glands, which are noticeably absent in lizards.
The full point descriptors are shown in Table 6.1.

Table 6.1   Derived features used in constructing the cladogram in Figure 6.3

Beyond point	Derived features
1	Hair and mammary glands
2	Cheek teeth for shearing and crushing meat
	Teeth of horses are adapted only to plant material
3	Retractable claws
4	Ability to purr: lions can roar but not purr, cats are unable to roar

Cladistics has some advantages over classical taxonomy in that, by concentrating on branching points, it relies less on morphological similarities, which may be open to subjective weighting. This leads, however, to tension between different modes of classification. Most traditional classical taxonomists, for example, would argue that, on the basis of the unique features of the human body, mind and behaviour, only humans now belong to the family Hominidae. Traditionally, the superfamily of hominoids Hominoidea was divided as follows:

Hylobatidae	(gibbons and siamangs)
Pongidae	(orang-utans, gorillas and chimps)
Hominidae	(hominids, that is, humans)

It is now recognised from molecular studies that humans are closer to gorillas and chimps (the African apes) than either of these are to orang-utans (the Asian ape). On this basis, a revised cladistic classification would be (see Figure 6.4):

Hylobatidae	(gibbons and siamangs)
Pongidae	(orang-utans)
Hominidae	(chimps, gorillas and humans)

Cladistic taxonomy on the basis of molecular similarities would therefore place chimps, gorillas and humans in the family Hominidae; humans would then be called hominines. We currently have the rather unsatisfactory situation in the literature that whereas the term 'hominoids' is taken to mean humans and the great apes (since we belong to the same superfamily), the term 'hominids' can either mean humans and their ancestors or, following cladistics, the African apes and humans. The problem is that the term 'hominid' has been traditionally and extensively used to describe humans and their ancestors in sharp distinction to the great apes.

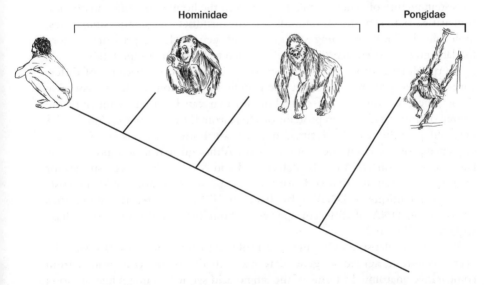

Figure 6.4   Cladistic classification of hominids and pongids

In this book, we present the cladistic classification as being the most valuable way of considering the relationship between humans and other species of apes, but we will use the term 'hominid' to mean the ancestors of humans after the split from our common ancestor with the apes up to and including modern humans (Figure 6.4). We will now examine the evidence that leads us to adhere to this classification.

6.1.2 *The human–primate gap: the 1.6 per cent advantage*

Early hominids are gone, and there is now only one species left: ourselves, *Homo sapiens*. However, several species of great ape still exist today, and we once shared a common ancestor with one of them that died out about 7 million years ago. It may be possible then to infer something about human behaviour by looking at the behaviour of the contemporary great apes.

This raises the problem of how similar ancestral hominids were to today's great apes – bearing in mind that the great apes have evolved since the early hominid species became extinct. To help with this, we need a phylogenetic tree detailing the evolution of the primates. There are two basic disciplines that can help us construct such a tree: comparative anatomy and molecular biology. Comparative anatomy examines the similarities between primates in terms of their basic body plan. Since humans more closely resemble chimpanzees than they do ring-tailed lemurs, it seems reasonable to suppose that we are more closely related to chimps than lemurs. By 'more closely related' we mean that we more recently shared a common ancestor. By such methods, humans and the four species of great ape (gorillas, chimpanzees, bonobos and orang-utans) were, by 1943, placed in the same superfamily called Hominoidea (Box 6.1). Morphological evidence, however, tends to be qualitative, and it has proved difficult to push the method further to establish unambiguously a branching order among the hominoids. More recently, however, molecular biology has provided a clearer and more consistent picture of genetic similarities between the hominoids. Since anatomical similarities between the hominoids are the product of genetic similarities, molecular biology, by going straight to the genetic level, provides a more fundamental approach that should be less open to biases in interpretation.

Morphological differences between species come about because of differences between proteins and hence the genetic information needed to assemble them. There is now a range of techniques that can be used to measure the degree of similarity between proteins or DNA from different species. Similarities between proteins can be estimated using antibody reactions or by direct sequencing of the component amino acids. Although it is a slow process, the base sequences on DNA can be determined and compared between species for regions of the genome that code for the same proteins. A more rapid and cost-effective technique is **DNA hybridisation**. This compares the similarities between the DNA of different species without the lengthy process of base sequencing (Box 6.2).

When molecular studies began in the 1960s, the exiting news was that virtually every technique agreed in general terms with the broad conclusions from comparative anatomy. In terms of the amino acid sequences in a range of blood proteins, and hence the genes responsible, we are virtually identical to

BOX 6.1

The great apes (Hominoidea): our nearest relatives among the primates

Name	Location, ecology and social organisation

Common chimp
(*Pan troglodytes*)

Tropical Africa. Forest, woodland and open savannah. Troglodytes means 'cave-dwelling', which reflects an early European misunderstanding. Chimps are arboreal and terrestrial but spend over 50 per cent of their time in trees. Mixed diet of fruit, vegetation and some meat. Multimale, multifemale groups, 'promiscuous' mating system but with dominant males. Most copulations are opportunistic, with little competition. Female exogamy, that is, females leave their native group at puberty. Male–male grooming accounts for nearly 50 per cent of all adult interactions. Tool use common, for example termite sticks to extract termites from nests to eat, stone hammers, and munched leaves to act as a sponge to soak up water from otherwise inaccessible places

Bonobo (*Pan paniscus*)

Tropical Africa south of Zaire river. Lowland forests. Diet similar to common chimps – fruit, shoots, buds, insects and some mammals. Multimale, multifemale groups. Females sexually receptive through most of the oestrus cycle. Less overall aggression in groups compared with common chimps. Despite their name, bonobos are only fractionally smaller than common chimps

Gorilla (*Gorilla gorilla*)

Distributed in Central and West Africa in three subspecies: the eastern lowland gorilla (*Gorilla gorilla graueri*), the western lowland gorilla (*Gorilla gorilla gorilla*) and the mountain gorilla (*Gorilla gorilla beringei*). Ground-dwelling shy vegetarians, utilising over 100 species of plant. Polygynous harems of one dominant male (the so-called 'silver back'), several females and infants. Groups are sometimes raided for females, and invading males can kill infants. Large degree of tolerance within groups. Copulation rate is low: once about every 1–2 years per female

Orang-utan (*Pongo pygmaeus*)

Forest ape found in Borneo and Sumatra. Largely an arboreal species but will travel on the ground. Mostly eat fruit, with some leaves and bark. Dominant male has a territory overlapping with those of a few females with whom he mates. Lower-ranked males are solitary and migratory, probably reflecting the wide dispersal of food. Males are intolerant of other males. Males have large cheek pads of subcutaneous tissue between the ears and eyes

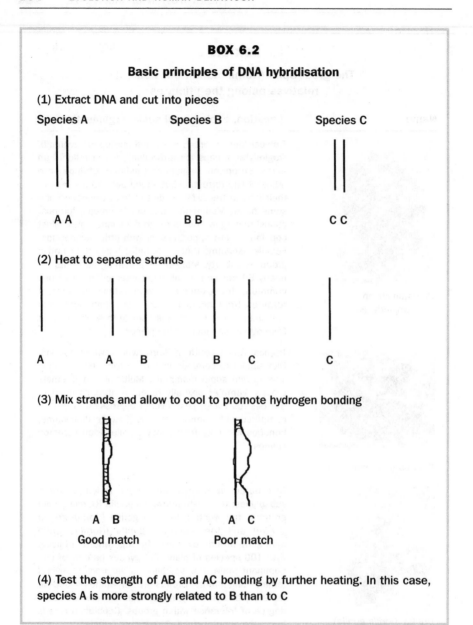

BOX 6.2

Basic principles of DNA hybridisation

(1) Extract DNA and cut into pieces

Species A   Species B   Species C

A A   B B   C C

(2) Heat to separate strands

A   A   B   B   C   C

(3) Mix strands and allow to cool to promote hydrogen bonding

A B   A C
Good match   Poor match

(4) Test the strength of AB and AC bonding by further heating. In this case, species A is more strongly related to B than to C

chimpanzees, slightly different from gibbons and different in many amino acids from the Old World monkeys. To construct fully a phylogenetic tree based on molecular differences, however, we need to introduce two assumptions:

1. That quantitative differences between the amino acid sequences of proteins or base sequences in complementary regions of DNA represent relative differences in terms of evolutionary divergence of the species involved

2. That differences in molecular (protein or DNA) structures can be translated into absolute times of divergence by the introduction of a molecular clock.

These two assumptions seem reasonable enough. When two species diverge from a common ancestor, mutations begin to accumulate in each species. Since it would be extremely unlikely for the same mutation to occur in both species, differences in DNA sequences accumulate as time passes. Controversy in this field arises when we try to relate the degree of mutational differences to a 'mutation clock'. If we take the simplest assumption of linearity, if species A differs from B by 2 per cent and from species C by 4 per cent, then A and C shared a common ancestor twice as long ago as A and B shared one (Figure 6.5).

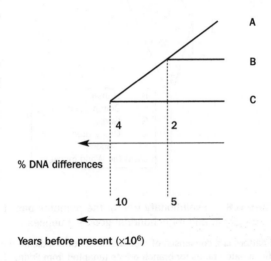

Figure 6.5   Hypothetical relationship between three species A, B and C

It then remains to translate this relative scale into an absolute scale using palaeontological evidence. If fossils show that A and B shared a common ancestor 5 million years ago, it would seem to follow that the ancestor to A, B and C lived 10 million years ago. In principle, it all looks very easy, but it is now appreciated that molecular clocks do not run so regularly, much effort is needed, and uncertainties remain in translating relative differences into absolute time.

6.1.4 *Phylogeny of the Hominoidea*

Bearing in mind the promise and difficulties of molecular comparisons, we can now construct a **phylogeny** for the hominoids based on molecular clockwork calibrated using fossil evidence. There are in fact two independent fossil calibrations available for the primates: the divergence between monkeys and apes (about 7 per cent DNA difference), which occurred between 24 and 34 million years ago, and that between orang-utans and gorillas (with about 4 per cent difference in DNA), which occurred about 12–18 million years ago. Figure 6.6 brings together data from a number of studies.

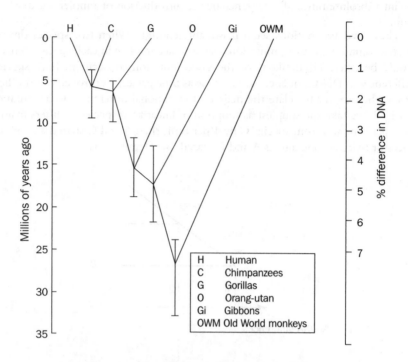

Figure 6.6   Evolutionary tree of the primates based
on globin and mitochondrial gene sequences

The pattern is a consensus of recent studies. Bars indicate the
range of estimated times for branch points (adapted from Friday, 1992)

The implications of Figure 6.6 are profound and take time to sink in. The first point to note is that chimpanzees, and not gorillas, are our closest relatives. In fact, chimpanzees are more related to us than they are to gorillas. Instead of talking about the huge gap between humans and the apes, we should, at least in DNA terms, talk about the small gap between the 'higher apes' – humans, chimps and gorillas – and the 'slightly lower apes': orang-utans and gibbons.

We differ in our DNA from the chimps by just 1.6 per cent, the remaining 98.4 per cent being identical. Our haemoglobin, for example, is the same in every one of the 287 amino acids units as chimpanzee haemoglobin; in terms of haemoglobin, we *are* chimps. Based on these findings, Jared Diamond, in his appropriately titled *The Rise and Fall of the Third Chimpanzee* (1991), makes the controversial suggestion that such small differences between humans and chimps suggest that they should be regarded as belonging to the same **genus**. The higher apes would then become *Homo troglodytes* (common chimps), *Homo paniscus* (pygmy chimps) and *Homo sapiens* (the wise chimp). Diamond's suggestion is unlikely to take root.

We must beware of making too much of the 1.6 per cent difference. Information coded on DNA is not a linear system; that 1.6 per cent could and has effected profound changes. As we shall explore more thoroughly later in this chapter, one obvious difference between chimps and ourselves is that we have a much larger brain. Even 4 million years ago, our ancestors had larger brains than modern-day chimps. Figure 6.7 shows the current distribution of the African and Asian apes.

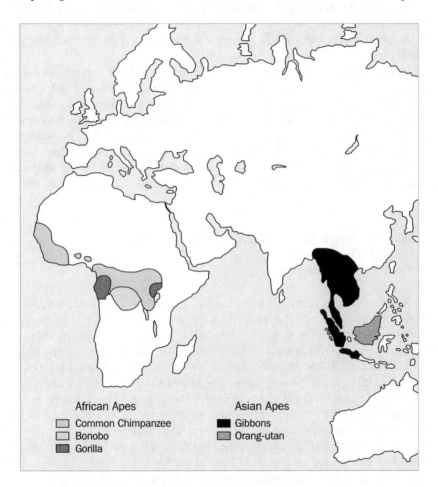

Figure 6.7 Current geographical distribution of the African and Asian apes (adapted from Boaz and Almquist, 1997)

6.2 Intelligence in humans and other primates

6.2.1 *The importance of being large*

When we visit zoos, we are so fascinated and impressed by large mammals, such as elephants, tigers, giraffes and rhinoceros, that we are apt to forget that we ourselves are among the largest mammals alive. The vast majority of mammals, including most non-human primates, are considerably smaller than humans. It has long been recognised that, as mammals evolve, there is a tendency for them to grow in size. The Eocene ancestor of the modern horse, for example, was only the size of a small dog about 40 million years ago. This trend, although it is not universal, is sometimes called Cope's law, and humans, as mammals, have been part of this trend towards increasing size. Estimating the body mass of early hominids is fraught with difficulties, but there is general agreement on the broad picture: early **Australopithecines** were small, *Australopithecus africanus*, for example, probably weighing between 18 and 43 kg. Body weight rose with *Homo erectus*, peaked with early *Homo sapiens* and has declined slightly over the past 80 000 years.

The causes of this gradual increase in body size are difficult to establish. It may be that a move from forests to a more terrestrial habitat relaxed the constraints on the size of arboreal species, or that in the more open environments that early hominids occupied, predation risks were greater and this selected for increased size (Foley, 1987). Whatever the causes, the ecological and evolutionary consequences were profound. One of the most significant effects relates to the fact that the metabolic rate of an animal rises with body weight in line with well-established physiological principles and according to the following equation:

$$M = KW^{0.75}$$

where: M = metabolic rate (energy used per unit of time)
W = body weight
K = some constant.

The effect of this relationship is that, as the size of hominids grew, so the absolute requirement for the intake of calories grew, but the rate of input of calories per unit of body mass fell. This effect can easily be seen in mice and humans. Mice consume absolutely less food per day than a human, but a mouse will typically eat half its own body mass of food in one day and spend most of the day finding it. Humans eat only about one twentieth of their body mass in food each day and thankfully have plenty of time left over to read books on evolution. The consequence for hominids of the absolutely larger amount of food that had to be found as size grew was that their home range increased. This was not the only option available of course. Gorillas, for example, grew in size but became adapted to eating large quantities of low-calorie foodstuffs, such as leaves. This requires a large body to contain enough gut to digest the plant material, and this is essentially the strategy pursued by large herbivores such as cattle. Chimps and early humans clearly opted for the high-calorie, mixed diet that required clever foraging to obtain.

Increases in size may in themselves have been a response to ecological factors, but it is important to note that such increases can have major effects on the lifestyle of an animal and, through complex feedback effects, force it to adapt further in other ways. An increase in size means that animals incur an increase in absolute metabolic costs. These could be met by increasing the size of the foraging range. Large animals also have a smaller surface area to volume ratio than small ones, which in tropical climates would lead to problems of overheating and consequently a greater reliance on water. The upright stance of hominids may have been a response to this, since standing upright exposes less surface area to the warming rays of the sun. Body fur loss would also have helped with temperature regulation. Larger animals take longer to mature sexually, so offspring become expensive to produce and require longer periods of care. The kin group and larger social groups now become important for care and protection (see Foley, 1987).

If a premium is placed on a large brain to cope with social exchanges or the demands of foraging, this requires a shift towards higher-calorie foodstuffs to support it. Meat can provide high-quality food, and meat-eating in turn reinforces co-operation, sharing and sociality. In addition, a foraging range tends to be correlated in primates with group size, so increasing the range would increase the group size.

To develop a large brain through evolutionary time, a fairly stable environment is needed in which energy rich foods can reliably be obtained. The most remarkable feature of the period between the Australopithecines and *Homo habilis*, however, is that brains grew larger than expected from body size increases alone. It is noteworthy that our brains are expensive to run. A chimp devotes 8 per cent of its basal metabolic rate to maintaining a healthy brain, whereas for humans the figure is 22 per cent even though the human brain represents only about 2 per cent of body mass. Larger brains require better sources of nourishment. In fact, the initial increase in brain size about 2 million years ago seems to correlate with a switch from a largely vegetation-based diet of Australopithecines to a diet with a higher percentage of meat, as found with *Homo habilis*. Figure 6.8 shows a time chart for the hominids.

Exactly why we developed such large brains is a disputed subject. The rapid growth of the human brain, which for about 1.5 million years remained at about 750 cm^3 and then in the past 0.5 million years doubled to its present volume, has led some, such as Geoff Miller (1996), to suggest that a runaway sexual selection process must have been at work. We have seen, in the case of the peacock's tail, how sexual selection can exert a powerful force and bring about rapid change that flies in the face of natural selection. Miller's hypothesis is that females were initially attracted to an intelligent use of language by males, and once this shift in a preferred trait took root, it led rapidly (by sexual selection) to an increase in brain power to keep potential mates amused.

Whatever the cause, an increase in brain size posed at least two problems for early hominids: how to obtain enough nourishment to support energetically expensive neural tissue and how to give birth to human babies with large heads. The first of these problems, as noted above, was probably solved by an earlier switch to a meat-eating diet about 2 million years ago.

The second problem was solved by what is, in effect, a premature birth of all human babies. One way to squeeze a large-brained infant through a pelvic canal is to allow the brain to continue to grow after birth. In non-human primates, the rate of brain growth slows relative to body growth after birth. Non-human primate mothers have a relatively easy time, and birth is usually over in a few minutes. Human mothers suffer hours of childbirth pains, and the brain of the infant still continues to grow at prebirth rates for about another 13 months. Measured in terms of brain weight development, a full term for a human pregnancy would, if we were like other primates, be about 21 months, by which time the head of the infant would be too large to pass through the pelvic canal. As in so many other ways,

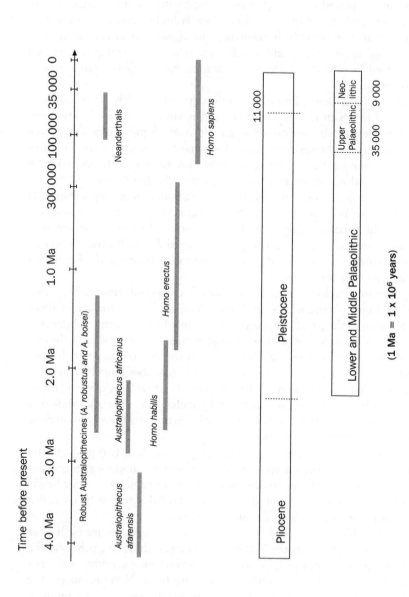

Figure 6.8 Time chart of early hominids

Note that, because of the uncertainties of
hominid phylogeny, no branching sequences are given to link hominids

natural selection has forced a compromise between the benefits of bipedalism (requiring a small pelvis) and the risks to mother and child during and after childbirth. Human infants are born, effectively, 12 months premature.

The premature birth of human infants required a different social system for its support than the unimale groups of our distant Australopithecine ancestors. As brain size grew, so infants became more dependent on parental care. Women would have used strategies to ensure that care was extracted from males. This would lead to the emergence of a more monogamous mating pattern since a single male could not provision many females. It is significant that the body size sexual dimorphism of hominids during the Australopithecine phase was such that males were sometimes 50 per cent larger than females. This dimorphism was probably selected by intrasexual selection as males fought with males to control sizeable harems. By the time of *Homo sapiens*, this figure had reduced to 10–20 per cent, signalling a move away from polygyny towards monogamy. Women probably ensured male care and provisioning for their offspring by the evolution of concealed ovulation. The continual sexual receptivity of the female and the low probability of conception per act of intercourse ensured that males remained attentive. This is a subject explored further in Chapter 8.

We can finally couple this reasoning with a knowledge of the environment of Africa during the Pliocene era to provide a tentative construction of what may be some significant changes in brain size and social structure (Figure 6.9). We must note, however, that even the broad generalisations offered here are far from certain. Australopithecines present a special puzzle because, although body size dimorphism is marked, with male to female body weight ratios as high as 1.7, there is very little dimorphism in the size of the canine teeth. The high level of body size dimorphism indicates intense male–male competition, probably for females, but the low canine dimorphism suggests little male–male competition. Darwin himself faced this same problem and suggested a 'weapons replacement' hypothesis. The argument is that, as hominids learnt to use weapons to fight each other, so tooth size became less important, but physical strength remained crucial. Put crudely, male Australopithecines began to throw stones at each other instead of biting. The problem here is that stone tools only show up a million years after the appearance of *A. afarensis*. Over 100 years after Darwin tackled this issue, it remains unresolved (Plavcan and van Schaik, 1997). In Figure 6.9, we accept the traditional interpretation that Australopithecines were polygynous with unimale, multifemale groups.

6.2.2 *Brain size in humans and other mammals*

If one were to list four obvious biological differences between humans and the great apes, they would probably be that humans walk with an upright gait, have relatively hairless bodies, use language naturally and enthusiastically, and have larger brains. Our hairlessness and upright gait would by themselves be incapable of accounting for the gulf that separates humans and apes. Given the fact that language is controlled by the brain, it is clear that it is important to examine exactly what are the differences between our brains and those of other primates.

Time before present

4.0 Ma   200 000 years

1 Ma = 1 x 10⁶ years

Figure 6.9   Conjectured evolution of hominid social systems
in relation to group size and brain size changes

Is there a special area of the brain unique to humans? In the 19th century, Richard Owen (*c.* 1858), an anatomist who vigorously opposed the application of Darwinism to humans, thought that there was. He claimed that humans have a special area of the brain called the 'hippocampus minor' that is not found in apes. This, he argued, was clear evidence that we could not have descended from the apes; here was the seat of human distinctiveness. His hopes for a special status for humans were, however, short lived: Darwin's 'bulldog' champion, Thomas Henry Huxley, rushed to the fray (*c.* 1863) and conclusively demonstrated that apes possessed the same structure that Owen had identified. The debate was parodied in popular culture. In Charles Kingsley's *Water Babies*, published in 1863, for example, there is much talk of 'hippopotamus majors'.

Since the time of Owen and Huxley, there have been numerous attempts to establish which features of the human brain, if any, confer upon humans their unique qualities. It is tempting to think that we simply have bigger brains than other mammals, but even a cursory examination of the evidence rules this out. Elephants have brains four times the size of our own, and there are species of whale with brains five times larger than the average human brain. We should expect this of course – larger bodies need larger brains to operate them. The next step would be to compare the relative size of brains among mammals (that is, the ratio of brain mass to body mass). The results are unedifying: we are now outclassed by such modest primates as the mouse lemur (*Microcebus murinus*), which has a relative brain size of 3 per cent compared with 2 per cent for humans.

We can find some reassurance, however, in the phenomenon of **allometry**: As an organism increases in size, there is no reason to expect the dimensions of its parts, such as limbs or internal organs, to increase in proportion to mass or volume. If we simply magnified a mouse to the size of an elephant, its legs would still be thinner in proportion to its body than those of an elephant. This happens in primates too: the bones of large primates are thicker, relatively speaking, than the bones of smaller primates. In fact, there is a fairly predictable relationship between brain and body size in mammals:

$$\text{Brain size} = C\,(\text{Body size})^k$$

where C and k are constants.   (Equation 1)

The constant C represents the brain weight of a hypothetical adult animal weighing 1 g. The constant k indicates how the brain scales with increasing body size and seems to depend upon the taxonomic group in question. Much of the pioneering work in developing these equations was carried out by Jerison (1973), who concluded that, for the entire class of mammals, k was about 0.67 and C about 0.12. There is much discussion about the precise values for these constants, and even within primate groups k varies from 0.66 to 0.88. Later revisions of Jerison's works suggest that k may be 0.75 for all mammals.

If we plot a graph of brain size against body weight for mammals on linear scales, a curve results, showing that brain size grows more slowly than body size (Figure 6.10).

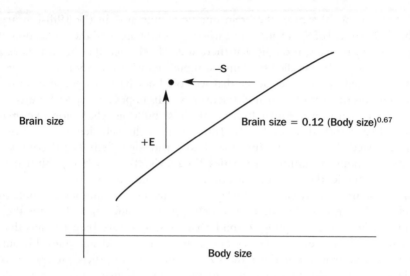

Figure 6.10 Growth of brain size in relation to body size for mammals

An animal occupying a point above the line is said to be encephalised, that is, it has a brain larger than expected for an animal of its body mass. This could be the result of a relative growth in brain size (positive encephalisation, +E) or a diminution of body size relative to brain mass (negative somatisation, –S) (see Deacon, 1997)

If we take log values of both sides of equation 1 with the constants for mammals inserted, then:

$$\log (\text{Brain size}) = 0.67 \log (\text{Body size}) + \log 0.12$$

Thus, a plot of the logs of brain and body size, or a plot on a logarithmic scale, should give a straight line of slope 0.67 (Figure 6.11).

Figure 6.11 starts to give an indication of what makes humans so special: we lie well above the allometric line seen for other mammals. If we insert a value of 60 kg as a typical body mass for humans into equation 1, our brains should weigh about 191 g. If we use the equation for primates and take k as 0.75, we calculate a prediction of 460 g. The real figure is in fact nearly 1300 g. Our brains are at least seven times larger than expected for a mammal of our size and about three times larger than that expected for a primate of our size. Despite our 1.6 per cent difference from chimps in terms of DNA, we can take some comfort from the fact that our brains are vastly different even in size.

Ancestral brains

A reasonable estimation of the size of the brains of our early ancestors can be obtained by taking endocasts of the cranial cavity of fossil skulls. There is some debate over how to interpret the fine detail of these casts (such as evidence for

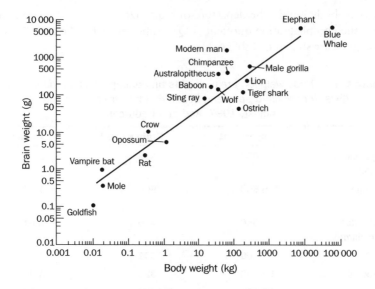

Figure 6.11   Logarithmic plot of brain size against
body size (adapted from Young, 1981)

The line of best fit is drawn as: Brain size = 0.12 (Body size)$^{0.67}$

folding), but there is consensus on the general trend: about 2 million years ago, the brains of hominids underwent a rapid expansion (Figure 6.12). Australopithecines possessed brains of a size to be expected from typical primates of their stature, but *Homo sapiens* now have brains about three times larger than a primate

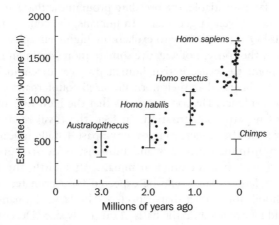

Figure 6.12   Growth in human brain volume during
human evolution (from Deacon, 1992)

Note that, during this period, body weight also grew but at a
slower rate than brain volume

of equivalent body build. The departure of brain size from the allometric line is known as the encephalisation quotient (EQ). Some values for the great apes and early hominids are shown in Table 6.2.

Table 6.2   Body weights, brain weights and encephalisation quotients (EQs) for selected apes and hominids (data from Boaz and Almquist, 1997, and other sources)

Species	Body weight (g)	Brain weight (g)	Jerison EQ*
Pongo pygmaeus (orang-utan)	53 000	413	2.35
Gorilla gorilla (gorilla)	126 500	506	1.61
Pan troglodytes (common chimp)	36 350	410	3.01
Homo habilis	40 500	631	4.30
Homo erectus	58 600	826	4.40
Homo sapiens	60 000	1 250	6.55

* Calculated from: $EQ = \dfrac{\text{Actual brain weight}}{\text{Predicted brain weight from } 0.12 \text{ (Body weight)}^{0.67}}$

The interpretation of encephalisation remains controversial. Intelligence is likely to be far too complex to have a simple relationship with the EQ. This is illustrated by what Deacon (1997) has called the 'chihuahua fallacy'. Small dogs such as chihuahua and Pekinese are highly encephalised, that is, they lie to the left of an allometric line of brain weight against body weight for carnivores. The reason is that they have been deliberately bred for smallness in body size, but since brain size is far less variable, the breeding programme that led to small dogs has left them with relatively larger brains. In humans, the condition of dwarfism also yields a high EQ. The best way to explain the highly encephalised condition of chihuahuas is by the concept of negative somatisation (see Figure 6.10 above). The important point to realise is that human dwarves or chihuahuas are not noticeably more intelligent than their normally sized counterparts.

Reviewing the evidence, Deacon concludes that the high EQs of humans are not the result of negative somatisation. In fact, the fossil record shows that hominid body size has been increasing over the past 4 million years. Instead, it seems that, in non-human primates and humans, body size growth during development is slower than for other mammals. After birth, the human brain continues to grow larger than would be expected for a primate, and body size growth in adulthood stops earlier than expected. This leaves humans with a brain larger than would be expected on the basis of our body size (Deacon, 1997).

As intelligent humans, we tend to take the value of high intelligence and large brains for granted and assume that it had self-evident survival value. It may seem obvious that natural selection would eventually deliver intelligent creatures capable of contemplating their own origins. If we think it inevitable, we are probably slipping, as is all too easy, into a progressive view of evolution. Even

Darwin was not immune from this frame of mind. Having demolished teleology and given a naturalistic explanation of the origin of species, at the end of *Origin* he says:

> And as natural selection works solely by and for the good of each being, all corporeal and mental endowments will tend to progress towards perfection. (Darwin, 1859, p. 459)

But progress to large brains was never inevitable. Natural selection is not interested in brainpower *per se* – if small brains are sufficient for the purpose of genetic replication, all the better. There is a species of sea-squirt that uses its brain to find a suitable rock to cling to; once found, there is no need for a brain and it is absorbed back into the body. The dinosaurs managed for about 200 million years with relatively small brains, and it is not at all clear that a large brain would have been of any help when the supposed asteroid hit earth about 65 million years ago and triggered their extinction.

Brains are costly to produce and expensive to run. An adult brain accounts for only about 2 per cent of body mass but consumes about 20 per cent of all the energy ingested in the form of food. Unfortunately for thinkers who want to lose weight, this percentage hardly alters whether we think hard or little.

Now we have established that there is something unique about the human brain, we need to examine why such a risky organ should have evolved to the proportions that it has. There are two related questions that quickly arise, and answering the first helps with the second. The first is why did primates evolve larger brains than other mammals, the second, why, among our own ancestors, did brain size increase well beyond that typical of other primates? Current thinking on these issues is interesting and important since it points a way towards understanding not only the roots of human cognition, but also the origin and behavioural significance of language. But first, primate brain size.

6.2.3 *Origins of primate intelligence*

There are two current popular theories that address the problem of why primates are vastly more intelligent than most mammals. One is that the environment of primates poses special problems in terms of the mental capacity needed to gather food. The other is that group-living for primates requires considerable mental skill and complexity. We will describe the reasoning behind each theory in turn and then move on to examine how the theories can be tested.

Environmental factors: food and foraging

Most herbivores, such as cattle, tend to have specialised guts that enable them to digest and ferment low-calorie food. Such animals need a large gut to contain the bacteria needed to help with breaking down the cellulose, and they need to spend considerable time grazing. For this style of life, a large brain is not necessary (although it may help with predator evasion), and it may be significant that cattle and horses lie below the allometric line shown in Figure 6.11.

Carnivores, on the other hand, receive a balanced diet with every kill: the meat of another animal is ideally suited to building new flesh. What is needed here is speed, strength and the perceptual apparatus to catch prey. Apart from the great apes, most primates are too small to have sufficient gut to ferment large quantities of cellulose, so resort to a more varied diet. They are, in effect, unspecialised vegetarians, and, at least for some species, obtaining a balanced diet can be fairly intellectually demanding.

Obtaining food can be broken down into a series of stages: travelling to find food, locating food using a perceptual system and extracting food from its source. For primates that rely on high-calorie foods, such as fruits, there may only be a few fruit trees in season in a large patch of forest. Remembering where they are is difficult work, and there is ample evidence that primates employ efficient cognitive maps to remember the location of the fruit and the optimum route between the trees (Garber, 1989). Fruits are easily identified by their colours. Colour vision is common in reptiles and birds (all those showy plumage colours not being for human delight), so it may be thought that the intellectual demands of colour vision are slight. Certainly, reptiles are not renowned for their high intelligence. But most mammals are colour blind, and it is therefore extremely likely that primates had to rediscover colour vision at a later date in evolution after it was lost. Primates do in fact have a very well-developed visual system, way beyond that which is required for locating fruits.

When food is found, the work of a primate is not over. Most primates feed upon plant matter and, whereas many plant species are only too glad to allow animals to eat their fruits and spread their seeds, they are equally averse to having their leaves and stems destroyed, having evolved counterattack mechanisms in the form of stings, prickles and poisons to deter foragers. Processing plant material to overcome such deterrents requires some cognitive skill and could have helped to drive up primate intelligence.

One obvious feature of food extraction and processing is the use of tools such as shaped twigs to catch insects, or stones to break nuts. Only chimps, however, regularly use tools in the wild, so the suggestion (often applied to humans) that tool use stimulated an increase in brain size can only be applied to chimps and not to other primates. Reviewing the evidence on the comparison of tool complexity between chimps and early hominids, Byrne (1995) concludes that the tools used by Neanderthals and *Homo erectus* were only marginally more sophisticated than those used by chimps today – hardly a credit to our forebears, yet the brains of these hominids were huge by primate standards. It seems unlikely then that tool use could explain the increase in brain size in most primates and the sudden increase in human brain size about 2 million years ago.

Social factors: Machiavellian intelligence and the theory of mind

In recent years, there have emerged several related hypotheses suggesting that it may be the demands on the social world that have been the main determinant of the growth in primate intelligence. Byrne and Whitten (1988) developed these theories and labelled them together as the **Machiavellian intelligence** hypothesis, named after the Renaissance politician and author Nicolo

Machiavelli who epitomised the cunning and intrigue of political life in Italy in the early 16th century. The essence of the Machiavellian intelligence hypothesis is that primate intelligence allows an individual to serve his or her own interest by interacting with others either co-operatively or manipulatively but without disturbing the overall social cohesion of the group. The analogy with human politics is clear: the successful and cynical politician uses his position to further his own ends while to all appearances serving the people, and without disrupting or bringing into disrepute the elective system. To investigate whether primates really employ such tactics, we will consider the social groupings of primates and the nature of their intelligence.

The size of a primate group is determined by a number of factors. The minimum group size is determined largely by the need to defend against predators: there is a security in numbers that benefits each individual. The maximum group size is probably determined by both ecological and social factors. The larger the group, the longer the time taken to move *en masse* from place to place, and the less food there is per individual when it is found. In addition, if the group becomes too large, conflicts over food and status are set up, and the group may split (Dunbar, 1996b). It follows that the benefits of group living must be weighed against its cost.

Social conflicts and costs result from the fact that the neighbour of any individual is a potential competitor for food and for mates. Some sort of order is often maintained by a linear dominance hierarchy, those at the top reaping the most rewards. The alpha male in chimpanzees, for example, will often take preference at a new feeding site and will have a good chance of fathering any offspring from a female in oestrus in the group. Beyond these simple predictions, however, social life is complicated, and the success of any individual chimp is the product of a network of coalitions and alliances. This can be seen even in the dominance

Figure 6.13   Pygmy chimps (*Pan paniscus*) grooming each other

A number of researchers have suggested that grooming enables individuals to perform effectively in social situations by forming alliances and reconciling conflict

hierarchy. The alpha male is not simply the chimp with the greatest physical strength: his status may also reflect the status of relatives and allies. His power is thus based on his 'connections'.

We can picture a primate group as a product of centripetal forces resulting from predatory pressures tending to keep the group together, and centrifugal forces emanating from tension and conflict in the group tending to push the group apart. Predation from without and conflict from within both act negatively on the reproductive fitness of an individual, and we can expect evolution to have come up with ways of mitigating these. With regard to conflict, chimps and other primates seem to have hit upon **grooming** as an effective mechanism to reduce intragroup tensions and thus enhance group cohesion.

Most visitors to zoos will have noticed the engaging way in which chimps and other primates spend time picking through each other's fur and removing bits of plant material, fleas and scabs. Such activity probably began as a form of reciprocal altruism: one individual could help to maintain the hygiene of another in return for the same 'back-scratching' favour later. However, the time that some primates, especially baboons, macaques, vervet monkeys and chimps, spend in grooming seems excessive for the demands of simple hygiene. Even the nutritional benefit to the groomer when a tasty flea or scab is found is low relative to the time expended. There is now general agreement that grooming serves a more subtle and sophisticated function than simple fur hygiene. A pair of primates that regularly groom each other are more likely to provide assistance to each other when one is threatened than are non-grooming partners. Grooming seems to serve to maintain friendships, cements alliances and is used to effect reconciliation after a fight.

But Machiavellian intelligence is more than just grooming, and the ambitious primate has other devices to help it to navigate the complex currents of social life. To really succeed in primate politics, deception is needed, and numerous observers have noted that primates will send out signals, such as false warning cries, that can only really be satisfactorily interpreted as being designed to mislead others. One example from the observations of Byrne on baboons, *Papio ursinus*, will suffice. Byrne noticed that a juvenile male, Paul (A), encountered an adult female, Mel (T), who had just finished the difficult task of digging up a nutritious corm (Figure 6.14). These are desirable food items, and the hard ground probably meant that Paul would be unable to dig his own. Paul looked around and, when assured that no other baboon was watching, let out a scream. Paul's mother (Tool), who was higher ranking than Mel, ran to the rescue and chased Mel away. Paul was left by himself, whereupon he enjoyed eating the abandoned corm (Byrne, 1995).

The significance of this is that the deployment of deceptive tactics demonstrates the ability of some primates to imagine the perspective of others. The interpretation of such observations must be made with great care since, for human observers, it is all too easy to impute intentions that are not there. Nevertheless, a body of evidence is beginning to suggest that some animals are capable of practising deception. This may not sound like a great intellectual feat but, in the animal kingdom, only humans and a few other primate species seem to have this ability. Grooming and deception make great demands on brain power. Social primates must be able to recognise one another, remember who gave favours to whom, who is related to whom and, most demanding of all, consider how a situation would look to another. No wonder we are fascinated by the social behaviour of such primates as chimpanzees, with their dominance hierarchies, old boy networks, nepotism and manipulation of others for selfish ends; it all looks very familiar.

Figure 6.14   An interpretation of deceptive behaviour in baboons, *Papio ursinus*, as interpreted by Byrne (1995) (drawing by D. Bygott)

Deception involves penetrating the mind of others. Consider an old trick used by people seasoned in the arts of institutional politics. It is often said that the best way to persuade someone else, especially a line manager, to support a particular idea is to convince her that the idea is in fact her own. It is a trick of which Machiavelli would have approved. Inspection of the ruse shows us that we are doing something quite complex. We may know that an idea is our own, but we must behave as if it were not. This requires that we must monitor our own words and body language, and imagine how we must appear to our interlocutor. We then need some technical skill of acting and mind manipulation to convince the other person that she thought of the idea; this requires some understanding of the way in which other minds work and an appreciation of the beguiling powers of vanity. To cap the deception, we may feign admiration for the originality of the idea and compliment our partner on her brilliance. To attempt this feat (the fact that you can understand it is enough) requires what psychologists call a **'theory of mind'**, with, in this case, third-order **intensionality** (the use of 's' rather than 't' in the spelling often being used to distinguish the concept from intention).

'Theory of mind' was a term first used by primatologists when they realised that chimps could solve problems that depended on their appreciating the intentions of another individual, in other words, realising that other objects out there in the world have minds complete with beliefs, intentions and mental states that can be predicted. We can conceive of this appreciation of other minds in terms of orders of intensionality. A dandelion probably has zero-order intensionality: it is not aware of its own existence; there is no one at home. Self-awareness indicates first-order intensionality. When Descartes began his famous train of sceptical reasoning and pushed doubting to its limit, he reached that fact that if he doubted his own existence, he thereby proved it since somebody must be doing the doubting, hence *cogito ergo sum*: 'I think therefore I am'. Second-order intensionality involves self-awareness and the realisation that others are similarly aware. From here on, we can posit an infinite sequence: 'I think' is first order, 'I think, you think' is second, 'I think that you think that I think' is third order and so on. Children acquire second-order intensionality between 3 and 4 years of age. Most adults can keep track of about five or six orders of intensionality before they forget who is thinking what. In the Machiavellian trick we noted earlier, we effectively want someone to believe that we think that she had a good idea ('I think that you think that I think that it was your idea'), which is third-order intensionality.

It is easy to ascribe zero-order intensionality to plants and machines but much harder to decide what has self-awareness, or first-order intensionality. Behaviourism faced this difficulty by treating all animals as machines and thus assuming zero-order intensionality. Some even adopted this approach to humans, but without much success. There are a number of problems in describing awareness in others. There is probably a natural bias in human thinking towards anthropomorphism. Humans have an acute sense of their own existence, our lives are dominated by goals and motives, and it is natural for us, and probably rightly so, to interpret the behaviour of other humans in our own terms. It is all too easy, however, to transpose this framework onto the behaviour of other animals. When a cat offers affection to humans, we are bright enough to spot that this may be just 'cupboard love' and more often than not the cat simply wants feeding. It is easy to interpret this as second-order intensionality: the cat is self-aware, aware that it is hungry and knows that it can convey this to us and so manipulate our minds to set about feeding it. In fact, we are probably far too generous in our attribution of intensionality here. The action of the cat may simply be a learned response. A cat may have first- or even zero-order intensionality, but how can we know?

One ingenious method occurred to Gordon Gallup who, in the 1960s, was a psychologist at the State University of New York. While shaving, Gallup realised that using a mirror indicates self-awareness. With very little training, humans realise that the image in the mirror is of themselves and can be used to judge and alter their appearance. Most animals, it seems, never appreciate the significance of their own image. Domestic kittens and puppies react as if the image were another individual and then gradually lose interest. Monkeys can use mirrors as a tool to see round corners to solve puzzles presented to them in captivity but never react to their image in a way that indicates self-awareness. One clever test of self-awareness is to place a spot of odourless paint on the hand and forehead of a monkey or ape while it is asleep. When the monkey or ape recovers consciousness, it typically notices the paint on its hand and attempts to remove it. When a monkey is presented with a mirror, it never makes the connection between the spot in the image and the fact that it is on themselves. In contrast, chimpanzees and orang-utans correctly grasp the significance of the image and use the mirror to help to remove the spot of paint (Gallup, 1970). Some gorillas fail the test, but one captive, called Koko, passed easily (see Tomasello and Call, 1997, for a review of the evidence on mirror self-recognition).

Moving up the ladder of intensionality, we can now ask whether self-aware animals such as chimps are also aware of other minds, that is, whether they have a theory that other minds exist. As noted earlier, Byrne and Whitten have concluded that the observational evidence demonstrates that only chimpanzees, orang-utans and gorillas practise intentional tactical deception, behaviour, that is, that can best and parsimoniously be explained by one animal deliberately manipulating the mind of another animal into a false set of beliefs (Byrne, 1995). Evidence on deception in cats and dogs was ruled out as being probably a result of trial and error learning. The evidence from tactical deception suggests, therefore, that only the great apes, some species of baboons and humans are capable of first- and/or second-order intensionality.

Theory of mind was a profound breakthrough for the apes and early hominids. In the case of humans, it has, coupled with language, given us science, literature and religion. All of these activities require a distinction between the self and the world, the realisation that other sentient creatures exist with their own views of the world that may be different to our own. All involve the realisation that words, evidence and ritual can influence the minds of others, and all of course require a standing back from phenomena and an appreciation that appearances may be different from reality. This third-party perspective lies behind the strange claim that some scientists (usually physicists in a metaphysical mood) make about 'knowing the mind of God'.

The theory of mind may have arisen from the complex social world of early hominids and itself promoted encephalisation, or it may have been a product of relative brain enlargement that developed anyway in relation to ecological factors. Against the promise held out by research into the social world of primates, we must balance the fact that not everyone is convinced by the idea that some primates have second-order intensionality. Tomasello and Call, for example, conclude that 'there is no solid evidence that non-human primates understand the intentionality or mental states of others' (Tomasello and Call, 1997, p. 340). These reservations do not, however, invalidate the whole social complexity hypothesis. However chimps represent their social world, it is clear that it makes significant cognitive demands in addition to the demands of foraging and physical survival. These two sets of factors, environment and sociality, may have been inextricably linked in the causation of hominid encephalisation and human intelligence. Some recent work has tested the competing claims of the two theories and does suggest one set of factors may have been crucially important. The next section examines this issue.

6.2.4 *Food or sociality: testing the theories*

To test these two competing theories of brain enlargement, we obviously need some way of measuring three things:

1. The level of environmental complexity associated with different foraging strategies
2. The level of social complexity set by group size and group dynamics

BOX 6.3

Relationship between group size and the number of possible pair interactions

Number in group:	2	3	4	5	6
Possible interactions					
Relative complexity indicated by number of interactions	1	3	6	10	15

If we assume that complexity is indicated by the number of possible pairs of interactions to take into account, it can be readily proven that the number of possible interacting pairs is given by:

$$\text{Interactions} = \frac{N^2 - N}{2}$$

3. The level of intelligence possessed by species that forage and live in groups.

It turns out that there are problems in tackling each of these measurements, and we must be aware of the assumptions that need to be made.

Measuring environmental complexity

The cognitive demands of feeding are obviously related to the type of food consumed, especially its spatial and temporal distribution, its ease of identification and the processing needed before it can be eaten. Most of these factors largely translate into a foraging range that can be measured in terms of distance, area or journey time. Such measures tend to ignore processing difficulty, but they should, roughly at least, correlate with environmental complexity.

Measuring social complexity

The social complexity of a group is to some degree indicated by the mean size of the group: the larger the group, the more relationships there are to keep track of, the higher the levels of stress and the greater the all-round level of harassment.

Measuring group size is also fairly easy and reliable data exist for a range of primate species. We need to be careful, however, since it does not follow that stress is linearly related to group size. In fact, in terms of the number of possible relationships of which to be aware, the complexity of a group rises rapidly with increasing group size, such that the number of possible two-way relationships in any group of size N is given by $(N^2-N)/2$ (Box 6.3). If complexity is indeed indicated by this, then it rises rapidly with group size.

Measuring intelligence

The assumptions made in measuring social and environmental complexity begin to look reasonable compared with those we face when estimating the intelligence of an animal. It is all too easy to be anthropomorphic and assess the cleverness of an animal by how well it performs tasks that humans have designated as clever. Trying to teach a chimp or some other ape to speak, for example, would be doomed to failure since the anatomy of their windpipes is not suited to the task, however intelligent they may be. Hence the trend now is to use sign language, although progress has been slow. Given the controversy surrounding the construction of a fair and culture-neutral IQ test for humans, it is not surprising that there is considerable disagreement over the creation of a 'species-fair' behavioural measure of intelligence.

Faced with the problems of measuring and interpreting intelligent behaviour, we could employ indirect methods. In fact, we already resorted to this when we concluded above that the high intelligence of primates is something to do with their deviation above the allometric line of body weight and relative brain size for mammals. There are reasons, however, for suspecting that brain size relative to body mass may be only a rough measure of animal intelligence and that we need something more precise.

Psychologists have long suspected that the brain is not an all-purpose learning device but instead consists of specific modules designated for particular tasks, rather like a Swiss army knife that unfolds to reveal a whole cluster of useful tools (Tooby and Cosmides, 1992). The basic principles of evolution support this model. As a body or organ develops through evolutionary time, it cannot at any stage be rebuilt from scratch. Even if the conditions in which an organism finds itself are very different from those of its ancestry, evolution must make do with what it already has to work on. Not that great feats cannot be achieved by this – a fin can turn into a leg, a leg into a wing and a wing back again into a fin – but evolution is essentially opportunistic, and as a consequence organisms often bear the scars of their past. The human forearm, the wing of a bat and the hand of a frog, for example, are all based on a five-digit plan reflecting an adaptation from some remote five-digit ancestor. As an analogy for this process, imagine that you must construct a boat using only items salvaged from a local scrapyard. You may succeed, but unless you were lucky enough to find a complete boat, your finished product would probably carry vestiges of the previous functions of its component parts: tyres, cans, car bodies and so on.

It may be fruitful then to consider the human brain in this light. As long ago as 1970, Maclean argued that the human brain can be divided into three main sections: a primitive core that we have inherited from our reptile-like ancestors, a

Reptilian core (striatal complex: corpus striatum and globus pallidus)
Area responsible for basic drives, repetitive and ritualistic forms of behaviour. Involved in 'innate' disposition to establish hierarchies and also possibly in storage of learnt forms of behaviour.

Old mammalian (limbic system: limbic cortex and primary nuclear connections)
Contains a number of areas concerned with fighting, feeding, self-preservation, sociability and affection for offspring

Neomammalian (neocortex)
Relatively recent in evolutionary time. Well-developed neocortex found only in higher mammals. Receives information from eyes, ears and body wall. Responsible for higher mental functions. Well developed in primates, especially humans

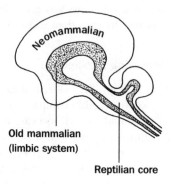

Old mammalian
(limbic system)

Reptilian core

Figure 6.15 Triune model of the brain as proposed by MacLean (adapted from MacLean, 1972)

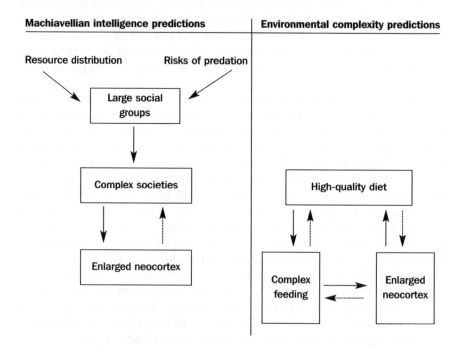

Figure 6.16 Comparison of rival theories to account for the increase in intelligence in primate ancestry (adapted from Byrne, 1995)

mid section that contains areas concerned with sensory perception and integrating bodily functions, and finally an outer layer or **cerebral cortex** that is distinctive to mammals (Figure 6.15). The word 'cortex' comes from the Latin for bark, and it is this crinkly outer layer which lies like a sheet over the cerebrum. It consists largely of nerve cell bodies and unmyelinated fibres (that is, fibres without a white myelin sheath), giving it a grey appearance – hence the phrase 'grey matter' to distinguish it from the white matter beneath. The cortex is only about 3 mm deep in humans. In non-primate mammals, it accounts for about 35 per cent of the total brain volume. In primates, this proportion rises to about 50 per cent for prosimians and to about 80 per cent for humans. If we desire some objective measure of animal intelligence, it could be the cortex that we need to focus on. The cortex surrounding the cerebellum is often more specifically referred to as the neocortex to distinguish it from other cortical areas of the brain, such as the pyriform cortex and the hippocampal cortex.

Environmental and social complexity and neocortex volume

If we accept that it may be the neocortex that is the advanced region of the brain concerned with consciousness and thought, it is this region of the brain that should correlate with whatever feature has driven the increase in intelligence in humans and other primates. Figure 6.16 shows how these two theories now stand with regard to neocortical enlargement.

To test these competing theories, Robin Dunbar (1993) of the University of Liverpool plotted the ratio of the volume of neocortex to the rest of the brain against various measures of environmental complexity and also against group size. The results were fairly conclusive. He found no relationship between neocortex volume and environmental complexity, but a strong correlation between the size of the neocortex and group size (Figure 6.17).

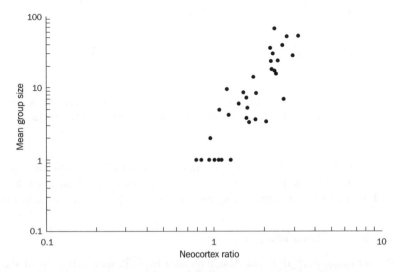

Figure 6.17   Plot of group size against neocortex ratio for various species of primate (adapted from Dunbar, 1993)

Neocortex and intelligence

The correlation observed in Figure 6.17 looks promising for the Machiavellian hypothesis, but neocortex volume is still an indirect measure of intelligence. In an attempt to establish whether the neocortex ratio correlates with Machiavellian intelligence in a more direct way, Byrne and Whiten (1988) collected data on actual observed instances of Machiavellian intelligence in action. If one primate deceives another in such a way that it shows some appreciation of the other's mental state, this is taken as an example of Machiavellian intelligence. Some primates have obviously been studied more than others, and this will tend to increase the number of tactical deceptions observed. Byrne and Whiten allowed for this effect by calculating a tactical deception index based on the number of studies undertaken. In addition, they were rigorous in excluding episodes that could be reasonably interpreted in other ways (Byrne, 1995). The result is shown in Figure 6.18.

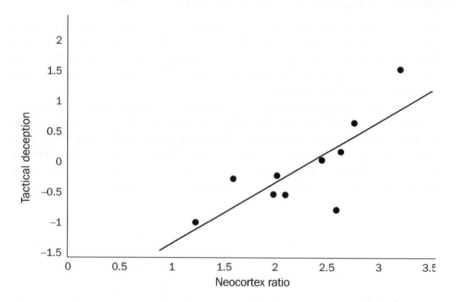

Figure 6.18   Relationship between neocortical ratio and index of tactical deception for a variety of primates (from Byrne, 1995)

The correlation of 0.77 found in Figure 6.18 is highly significant and offers further support to the idea that neocortical enlargement may have been driven by the advantages to be had from processing and utilising socially useful information.

Grooming coalitions and group size

The encouragement that the Machiavellian hypothesis receives from data on neocortical enlargements prompts a deeper inquiry into why large groups should make greater cognitive demands. We have noted that the number of possible relationships to be aware of in a group increases as a quadratic function of group size, $(N^2-N)/2$, but it is probable that the quality of relationships is just as important as their number.

As humans, we are acutely aware that large groups quickly resolve into 'cliques' or small subgroups that trade information and help. A similar phenomenon is observed in other primates (but without the mediation of an aural language) when they spend time in fairly stable subgroups grooming one another. We have already noted that grooming is both a form of reciprocal altruism and a signal of friendship and willingness to help and to expect help in future encounters. If grooming serves as some sort of social cement that binds groups together, we would expect the proportion of time that a primate spends on grooming to be related to group size. Dunbar (1993) tested this prediction using 22 species of primate that live in stable groups. The correlation, shown in Figure 6.19, is at least consistent with the idea that grooming is needed to maintain the cohesion of primate groups. It is important to note that grooming takes place among special subgroups or cliques. Any individual does not distribute grooming time equally among all other members of the total group.

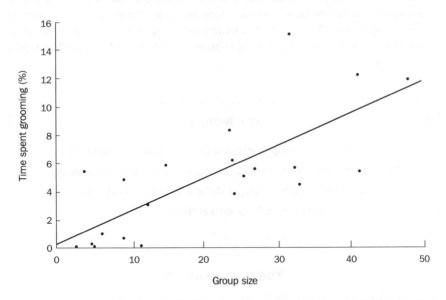

Figure 6.19   Mean percentage of time spent grooming versus mean group size for selected species of apes and Old World monkeys (from Dunbar, 1993)

The equation of the regression line is:

$$G = -0.772 + 0.287\,N$$

The work of such people as Dunbar, Byrne, Whiten and Aiello suggests that it is the social complexity of primate life that may have demanded an increase in brain size, and we may reasonably infer that this was also probably a powerful factor in driving up the size of brains in early hominids. From a gene's eye view, brain tissue is expensive and risky but probably worth the effort given the potential benefits. One of the benefits must surely have been language. Language is probably the greatest invention humans ever made, or alternatively at least the greatest endowment that natural selection ever gave us. We turn to the relationship between brains and language in the next chapter.

SUMMARY

- Cladistic classification, reinforced by molecular biology, places modern humans in the family Hominidae along with chimps and gorillas. DNA hybridisation techniques and palaeontology suggest that we last shared a common ancestor with chimps about 7 million years ago. Since then, human evolution has been characterised by increasing bipedality, a slight overall increase in body size and, most importantly, over the past million years, a rapid increase in brain size.

- Large brains, their metabolic requirements and the consequent need for prolonged infant care would have favoured increasingly monogamous sexual relationships between our ancestors in the Pleistocene era. It is probable that environmental and social factors were responsible for the selective pressure that led to large brains. The role of social factors in the evolution of primate and human intelligence is increasingly receiving support as part of the Machiavellian intelligence hypothesis.

KEY WORDS

Allometry ■ Australopithecines ■ Cerebral cortex ■ Cladistics
DNA hybridisation ■ Genus ■ Grooming ■ Hominids ■ Intensionality
Machiavellian intelligence ■ Order ■ Phylum ■ Phylogeny
Speciation ■ Taxonomy ■ Theory of Mind

FURTHER READING

Byrne, R. (1995) *The Thinking Ape*. Oxford, Oxford University Press.
A clear exposition of the Machiavellian intelligence hypothesis.

Dunbar, R. I. M. (1996) *Grooming, Gossip and the Evolution of Language*. London, Faber & Faber.
A readable and popular account of the evolution of brain size and its possible causes.

Geary, D. C. (1998) *Male, Female: The Evolution of Human Sex Differences*. Washington DC, American Psychological Association.

Geary explains the principles of sexual selection and how these can be used to understand differences between males and females. Provides a good discussion of the evidence for real cognitive differences between males and females.

Harvey, P. H. and Pagel, M. D. (1991) *The Comparative Method in Evolutionary Biology*. Oxford, Oxford University Press.
An excellent book for dealing with the theoretical and methodological problems of constructing phylogenetic trees. Gives a good treatment of allometric equations.

Jones, S., Robert, M. and Pilbeam, D. (1992) *The Cambridge Encyclopedia of Human Evolution*. Cambridge, Cambridge University Press.
Numerous experts have contributed to this book. Thorough and well illustrated with plates and diagrams. Excellent for comparing human and primate evolution.

Tomasello, M. and Call, J. (1997) *Primate Cognition*. Oxford, Oxford University Press. Although only a small part of this book deals with human cognition, it offers a valuable digest and review of recent work on non-human primate intelligence and its relevance to humans. The authors adopt an ecological approach and offer useful advice on why the theory of mind data should be interpreted with caution.

Language and the
Modular Mind

Plato... says in Phaedo that our 'imaginary ideas' arise from the
pre-existence of the soul, are not derivable from experience –
read monkeys for pre-existence

(Darwin, 1838, quoted in Gruber, 1974, p. 324)

It was noted in Chapter 1 how Darwin and Spencer cut through a philosophical problem that had entertained philosophers since the time of Plato, namely how the mind is able to obtain knowledge of the external world. Both Plato and Kant viewed the mind as an entity born pre-equipped with categories or ways of structuring experience. For Locke, on the other hand, the mind started out as a formless mass and was given structure only by sensory impressions. In a note to himself, Darwin dismissed Locke in the same way as he dispatched Plato's pre-existing soul: 'He who understand baboon would do more toward metaphysics than Locke' (Darwin, 1838, quoted in Gruber, 1981, p. 243).

Darwin was probably right: the brain at birth is not a formless heap of tissue, nor does it carry a recollection of eternal verities associated with an immortal soul. The brain enters the world already structured by the effects of a few million years of natural selection having acted upon our primate and hominid ancestors. It is therefore born ready shaped, but also eager for experience for calibration and fine-tuning to the world in which it finds itself. Eibl-Eibesfeldt expressed this succinctly:

> The ability to reconstruct a real world from sensory data presupposes a knowledge about this world. This knowledge is based in part on individual experience and, in part, on the achievements of data processing mechanisms, which we inherited as part of phylogenetic adaptations. Knowledge about the world in the latter instance was acquired during the course of evolution. It is so to speak, *a priori* – prior to all individual experience – but certainly not prior to all experience. (Eibl-Eibesfeldt, 1989, p. 6)

It has been known for many years that regions of the brain are associated with distinct functions. Vision and language, for example, can both be identified with specific parts of the brain. The origins of language and the **modularity** of the

192

mind are currently two controversial topics both in linguistics and in psychology. Many evolutionary psychologists have committed themselves to a view of the human brain as an organ consisting of a number of specific problem-solving areas or modules. The fact that language itself seems to be associated with one or two distinct regions of the brain has been used as evidence that a language acquisition facility is itself one such module.

In this chapter, the concept of the modular mind is discussed and one well-established feature of human thinking, called the Wason's selection task, is considered as evidence for modularity. The chapter also draws upon work explored in Chapter 6, on the social factors implicated in the evolution of brain size, to examine the function and evolution of language.

7.1 The modular mind

7.1.1 The issue of modularity

The evolutionary approach to psychology has been much influenced by the work of Tooby and Cosmides and the powerful manifesto on evolutionary psychology that they issued in *The Adapted Mind*. Many of the issues, in particular the extent to which the brain can be thought of as a set of discrete problem-solving modules, remain controversial, but the approach has met with considerable success as a heuristic model. For Tooby and Cosmides, psychology is to be seen as a branch of biology that studies the structure of brains, how brains process information and how the brain's information-processing mechanisms generate behaviour. The key principles of this **paradigm** are as follows:

1. The human mind is what the brain does. It is an information-processing device that receives inputs and generates outputs in a manner directly analogous to a computer. In this view, both thought and behaviour are cognitive processes. Consider the example of how to act towards kin. Kin selection theory suggests a number of factors, such as the degree of relatedness of kin to the self, their reproductive value and the costs and benefits of any action, that must be taken into account. To behave appropriately requires a cognitive computational program that factors these parameters into the decision-making. Simple instincts will not suffice.

2. The neural circuits that make up the brain were 'designed' by natural selection to solve problems that our ancestors faced in their environment of evolutionary adaptation (EEA). The most important feature of the problems solved is that they were adaptive: such problems were repeatedly encountered and, more importantly, impinged on the survivability of the organism concerned. In fact, the only problems that natural selection can solve are adaptive ones; problems to do with growth, survival, harvesting resources, avoiding predators, finding mates, reproducing and so on.

3. We are conscious of only a very small part of the working of our brain, most of its mechanisms being hidden from view. Intuition grossly oversimplifies the processes at work and can be misleading.

4. The way in which brains solved the vast array of adaptive problems was not through some general problem-solving device, which would probably be highly inefficient, but through the construction of a set of discrete and functionally specialised problem-solving modules. Each module is capable of responding to and solving a problem only over a restricted domain; hence they are called 'domain-specific modules'. Tooby and Cosmides use the analogy of a Swiss army knife with numerous blades and attachments for specific purposes.

5. Cognitive mechanisms that were sculpted during the hundreds of thousands of years that humans spent in a hunter-gatherer lifestyle will not necessarily appear adaptive today: we carry Stone Age minds in modern skulls.

6. Because humanity belongs to one species, all members of which can pool genes with any other member of the opposite sex to create viable offspring, so the variability between mental organs must be limited. These **domain-specific mental modules** are therefore common to all people, with only superficial intergroup variation. Tooby and Cosmides use an analogy with *Gray's Anatomy*.

'Just as one can now flip open *Gray's Anatomy* to any page and find an intricately detailed depiction of some part of our evolved species-typical morphology, we anticipate that in 50 or 100 years one will be able to pick up an equivalent reference work for psychology and find in it detailed information-processing descriptions of the multitude of evolved species-typical adaptations of the human mind. (Tooby and Cosmides, 1992, p. 68)

What follows from this is the fact that the genetic variation that exists between people and peoples will inevitably be minor and have very little effect on the cognitive architecture common to all that constitutes a universal human nature:

In actuality, adaptationist approaches offer the explanation for why the psychic unity of mankind is genuine and not just an ideological fiction. (Tooby and Cosmides, 1992, p. 79)

7. As with all manifestos, there is an enemy, the enemy in this case being the 'standard social science model'. The only feature of this model that is accepted by Tooby and Cosmides is the idea that genetic variation between racial groups is trivial and insufficient to explain any observed difference in behaviour. From thereon, there is fundamental disagreement. The social science approach, which they admit for this purpose is a conflation of many schools of thought, is taken to suggest that the mental organisation of adults is determined by their culture. As such, the human mind has as its main property merely a capacity for culture. Culture itself rides free from any strong influence from the lives of any specific individuals and certainly free from human nature. The blank slate approach is of course an oversimplification, but Tooby and Cosmides detect in the social science literature an assumption that the human mind is akin to a computer without programs: it is structured to learn but obtains its programs from an exterior culture rather than an interior nature.

Some potential candidates for domain-specific modules

Some modules that might be expected to form part of the human mental tool kit are mechanisms for co-operative engagements with kin and non-kin, means by which to detect cheats, parenting, disease avoidance, object permanence and movement, face recognition, learning a language, anticipating the reactions and emotional states of others (theory of mind), self-concept and optimal foraging – to name but a few. A central problem with all of this is of course knowing when to stop. At what level of discrimination and finesse have we reached an indivisible module? A useful, albeit conjectured, hierarchical organisation of modules is provided by Geary (1998) (Figure 7.1).

The importance of the social group of modules can be gleaned from the fact that nearly all primates live in complex social groups, the orang-utan being one notable exception. Furthermore, social living has almost certainly been a feature of the evolutionary ancestry common to humans and primates over the past 30 million years. The group-level modules enable individuals to function in groups in mutually beneficial ways. The recognition and discrimination of kin provides an important facet of increasing inclusive fitness, while a disposition to accept social ideologies would provide a much-needed cement to bind together non-kin for reciprocal exchanges. If early hominids were engaged in intergroup competition, mechanisms to bind groups together, and moreover treat in-group and out-group members differently, would provide a competitive advantage.

Within the category of individual modules, we can see that, in this schema, language acquisition is expected to be facilitated by specific language centres of the brain, something that is in fact the case. Similarly, all normally functioning humans can be expected to have a theory of mind centre to enable predictions about the behaviour of others to be made. Facial processing is essential for anticipating the mood of others as well as making choices about mating and remembering individuals with whom one may have made social contracts.

Under ecological modules, we note an array of specialised mechanisms for helping the human organism to exploit features of the biological and physical landscape to assist in survival and reproduction. One overall point stressed strongly by Geary is that such modules do not constrain people to behave in fixed patterns along the lines of stimulus–response thinking; predators would soon home in on organisms that behaved in predictable ways. We should instead regard the modules as conferring a flexible response according to the context.

In this view, one can see the purpose of development as enabling the calibration of these modules (setting the start-up conditions) to local social, biological and physical conditions. The modules constrain and bias the type of experiences to which a growing child should attend, but then use the result of the experiences gained to fine-tune the inherent functional mechanisms. Thus, for example, we are born with a disposition to attend to verbal sounds and organise utterances. The language that we finally speak is a result of this cognitive bias coupled with the actual evidence obtained about syntax and vocabulary. This ontogenetic development of modules in relation to the local environment can be thought of as an 'open genetic program', an open program being one that takes on board instructions from the environment (Geary, 1998).

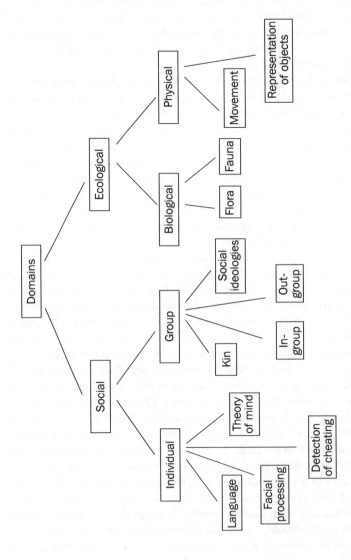

Figure 7.1 Some possible evolved domains of the mind (modified from Geary, 1998)

Problems with the modular approach

The rallying call of Tooby and Cosmides has been both effective and contro-versial, and has sparked off numerous debates. Some of the topics that are still the subject of vigorous questioning are:

- *The nature of the EEA.* Was there a single environment that shaped the human brain, and what were its features? How can we establish what the environment was like?

- *Domain specificity.* Must problem-solving modules be domain specific? Might there not be room for cognitive processes that can be brought to bear on a range of adaptive problems? Might not domain-general processes also have some evolutionary plausibility?

- *Correspondence with neurophysiological structures.* It would be unlikely that modules discovered and investigated by evolutionary psychologists would have a simple mapping structure with neurological structures, but what is the relationship between domain-specific modules and the structure of the brain?

- *Behaviour as a result of a cognitive process.* Must all adaptive behaviour be seen as the result of cognitive processes? Because behaviour has to solve a complex problem and provide an adaptive solution, this does not imply that it is driven entirely by mental processes. A physiological system might suffice. The heart, for example, is a highly adaptive organ that has to solve numerous problems to do with pumping rates and the co-ordination of pressure changes according to the activity of the organism, yet its behaviour is hardly a product of cognition.

- *The superficiality of genetic variation.* Tooby and Cosmides offer a very restricted scope to evolutionary psychology. They posit a world shared by humans with an essentially similar genetic make-up in which differences between humans are not a significant part of the adaptationist paradigm. No sooner is this thrown down than exceptions accumulate.

The first and obvious one is that there must be differences between the architecture of males and females; they must solve, for example, the problem of mate choice in different ways. So now we have two universal human natures. Tooby and Cosmides also concede some, albeit minor, role to variation, noting that there 'may also be some thin film of population-specific or frequency dependent adaptive variation in this intricate universal structure' (Tooby and Cosmides, 1992, p. 80).

David Wilson offers a constructive criticism of this position and argues the case for incorporating adaptive genetic variation within the scope of evolutionary psychology (Wilson, 1994). Wilson makes the point that, in any population, there may exist a genetic **polymorphism** such that populations within a single species may have substantial genetic differences that dispose them to behave differently but nevertheless adaptively in their local environ-ment. Many fish species seem to display just this sort of adaptive variation. Males of the blue-gill sunfish (*Lepomis macrochirus*), for example, exist in three forms: a 'a parental' form that grows to a large size and defends a nest, a smaller 'sneaker' form and a 'mimic' form that is also small but resembles a female. The sneakers and mimics thrive by dashing into the nest of the larger form at the moment at which he deposits his sperm. Here we have a genetic variation that underlies the shape and behaviour of the phenotype that gives rise to different adaptive solutions (see Wilson, 1994). The very existence of the parental form enables the sneakers and mimics to parasitise their behaviour.

Adaptive variation between individuals that is significant could also be achieved by **phenotypic plasticity**. A single genotype might produce a range of phenotypes differing as a result of the environmental cues received during development. It would in fact be a highly efficient and adaptive mechanism that allowed development to be structured by experience in ways to achieve an adaptive fit to local conditions.

Despite these reservations, many now acknowledge that the vision outlined by Tooby and Cosmides has encapsulated a major advance in thinking about the human mind. There remains a lively debate concerning different components of the programme outlined by Tooby and Cosmides, most of which is outside the scope of this book. A critique of the reliance on the EEA is provided by Betzig (1998). While broadly sympathetic, Wilson calls for a more flexible approach to allow the incorporation of genetic variation (Wilson, 1994). The domain specificity of mental modules and the assertion that behaviour must be seen as the product of cognitive processes are rejected by Shapiro and Epstein (1998). Samuels (1998) also rejects the proliferation of mental modules (the 'massive modularity hypothesis') for specific tasks.

With these reservations in mind, we will examine one of the best-documented examples of how the modular model of the human mind may help to explain puzzling features of human cognition.

7.1.2 *The modular mind in action: cognitive adaptations for social exchange*

> Blow, blow, thou winter wind,
> Thou art not so unkind
> As man's ingratitude. (Shakespeare, *As You Like It*)

This quotation reminds us that our ancestors had to cope with a difficult physical environment and a social world that contained difficult unreciprocating people. We have in-built mechanisms to cope with the cold: we shiver, reduce our exposed surface area and raise hairs from 'goose bumps' to improve insulation. So do we have in-built mechanisms to cope with cheaters? Tooby and Cosmides suggest that we do and that this is reflected in our powers of reasoning. They contend that since hominids have engaged in social interactions over a few hundred thousand years, our brains should have evolved a constellation of **cognitive adaptations** to social life. If interactions with other humans in our EEA involved exchanges of help and favours (reciprocating altruism), our cognitive algorithms (sequence of thought processes) should be adapted to possess the following abilities:

1. To estimate the costs and benefits of various actions to oneself and to others
2. To store information about the history of past exchanges with other individuals
3. To detect cheaters and be motivated to punish them.

To investigate human cognition, Tooby and Cosmides used a technique called the Wason selection task (Wason, 1966). Wason was interested in Popper's view of science that identified the hallmark of scientific reasoning to be the hypothetico-deductive method. In particular, scientists should test hypotheses by looking for the evidence that would falsify them. Box 7.1 shows the structure of a typical Wason selection task.

BOX 7.1

Basic form of Wason's selection task

Context: Part of your new clerical job in the Registry of your university is to check that student documents have been processed properly by your previous colleague. The document files of each student have a letter code on the front and a numerical code on the back. One basic and important rule for you to check is:

Rule: If a person has a D code on the front, the numerical code on the back must be a '3'.

You suspect that the person you have replaced did not label the files accurately. Examine the four documents below (some showing the front of the file and some the back).

D	F	3	7

Which document(s) would you turn over to test whether any file violates the rule?

The logical structure to problem in Box 7.1 can be written as:

D	F	3	7
P	not P	Q	not Q

The rule takes the form: 'If P then Q' (if a D on the front of the file, then a 3 on the back). The rule is violated if there is a D on the front but not a 3 on the back, or 'If P and not Q'. Thus we need only examine files D and 7. Cosmides then applied the problem to a context that involved social exchange and hence the recognition of benefits and the payment of costs. In such situations, the potential for cheating is exposed. Box 7.2 shows the logical structure of this new setting.

BOX 7.2

Wason's selection task in a social exchange context

Context: You are a bouncer in a Boston bar. You will lose your job unless you enforce the following rule:

Rule: If a person is drinking beer, he/she must be over 20 years old.

Information: The cards below represent the details of four people in the bar. One side indicates what they are drinking, the other side their age.

Instruction: Indicate only the card(s) you would definitely turn over to see whether any of the people is breaking the law.

| Beer | Coke | 25 | 16 |

In this new context, the proportion who chose 'Beer' and '16' (the correct answers) rose to 75 per cent. Tooby and Cosmides explain this improvement in performance by suggesting that the social context evokes a 'search for cheats procedure' in the human mind.

A rival explanation might be that people are simply better at reasoning in a non-abstract context of which they have some experience. Most of us are familiar with the illegality of under-age drinking but not with strange rules about student files. To test this Tooby and Cosmides varied the reasoning tasks as follows:

1. A task that has the same formal structure as the drinking problem but in a totally alien cultural setting
2. A task in which a concern to detect cheating would in fact lead to logical errors.

With regard to the second task, Tooby and Cosmides found that people were actually led into errors of reasoning by their propensity to look for cheats. With regard to the first challenge, they found that, even if it were couched in unfamiliar cultural terms (Box 7.3), the Wason problem was solved better when the problem entailed costs and benefits. Despite the unfamiliarity of the context, over 70 per cent of subjects were still able to reason correctly and choose 'P' and 'not Q' (Figure 7.2).

BOX 7.3

Wason's selection task in unfamiliar context of social exchange

Context and Rule: You are part of a tribe where a fundamental rule is that only married men are allowed to eat the aphrodisiac cassava root. Married men are always given a tattoo. All men (married or not) may eat molo nuts – a foodstuff less desirable than cassava root both in terms of taste and effects.

Instruction: Which cards would you turn over to test whether the rule has been violated?

| Eats cassava | Eats molo nuts | Has tattoo | No tattoo |

The formal expression of this is:

| P = Benefit | Not P = Benefit not drawn | Q= Has paid cost | Not Q = Not paid cost |

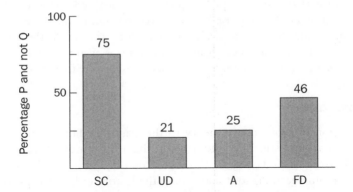

The y axis represents the percentage of correct responses

SC = task in the form of a social contract in an unfamiliar setting involving the detection of cheating
UD = unfamiliar context but task descriptive, that is, no cheating connotations
A = task in the form of an abstract logical question (for example, P and not Q)
FD = familiar context but descriptive without cheating implications

Figure 7.2 Percentage of correct responses in Wason's selection task according to context (after Tooby and Cosmides, 1992)

If the general thrust of Tooby and Cosmides' work is supported, the implications may be profound. Instead of thinking of the human mind as containing a single 'reasoning faculty', it may indeed be better to think of it as a cluster of mechanisms designed to cope with problems that we commonly encountered in the Pleistocene period of our biological evolution. One of these problems was how to reason accurately enough to detect cheating. The critical response to this work has been mixed. For methodological criticisms that still support the social contract hypothesis see Gigerenzer and Hug (1992). For a more severe criticism of this view, see Davies *et al.* (1995).

If the detection of cheating were so important in the social environment of our evolution, we might expect that other components of our mental apparatus would be finely tuned to detect cheating. Mealey *et al.* (1996) investigated whether our memory of faces is enhanced by a knowledge that the face belongs to a cheater. They presented a sample of 124 college students with facial photographs of 36 Caucasian males. Each photograph was supplied with a brief (fictitious) description of the individual, giving details of status and a past history of trustworthy or cheating behaviour. Students were allowed about 10 seconds to inspect each face. Of the 36 pictures seen, 12 were described in the category of trustworthy, 12 as neutral and 12 as threatening or likely to cheat. One week later, the subjects were shown the pictures again (together with new ones) and asked whether they remembered the faces.

The overall finding was that both male and females were more likely to remember a cheating rather than a trustworthy face. The effect was significant for males and females but stronger for males ($p = 0.0261$). This work supports the general notion that our perceptual apparatus is adapted to be efficient at recognising cheaters. Puzzling features, however, remain. In the same study, the authors found that if pictures of high-status males were used, the enhanced recognition of cheaters disappeared for males and was even reversed for female subjects, in that they were now able to recollect trustworthy faces more reliably.

7.2 Language

The application of evolutionary theory to the origins of human language has always been controversial. The arguments in the 19th century were so heated that, in 1866, one scientific society, the Societé de Linguistique de Paris, banned all further communications to it concerning the history of language. Over 130 years later, there are still wide-ranging disagreements over such basic issues as the probable timing of the start of human language and even whether language is a product of natural selection or is merely some emergent property of an increase in brain size. There are those who maintain that language first appeared in the Upper Palaeolithic period about 35 000 years ago, and those who suggest that language arrived with the appearance of *Homo erectus* about 2 million years ago. There is only room here to glance briefly at some of the main arguments in these debates. We will, however, explore the implications of some recent work on **grooming** and language.

7.2.1 *Natural selection and the evolution of language*

There has always been a strong anti-adaptationist tradition in linguistics. Noam Chomsky, one of the world's leading linguists, and Stephen Jay Gould, a prolific and widely read evolutionary theorist, have both repeatedly argued that language is probably not the result of natural selection. Gould's position seems to stem from a general concern about the encroachment of adaptive explanations into the territory of human behaviour. We saw in Chapter 1 that he has used the term 'Panglossianism' to deride those who see the products of natural selection in every biological feature. Gould seems to have a view of the brain as a general purpose computer that, being flexible, can readily and quickly acquire language from culture without needing any hard wiring. Gould's output and influence have been great but one cannot help but feel that his scepticism towards an evolutionary basis for language stems in part from a political agenda that may be well intentioned but unreasonably resistant to any claims for a biological underpinning of human nature.

Chomsky takes the view that language could have appeared as an emergent property from an increase in brain size without being the product of selective forces. He argues that when 10^{10} neurones are put in close proximity inside a space smaller than a football, language may emerge as a result of new physical properties. Chomsky's position is all the more surprising since he has battled long and hard to show that a language facility is something we are born with and not something that the unstructured brain simply acquires by cultural transmission.

The leading exponent of language as a product of natural selection is probably Steven Pinker, a linguist at the Massachusetts Institute of Technology in the United States. Pinker advances a number of arguments tending to suggest that language has been the outcome of a selective force (Pinker, 1994). In summary, these are:

- Some people are born with a condition in which they make grammatical errors of speech. These disorders are inherited (see also Gopnik *et al.*, 1996)
- Language is associated, although not in a simple way, with certain physical areas of the brain, such as Wernicke's and Broca's areas
- If language emerged as the product of a large brain (and only recently), what use was a large brain before that?
- Complex features of an organism that have been naturally selected, such as the eye of a mammal or the wing of a bird, bear signs of apparent design for specialised functions. Pinker argues that language bears these same types of design feature
- Children acquire language incredibly quickly. Parents provide children with only complete sentences and not rules, but children nevertheless infer rules from these and apply them automatically
- The human vocal tract has been physically tailored to meet the needs of speech. Specifically, humans, unlike chimps, have a larynx low in the throat (Figure 7.3). This allows humans to produce a greatly expanded range of sounds compared with a chimp
- The workings of the human ear indicate that auditory perception is specialised in a manner ideal for decoding speech
- If languages were products of culture, we would expect some correlation between the level of sophistication of a culture and the grammatical complexity of its language, yet no such correlation is found and even the language of hunter-gatherers is grammatically complex.

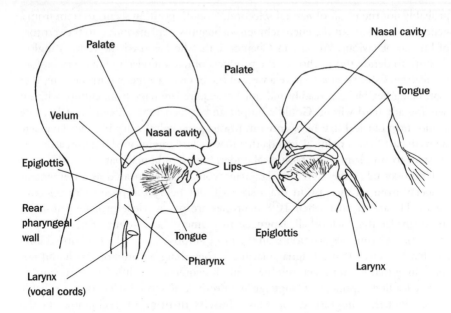

Figure 7.3   Comparison of head and neck of a human adult and an adult chimpanzee (based on Passingham, 1988)

The problem that all evolutionary accounts of the origin and development of a complex structure such as an eye or a wing or language have to face is the task of explaining the adaptive advantage of all the intermediate stages. Assuming that an eye, a wing or language did not suddenly fully emerge – and the probability of all the mutations appearing at once would truly be infinitesimal – we must ask what is the advantage of 5 per cent of a characteristic? There are in fact standard and convincing answers for the emergence of eyes and wings, proposed by Darwin himself. Five per cent of an eye in the form of, for example, a few light-sensitive cells could convey considerable advantages in the same way as proto-wings could be used for gliding or stabilising motion. The charge often levelled against adaptionist accounts of language is that 5 per cent of a language would be useless, and in any case the first language mutant would have no one to talk to. Pinker and Bloom respond to this by arguing that the early language-prone genetic variants would probably be close kin and could thus benefit from sharing language (Pinker and Bloom, 1990). Also, as is well known, 'pidgin' languages, the language of children and non-fluent tourists, show that a continuum of language skills and attributes can exist that still have use value. Other animals also show the value of even a limited vocabulary. Vervet monkeys, for example, have alarm calls that distinguish between leopards, eagles and snakes.

In relation to the Machiavellian intelligence hypothesis discussed earlier, we may also note that once hominids acquired language, it would not have taken long before it was realised that words could be used to convey half truths and untruths. The need to distinguish truth from falsehood would place a strong selective pressure on the ability to decode utterances carefully, attend to their nuances and assess the plausibility of what was on offer. For this reason, and other selective advantages accruing from sharing ideas and conveying information, the move from a simple language to the complex languages we see today could have been quite rapid.

Despite a vigorous exposition of their ideas, the work of Pinker and Bloom is likely to be controversial in the years ahead. Biologists tend not to study linguistics, and evolutionary theory is largely absent from the curricula in the cognitive and behavioural sciences. Perhaps disciplines, like the genome of individuals, have much to gain from cross-fertilisation and outcrossing.

7.2.2 *Dating the origin of human language: culture, anatomy and grooming*

Predictably, if there is debate about whether language is an instinctive evolved trait or a cultural achievement, there will be debate about when it began. If we agree with Pinker that language is a feature of human biology, it must have begun no later than about 200 000 years ago when *Homo sapiens* appeared, and could have been around in earlier hominid species. If we take the cultural achievement view, we must place the origin of language with the origin of symbolic culture, generally around 35 000 years ago. There is in fact a variety of evidence that has been used at some time or other to try to pinpoint the start of human language. We will briefly review some of the evidence but concentrate on recent work relating language to Machiavellian intelligence and grooming.

Cultural practices: the evidence from archaeology

Some, such as Randall White of New York University, suggest that language must have begun about 35 000 years ago at the end of the phase known as the Upper Palaeolithic (White, 1985). The argument is basically that it is at this time that we observe the first efflorescence of art and the practice of deliberately burying the dead, often with grave goods. Art and the burial of the dead imply a shared system of meanings and cultural beliefs. About this time, we also observe rapid technological change: materials other than stone, for example bone, are recruited for tools, and stone tools now appear as if order has been imposed on them according to some mental plan, instead of in earlier artefacts, which were opportunistic chippings. We also observe some regional differences in culture, increasingly large and complex living sites and evidence of trade. All of these achievements would, it is suggested, require a language.

There have been various criticisms of these arguments. Pinker (1994) objects that the whole premise is false: there is no single symbolic capacity that underlies art, technology and language. In any case, recent discoveries of the use of red ochre and engraved ostrich shells at some South African sites have pushed back evidence of symbolic culture to about 100 000 years before the present (Knight *et al.*, 1995).

Anatomy

A sudden expansion in brain size among our ancestors began about 2 million years ago with the first member of our genus, *Homo habilis*. By about 1 million years ago, *Homo erectus* had a brain capacity of about 1100 cm³, which is not far from our own of 1300 cm³. If language coincided with sudden brain expansion, this points to *Homo erectus* and *Homo habilis* as having had language capabilities. The problem is of course that we do not know whether there is a minimum brain size for language to start. If the minimum were 1300 cm³, only early *Homo sapiens* at 200 000 years ago could have had language. We know that language is associated with certain areas of the brain such as Broca's area and Wernicke's area. It has been claimed that an examination of endocasts of fossil skulls shows evidence of the existence of a Broca's area in the cranium of a *Homo habilis* specimen dated at about 2 million years old (Falk, 1983; Leakey, 1994). There have also been attempts to link asymmetries in fossil crania with the origin of language. Language in most people is associated with the left hemisphere, and, partly as a consequence of packing language circuitry into one half of the brain, the left hemisphere is slightly larger. Holloway (1983) suggests that this too can be detected in a *Homo habilis* specimen. Objections to this have included the fact that even specimens of Australopithecines show brain asymmetries, and the attribution of language to this small-brained ancestor seems improbable.

The most compelling anatomical evidence, however, comes from an examination of the throats of humans, apes and our ancestors (see Figure 7.3 above). The voice box of humans is called the larynx and contains a cartilage, called the Adam's apple, that bulges from the neck. If you feel your Adam's apple, you will notice that it is some distance down from your mouth. Humans have their larynx low in their throats, which allows the space for a large sound chamber, the pharynx, to exist above the vocal cords. In all other primates, the larynx is set high in the throat, which restricts the range of sounds that can be made but does at least allow the animal to breathe and drink at the same time. Humans are in fact born with their larynx high in the throat, which helps the infant to suckle while breathing. After 18 months, the larynx begins to move down the throat, finally reaching the adult position when the child is about 14.

In adult humans, the tongue forms the front wall of the pharangeal cavity, so movement of the tongue enables the size of the pharynx to be altered. In chimps and infants, the higher position of the larynx means that the tongue is unable to control the size of the air chamber immediately above it. The differences between humans and primates in the position of the larynx strongly suggests that we have evolved vocal apparatus to convey a wide range of sounds.

It would seem at first glance an easy task to determine the position of the larynx in ancestral fossils and thus work out when the enlarged speech-facilitating pharynx was established. The problem is that all this vocal apparatus is composed of soft tissue, which does not fossilise. Clues can, however, be found in the bottom of the skull, the basicranium, which in humans is arched as a result of our specialised vocal equipment and quite unlike the flat shape of other mammals. Following this approach, Laitman (1984) found that a fully fledged basicranium emerged about 350 000 years ago. The earliest *Homo erectus* specimen of about 2 million years ago has a slightly curved basicranium typical of the larynx position of a modern 6-year-old boy.

Grooming and the origin and function of language

Work by Dunbar referred to earlier provides another approach to the determination of the origin of language, which also suggests a function for the origination of language within Machiavellian theories of intelligence. There have been criticisms of Dunbar's methodology (see peer reviews in Dunbar, 1993), but the sheer originality of the reasoning makes it worth following. If we examine once again the plot of group size against neocortical ratio shown in Figure 6.17, we can use the equation of the regression line to predict something about the 'natural' size of human groups. The equation is:

$$\log N = 0.093 + 3.389 \log Cr$$

where: N = group size
Cr = neocortical ratio.

The neocortex in humans occupies about $1006 \, cm^3$ out of a total brain volume of $1252 \, cm^3$. This gives a Cr of 4.1, which, when inserted into the equation above, predicts a group size for humans of 148, with a 95 per cent confidence interval range of 100–231. We should note that this is the predicted size for a group of hominids living a hunter-gatherer lifestyle. Dunbar assembles evidence, including the size of early Neolithic villages, the number in contemporary hunter-gatherer bands and military companies over the past 300 years, and the number of people with which modern humans can have a genuinely social relationship, to suggest that this figure has some credence. It is difficult to determine what ecological factors drove the size of human groups up from what we must assume were smaller groups of our ancestors, which may have been the size of those of modern chimps, at about 50. It could be that the more open habitat colonised by early hominids was more vulnerable to predation, that early hominids practised a nomadic way of life or that competition existed between different hominid groups (Dunbar and Aiello, 1993).

If we now insert the figure for the size of a human group into the equation represented in Figure 6.19, $G = -0.772 + 0.287 N$, we arrive at a predicted grooming time of 42 per cent. Now, grooming is part of the overall time budget of an animal: time spent grooming is time lost for hunting, foraging, looking out for predators and attending to young. Dunbar suggests that the upper affordable limit for most primates is about 20 per cent, the highest recorded value for an individual species being 19 per cent for *Papio papio* baboons. Dunbar then proposes that, as the enlargement of the neocortex proceeded hand in hand with increasing group size and hence social complexity, there arose a point at which a more time-efficient means of grooming became essential. Dunbar's thesis is that language evolved as a cheap form of social grooming. Language enabled early *Homo sapiens* to exchange socially valuable information, not of the 'There is a beast down by the lake' sort as is usually supposed, but about each other: who is sleeping with whom, who can be trusted and who cannot. In short, language began as a device to facilitate the exchange of socially useful information, gossip as it is sometimes called.

In support of his thesis, Dunbar suggests that the subject matter of most conversations today is still predominantly of the gossip variety. In one study, conversations that were monitored in a university refectory were for over half the time concerned with social information, dealing with academic matters for only 20 per cent of the time. The crucial thing about language as an effective grooming device is that it can be used while other activities, such as walking, cooking and eating, are still taking place. Moreover, several groomers can be linked together in a conversation, unlike the pairwise interactions of physical grooming. In fact, if we assume that early humans could only afford the same time as chimps for grooming (about 15 per cent of the time available), then, since the ratio of human group size to chimp group size is about 148:54 or 2.7:1, early humans needed something about 2.7 times more effective than one-to-one grooming. Dunbar *et al.* (1994) suggest that it is no accident that typical human groups forming for conversation or gossip usually number about four. From an individual point of view, interacting with three other people is three times more effective in the use of time than interacting with one, and three is pretty close to 2.7.

Dunbar's analysis can also be used to date, very approximately, the start of human language. If we accept the view that unspecified ecological factors drove up the group size for early hominids, and that this in turn spurred the growth of the neocortex and thus the time spent grooming, there came a point at which the time needed for grooming called for language as a more time-efficient grooming device. Dunbar suggests the 'Rubicon' of grooming time that had to be crossed by language was about 30 per cent. Now, there is an ingenious way to predict grooming time for early hominids. Since grooming time is related to group size and group size is related to neocortical ratio, we can, even though we have no direct idea of the size of the neocortices of early hominids, use the fact that there is an allometric equation linking neocortex size to overall brain size to make some predictions. The results are that a 'Rubicon' of 30 per cent of the time budget translates to a group size of about 107. When this is plotted on a graph of predicted group sizes for various hominid species based on suggestions for the neocortical ratios, it provides a date for the start of language-based grooming for late members of *Homo erectus* and early *Homo sapiens* somewhere between 300 000 to 200 000 years ago. Table 7.1 shows data on predicted grooming times for fossil and modern hominids.

One problem with these data is the high grooming time predicted for Neanderthals. If language provided the answer to the demands for grooming, this tends to suggest that Neanderthals had language, yet evidence from the cultural achievements of Neanderthals would suggest they were not as advanced, at least in symbolic culture, as was *Homo sapiens*.

Dunbar's prediction is consistent with the idea that language is a distinguishing mark of *Homo sapiens*. It provides a rather later date than some of the anatomical evidence on cranial shapes and topographical features, and Dunbar's ideas are original and provocative. We will summarise a few areas in which the data or methodology has been criticised.

Association between grooming and group size

Figure 6.19 shows the relationship between primate group size and grooming time, which may be an association rather than a causal relationship. It could be

Table 7.1 Predicted percentage grooming times for fossil and modern hominids (data from Dunbar and Aiello, 1993)

Taxon	Number in sample	Mean % predicted grooming time
Australopithecus	16	18.44
Homo habilis rudolfensis	7	22.73
Homo erectus	23	30.97
Archaic *Homo sapiens*	18	37.88
Neanderthals	15	40.46
Modern *Homo sapiens* (female)	120	37.33
Modern *Homo sapiens* (male)	541	40.55

that living in larger groups leaves more free time for grooming. Much also depends on the choice of species used to obtain the relationship between grooming time and group size. The inclusion of species that form fission–fusion groups, such as chimpanzees, gives a different regression line. There is even some debate over whether grooming does function to maintain group cohesion.

There are particular problems when Dunbar's method is used to date the origin of language since there is a three-step process of inferring neocortex size from total brain size, then from neocortex size inferring group size, and from group size deducing grooming time. Any errors in the first or second stages (and there are wide confidence limits) are compounded by the time we get to the last.

Language as grooming

It is not obvious that language used to gossip is a direct equivalent of grooming: one would expect to find some structural similarities in the behaviours. Language appears too complex for grooming, and grammar, if anything, seems as much designed to describe the physical as the social world. It is also not obvious that a conversation group of four really is three times as efficient as speaking to individuals on a one-to-one basis. One-to-one conversations may be more intense and more productive in the sharing of valuable information compared with group gossip. To infer from gossip today an original function for language is a huge leap. We may gossip because our standard of living has given us time. The gossiping of academics and students in university refectories may not be typical of early hunter-gatherers.

Despite these reservations, Dunbar may have hit upon something very important. Language may have a role in supporting social intelligence. To function effectively in groups, we do need to know who are our relatives and friends. This may be obvious with close kin, but we need social information to recognise more distant relatives. In groups in which reciprocal altruism and especially indirect reciprocal altruism are found, we also need to keep track of the reputations of others and ensure that our own reputation is sound. Whether language started as gossip, and information about the physical world was a byproduct, or whether it was the other way around is still debatable.

Figure 7.4   Young people talking

Theorists such as Robin Dunbar have argued that language may have
originated as a means of exchanging socially useful information. Gossip, as it is
sometimes called, thereby functions as an efficient form of grooming

In the last analysis, the interactions between ecological factors and social complexity are extremely difficult to disentangle. We have dwelt upon theories dealing with social intelligence not because this is likely to have been the only route towards larger brains and the development of language – as ecological factors and bipedalism must have played a significant role – but because if the social intelligence hypothesis can be shown to be applicable to human evolutionary development, this whole area has a great potential to explain the way in which we use language now and in particular how we interact with others. The physical environment that our ancestors had to face would now be totally alien for most people, and the ecological dimensions of city life are not even apparent, but we may still carry around with us those mind modules that evolved to enable our species to cope with its social existence 300 000 years ago. Most of us in the wealthier countries are fortunate enough now not to worry about optimum foraging strategies, tool-making, keeping warm and predator evasion. The carnivores have gone and so have the physical exigencies of life, but we still have to negotiate our way through a social mix of relatives, friends, rivals and enemies.

SUMMARY

One approach to the theory of mind (Chapter 6) and the language facility is to regard them as domain-specific modules. Two American psychologists, Tooby and Cosmides, have been strong advocates of an approach to evolutionary psychology purporting to show that the human mind is constructed as a series of specialised problem-solving mechanisms or 'domain-specific modules'.

■ Such modules have been shaped by the thousands of years that early humans spent in their EEA and are adaptive in the sense of having been sculpted to solve ancient problems. There are expected to be modules to deal with group membership, the treatment of kin, facial processing and features of the biological and physical environment such as the distinction between flora and fauna and the properties of objects. There is much debate about the rigidity and specificity of these modules. The approach has been successful at least in one area – showing how human reasoning is biased by the need to detect cheating in social exchanges.

■ Although there is still widespread debate about the origin of language, and even about whether it is a product of natural selection, one intriguing set of ideas suggests that social complexity led to large brains. As group size grew in relation to ecological parameters, language evolved to serve as a grooming device in complex social groups. This line of thought places the origin of language somewhere between 300 000 and 200 000 years ago and identifies a surprisingly significant role for gossip.

KEY WORDS

Cognitive adaptations ■ Domain-specific mental modules ■ Grooming
Modularity ■ Paradigm ■ Phenotypic plasticity ■ Polymorphism

FURTHER READING

Barkow, J. H., Cosmides, L. and Tooby, J. (1995) *The Adapted Mind*. Oxford, Oxford University Press.
A highly influential work containing the manifesto of evolutionary psychology by Tooby and Cosmides. Other chapters of interest deal with cognitive adaptations for social exchange, the psychology of sex and language.

Deacon, T. (1997) *The Symbolic Species*. London, Penguin.
An evolutionary account on the growth of human brains that stresses the importance of the co-evolution of language and the brain. Deacon argues that the ability of the mind to construct and hold symbols is key.

Dunbar, R. I. M. (1996) *Grooming, Gossip and the Evolution of Language*, London, Faber and Faber.
A readable and popular account of the evolution of brain size and its possible causes.

Pinker, S. (1994) *The Language Instinct*. London, Penguin.
A popular account of the modular and evolutionary approach to language.

8

Understanding Human
Sexual Behaviour:
Anthropological Approaches

Oh wearisome condition of humanity!
Borne under one Law, to another bound;
Vainely begot, and yet forbidden vanity,
Created sick, commanded to be sound:
What meaneth Nature by these diverse Lawes?

Fulke Grenville, Lord Brooke, *Certaine Learned and Elegant Works* (1633)

Most people find the mating behaviour of our own species a source of endless fascination. The sexual antics of the rich and famous figure prominently in the millions of tabloid newspapers sold each day – and who among us, however cerebral, is immune to a bit of gossip about the affairs of our friends? The other end of the cultural spectrum is no exception: the themes of love, passion and jealousy have inspired some of the world's greatest art and literature. In terms of our daily activities, our conversations, high culture or low, sex infuses and dominates our lives. We are one of the sexiest primates alive.

This chapter examines the physical and historical evidence on human sexuality, employing perspectives often found in biological or physical anthropology. Its goal is to establish the species-typical mating strategies employed by human males and females. It has already been established in Chapter 4 that males and females will have different interests, so it should come as no surprise to find that these strategies are different. We should also expect these strategies to vary with local conditions.

8.1 Contemporary traditional or preindustrial societies

Humans living in today's industrial or more developed countries are living in conditions far removed from those prevailing in environments where the basic plan of the human genotype was forged. In addition, many such cultures are strongly influenced by relatively recent ideologies and belief systems such as Judaeo-Christianity, with its strong injunction in favour of **monogamy**. If we

212

want to ascertain the sexual behaviour of humans before our mode of life was transformed by industrialisation, or before our heads were filled with strict religious and political ideologies, it makes sense to look at traditional hunter-gatherer cultures that still exist, or cultures that have been relatively immune to Western influence and maintain their traditional patterns of life. This is not to suggest that such cultures are primitive or have no ideas of their own, but we could be assured that such cultures have not been subject to the mass persuasion systems of the Church and state found in the West.

8.1.1 *Cultural distribution of mating systems*

A broad sweep of different human societies reveals that, in many, the sexual behaviour observed departs from the monogamy advocated (at least in a legal sense) in most Western cultures (Figure 8.1).

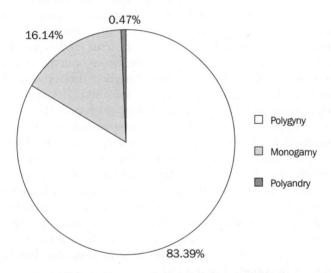

Figure 8.1 Human mating systems in traditional cultures prior to Western influence (from Smith, 1984)

It is difficult to jump to strong conclusions from such surveys, however, for at least two reasons. First, there is the problem of the independence of data points. If we identify two cultures that practise, for example, **polygyny**, this could quantitatively represent either two data points or one. It would count as two if the cultures were isolated and underwent independent social evolution but only one if they both descended from some ancestral culture that practised polygyny. Second, by focusing on traditional tribal societies relatively isolated from Western influence, we are making the questionable assumption that such cultures are 'throwbacks' or survivals of earlier stages in the social evolution of humans. Most anthropologists today see tribal societies not as frozen in time but as having developed in different ways from industrial cultures – but as having developed nonetheless.

With these reservations in mind, there is still some merit in examining some traditional cultures. If we look at hunter-gatherer societies, we at least know that people here face ecological problems similar to those of our ancestors 100 000 years ago. The often harsh existence of hunter-gatherers also places restrictions on the sort of social development that can take place. Patterns of mating behaviour cannot drift too far from that which is optimal for survival.

8.1.2 *Hunter-gatherer societies*

It is a sobering thought that at least 50 000 generations of differential survival and reproduction took place in a foraging way of life. No hominid subsisted in any other way until the invention of agriculture about 10 000 years ago. In studying hunter-gatherers today, however, we meet a set of problems different from those prevailing in the Pleistocene period. Although humans have been hunter-gatherers for 99 per cent of their time on the planet, contemporary hunter-gatherers are mostly confined to marginal environments such as deserts or sub-Arctic tundra rather than the more hospitable environments that have for obvious reasons been the subject of development and colonisation. Contemporary hunter-gatherers may therefore be 'survivors' from the inroads of agriculture, and we must be wary that they may not be entirely representative of what hunter-gathering was once like.

Bearing such problems in mind, we note that studies, such as those by Howell (1979) on the !Kung San people, tend to reveal a pattern of mild polygyny, males having a slightly higher variance in reproductive success than females as a result of serial polygyny resulting from remarriage or simultaneous polygyny.

Foley (1992) argues that the distribution of modern hunter-gatherers in tropical Africa is confined to areas of low and high rainfall where large mammals are not particularly abundant. It is probable therefore that contemporary hunter-gatherers live in environments depleted relative to those of our ancestors in large herbivores and carnivores, and we must bear in mind that hunting was once probably more important than it is today.

It turns out that food supply and the role of hunting in obtaining food are important factors in understanding the mating strategies of hunter-gatherers. The Ache people of Paraguay were hunter-gatherers until 1971, when they were enticed to live on a government reservation. Studies by Hill *et al.* (1988) showed that men would often donate meat to women in exchange for sex, high-ranking men gaining most from this practice. In this case, it seems that some Ache men achieved polygyny through meat-induced adulterous affairs. As a general rule, however, it is highly likely that a foraging way of life, especially where hunting produced an important part of the diet, never really sustained a high degree of polygyny.

The reasons for this are basically twofold and fairly simple. First, hunting large animals is risky and needs a combination of co-operation and luck. Given the prolonged period of gestation and nurturing for human infants, hunting is carried out by males, and the co-operation needed means that male rivalry must be kept within strict limits. Following a kill, the meat must be shared between all those who helped, and also with other unsuccessful groups along the lines of reciprocal altruism. If a high degree of polygyny prevailed in such groups, the sexual rivalry would militate against such altruism. In fact, the equitable sharing of hunted food is characteristic of hunter-gatherers and totally unlike that of other social hunting species where after a kill there is a free for all. Second, even if there were a surplus after sharing, meat is difficult to store. It is hard to see how, in a foraging culture, sufficient wealth or resources could ever be accumulated by one man to support a sizeable harem. Predictably, polygyny has been found to be pronounced or common in very few known hunter-gatherer societies. In most hunter-gatherer groups, men will have one or at most two wives.

The evidence above on the difficulty of accumulating resources and the need for co-operative hunting does not rule out the possibility that, if resources were more abundant in the late Pliocene period, and thus easier to accumulate by a few men, a higher degree of polygyny might have been found. Foley (1996) takes this view, arguing that the social structures we observe in today's hunter-gatherers are not vestigial but represent novel adaptations to a resource-depleted post-Pliocene environment. Foley argues that, as the Neolithic revolution left the hunter-gatherers outmoded, it was the agriculturists who would have maintained the ancestral social system of the Pliocene era consisting of 'polygynous family groups linked by alliances of male kin organised patrilineality' (Foley, 1996, p. 108). If Foley is right, there may be some merit in examining social systems of early agricultural communities, some of which now enter the historical record.

Figure 8.2   Yanomami Amerindians washing vegetables

The Yanomami, living in the rainforests of southern Venezuela and northern Brazil, are tribal people practising gardening and hunting. They number about 15 000 individuals and have no written language or formally coded laws. Yanomami males are often polygynous and this results in violent conflicts between males as they compete for wives

8.2 Sex and history

An inspection of historical societies allows us to examine how people have behaved under different cultural constraints and opportunities. This approach carries the advantage that we can have access to a diversity of legal and cultural norms. We will use the term 'historical societies' to refer to human groups living after the invention of agriculture and when written records became available for the first time. The problem with the historical approach is that evidence is often patchy, incomplete and biased by the recorder at the time, or by later selection on the part of historians. Moreover, historical arguments tend to adopt the narrative technique of juxtaposing and evaluating plausible assertions rather than the testing of hypotheses under controlled conditions that is the characteristic of science. Nevertheless, unless we want to subscribe to the Olympian notion that, lacking scientific rigour, historical knowledge has no value, we should concede that the study of historical societies might have something to offer.

8.2.1 *Power, wealth and sex in early civilisation*

In the United States, Mildred Dickemann, John Hartung and Laura Betzig have all pioneered the Darwinian approach to human history. Betzig (1982, 1986) examined six civilisations of early history: Babylon, Egypt, India, China, the Incas and the Aztecs. She found that, in all of them, an accumulation of power and wealth by a ruling élite coincided with the prodigious sexual activity of the rulers. All of these societies at some stage of their history were ruled by male despots or emperors who kept harems. The harems of these male rulers were vigorously defended and guarded by eunuchs, extreme penalties being meted out to any subject who had sexual relations with the ruler's concubines. This degree of polygyny, consisting of hundreds or even thousands of wives, would be unimaginable in hunter-gatherer societies. Economically, such harems were made possible by the Neolithic revolution that enabled the taxation of the majority to support the retinue of a minority. In short, harems were favoured by social inequality. As Betzig concludes: 'across space and time, polygyny has overlapped with despotism, monogamy with egalitarianism' (Betzig, 1992, p. 310).

In seeking to explain this phenomenon, it could be argued that such harems were displays of wealth – items of conspicuous consumption that gave pleasure to the ruler as well as broadcasting his power. Each of these reasons may offer a partial explanation, but several features of the regulation of the harems do not fit easily into such conventional accounts. Betzig shows how the structure of the harem seems to be designed to ensure the maximum fertility of the women concerned as well as ensuring absolute confidence of paternity for the despot. In some cases, wet nurses were employed, thereby allowing the harem women to resume **ovulation** soon after the birth of a child. In the civilisations of Peru, India and China, under the emperors Atahallpa, Udayania and Fei-ti respectively, great care was taken to procure only virgins for the harem. In the Tang Dynasty of China, careful records of the dates of conception and menstruation of the women were kept as a means of ascertaining their fertility. In this light, harems appear as breeding factories designed to maximise the propagation of an emperor's genes.

Betzig also studied the aristocracy of the early Roman Empire. Roman marriage was monogamous by law, but men still found ways to secure extrapair copulations. Historical sources such as Tacitus and Seutonius consistently speak of the voracious sexual appetite of the early emperors and how they were provided with virgins and concubines. Wealthy Romans also kept male and female slaves, even though few of the female slaves had real jobs around the household. Betzig rejects the suggestion that female slaves were employed to breed more slaves, since male slaves were forced to remain celibate and pregnant female slaves did not command a higher price. Betzig argues that female slaves were used by noblemen to breed their children. The fact that slaves born in a Roman household were (unlike slaves used in the mines) often freed with an endowment of wealth suggests that noblemen were freeing their own offspring.

We find the same pattern again and again: from the tyrants of old to the presidents of democracies today, when men become rich and powerful enough, they pursue and achieve polygyny. Exactly why extreme polygyny faded is not certain. The extreme polygyny of harems seems to represent an interlude between the end of hunter-gathering and the spread of democracy. One suggestion is that intense polygyny was destabilised by the inability of the ruling polygynists to command the loyalty of deprived and frustrated foot soldiers (Alexander, 1979). Ridley (1993) suggests that the rise of democracy allowed the ordinary man to express his resentment at the sexual excess of others. Even today in some countries, the sexual excesses of men elected to powerful positions can bring

Figure 8.3 'The Harem', by John Frederick Lewis (1805–1876)
Throughout history powerful men have employed harems and concubines. As well as serving as a status symbol for the male and providing him with sexual pleasure, they seem to be expressly designed to ensure the propagation of his genes

about their downfall. Whatever the historical reasons, no doubt complex, it is fairly clear that the extreme polygyny practised by ancient despots is not typical of the human condition for most of its history. Such evidence does show, however, how males can act opportunistically to achieve extreme polygyny in conditions favourable to their reproductive interests.

8.3 Physical comparisons between humans and other primates

> The next time a new species of primate is discovered we should be able to deduce its social behaviour by examining its testes and dimorphism in body and canine size. (Reynolds and Harvey, 1994, p. 66)

The above claim is indeed a remarkable one since it suggests that complex social characteristics can be inferred from simple measurable parameters. If the claim is reliable, the same technique should also, since humans have a shared ancestry with the primates, throw considerable light on ancestral human sexual behaviour. We will now examine the potential of this claim.

8.3.1 *Body size and canine size dimorphism*

Figure 8.4 shows how body size dimorphism and canine size dimorphism vary in relation to breeding system for primates. There are several hypotheses that could in principle account for the **sexual dimorphism** shown:

1. *Food competition between the sexes*
   Given that the two sexes of any species will tend to live in the same locality or ecological niche, it could be that, to avoid competition, the sexes have shifted to different diets or means of obtaining food. In monogamous birds in fact, dietary divergence does explain some of the diversity in body size and beak shape.

2. *Defence against predators*
   Large canines and increased male size could represent an adaptation for male defence of the young against predators. There is some support for this from Figure 8.4(b). Single-male groups would have more females to defend, so multimale groups might have become multimale precisely for defence purposes. Consistent with this hypothesis is the fact that, among multimale groups, male canine size is generally larger for terrestrial primates than for arboreal primates, reflecting perhaps the fact that predation pressure is usually larger for ground-living primates.

3. *Intrasexual selection*
   Male versus male contests over females are entirely consistent with the data in Figure 8.4. The large differences between measures of dimorphism in monogamous and polygynous (single male) contexts can be explained by the more intense competition between males for females in the latter. Even in multimale groups, some competition is observed to secure a place in dominance hierarchies.

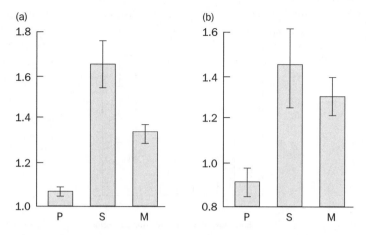

P = pair (monogamous); S = single-male (polygynous/harem); M = multimale (promiscuous)

Bars represent one standard error either side of the mean

Figure 8.4   (a) Body size dimorphism (adult body weight of male divided by adult body weight of female) versus mating system. (b) Relative canine size (a measure of canine dimorphism, equal to male canine size divided by female canine size) versus mating system (adapted from Harvey and Bradbury, 1991)

8.3.2 *Testis size*

The significance of **testis size** is that it indicates the degree of sperm competition (see Chapter 5) in the species. In the 1970s, the biologist R. V. Short suggested that the difference in testis size for primates could be understood in terms of the intensity of sperm competition. To obtain reliable indicators, testis size has to be controlled for body weight since larger mammals will generally have larger testes in order to produce enough testosterone for the larger volume of blood in the animal, and a larger volume of ejaculate to counteract the dilution effect of the larger reproductive tract of the female.

When these effects are controlled for, and relative testis size is measured, the results support the suggestion of Short that relatively larger testes are selected for in multimale groups where sperm competition will take place in the reproductive tract of the female (Figure 8.5). A single male in a harem does not need to produce as much sperm as a male in a multimale group since, for him, the battle has already been won through some combination of body size and canine size, and rival sperm are unlikely to be a threat. In contrast, in promiscuous multimale chimpanzee groups, females will mate with several males each day when in **oestrus**. It is possible that the sexual swellings that advertise oestrus in many female primates, and which males find irresistible, actually promote sperm competition since the best way for a female to produce a son who is a good sperm competitor is to encourage competition among his potential fathers.

A rival hypothesis to sperm competition is that of sperm depletion. The argument here is that large testes are needed by males who engage in frequent sexual activity to replenish depleted supplies of sperm. This hypothesis fails when

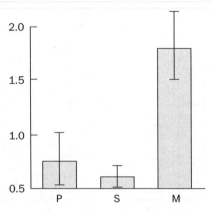

Error bars represent one standard error

Figure 8.5   Relative testis size versus mating system
(adapted from Harvey and Bradbury, 1991)

The y-axis refers to deviations from the allometric line of testis size against body
weight. Thus, a value of 1 would indicate a testis size expected for an average
primate of a given body weight. Values above 1 indicate that testes are larger than
would be expected. The values are for genera rather than individual species. The
difference between monogamous (P) and single-male (S) groups is not significant,
but both differ significantly from multimale (M) groups (see Harcourt *et al.*, 1981)

tested on bird species. From Figure 8.6 we can see that relative testis size in
polyandrous birds is higher than in leking species. Yet in **polyandry** the number of
females per male is low, but in leking species males will mate with many females.

A better explanation for these differences is that, in polyandry, high levels of
sperm competition are found since a female is impregnated by more than one
male. Leks entail low levels of sperm competition since females mate with only
the one successful male. We should also note from Figure 8.6 that male birds in
monogamous colonies have larger testes than those in solitary monogamous
breeding situations. Birkhead and Moller (1992) suggest that colonies provide
ample opportunities for 'extrapair copulation'. Males and females may be
ostensibly monogamous in terms of caring for offspring, but both are willing to
undertake adulterous affairs, hence the need for larger testicles in the male.

8.3.3 *Testis size and bodily dimorphism applied to humans*

Diamond (1991, p. 62) has called the theory of testis size and sperm competition
'one of the triumphs of modern physical anthropology'. As we have seen, the
theory has great explanatory power, and we will now apply it to humans.

Table 8.1 shows some key data on testis size and bodily dimorphism for the
hominids and pongids. The fact that men are slightly heavier than women could
reflect a number of features of our evolutionary ancestry. It could indicate the

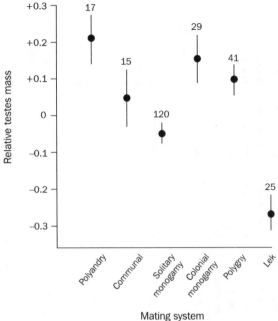

Mating system
Numbers represent the number of species in each set.
Vertical lines represent standard errors

Figure 8.6   Relative testes mass in relation to mating systems
for various species of birds (after Birkhead and Moller, 1992)

The vertical axis represents the deviation from the value expected for birds of a
given mass. A high plus value indicates that testes are heavier than would be
expected for birds of this body weight. A minus value indicates testes are smaller
than would be expected from birds of a given average body weight

protective role of men in open savannah environments, it could be the result of food
gathering specialisation whereby men hunted and women gathered, or it could
reflect male competition for females in unimale or multimale groups. The
dimorphism for humans is mild, however, compared with that for gorillas; this would
indicate that *Homo sapiens* did not evolve in a system of unimale harem mating.

Also, if early humans routinely competed to control groups of females, we
would not only expect a higher level of body size dimorphism but also smaller
testes. The testes of gorillas relative to their body size are less than half the average
for humans. On the other hand, if early humans had behaved like chimpanzees in
multimale groups, we would expect larger testes. In fact, if human males had the
same relative size of testis as chimps, their testes would be roughly as large as
medium-sized oranges. Figure 8.7 shows human, gorilla, orang-utan and chimp
males as females see them. The size of the large circle relative to the female shows
the degree of sexual dimorphism in the species. The length of the arrows and the
pair of dark shapes show the relative size of the penises and testes of the males.
One unexplained feature of male morphology is the large size of the human penis.

Table 8.1 Physical characteristics of the great apes in relation to mating and reproduction (data from Harcourt et al., 1981; Foley, 1989; Warner et al., 1974)

Species	Male body weight (kg)	Female body weight (kg)	Dimorphism: male to female	Mating system	Weight of testes (g)	Weight of testes as % of body weight	Approx. number of sperm per ejaculate ($\times 10^7$)	Estimated number of copulations per infant produced	Estimated global population
Human (*Homo sapiens*)	70	63	1.1	Monogamy and polygyny?	25–50	0.04–0.08	25	50–100	More than 6 000 000 000
Common chimp (*Pan troglodytes*)	40	30	1.3	Multimale in promiscuous groups	120	0.3	60	500–1000	Fewer than 110 000
Orang-utan (*Pongo pymaeus*)	84	38	2.2	Unimale Temporary liaisons	35	0.05	7	?	Fewer than 25 000
Gorilla (*Gorilla gorilla*)	160	89	1.8	Unimale Polygyny	30	0.02	5	10	Fewer than 120 000

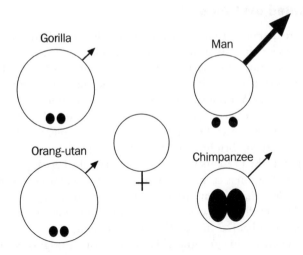

Figure 8.7   Body size dimorphism: female's view of males illustrating the relative size of the body (open circle), penis (arrow) and testes (dark ovals) (from Short and Balaban, 1994)

Harvey and May (1989) have taken ideas about testis size one step further and argued that interracial differences in testis size in human populations may be related to differences in mating behaviour and as such may be adaptive. Danish men, for example, have testes twice the size of Chinese men, which is a ratio much greater than expected from body size differences alone. Caucasians also produce twice the number of spermatozoa than Chinese (about 210×10^6 compared with 84×10^6). The implications of such findings are difficult to establish and interracial comparisons of testis size enter dangerous territory, not least because of the problem of defining race.

From a comparison of human testis size with other primates, Short concludes that we are not 'inherently monogamous... neither are we adapted to a multimale promiscuous mating system'. His view is that 'we are basically a polygynous primate in which the polygyny usually takes the form of serial monogamy' (Short, 1994, p. 13).

Patterns of mating behaviour may be revealed in the size of male testes and bodily dimorphism, but it could also be expected that the female body might hold some clues about the mating propensities of early *Homo sapiens*. Moreover, phylogenetic trees can be used to infer something about ancestral mating systems. One notable aspect of human female sexuality is that women are continually sexually receptive throughout the **ovarian cycle**. Another is that the precise moment of ovulation is concealed from both male and female. We will examine these topics further in the next section. Finally, we may note, from the last column in Table 8.1, the precarious state of the natural populations of our nearest relatives. The threatened extinction of these apes is nothing short of a tragedy and contrasts markedly with our own burgeoning population.

8.4 Concealed ovulation

Couples who are desperate to have a baby know only too well that it is difficult to pinpoint the precise time at which a woman ovulates. Ovulation is concealed from both the woman and the man. Clear testimony to this is the lucrative business of manufacturing contraceptives on the one hand and test kits for ovulation on the other. Yet humans are part of a minority among mammals in this respect. In most mammalian species, oestrus is announced with an explosive fanfare of signals. In the case of female baboons, for example, the skin around the vagina swells and turns bright red, she emits distinctive odours and, just to reinforce the point, she presents her rear to any male she happens to fancy.

The fact that ovulation is cryptic for human males and females is also borne out by the strange beliefs that humans have, until relatively recently, held about sexual fertilisation. Many traditional societies saw virtually no link between copulation and conception (Dunham *et al.*, 1991). Some thought that babies entered the mother from the environment, others that whole babies originated in the male. Aristotle supported this idea, and early microscopists such as Van Leeuwenhoek even thought that they could detect miniature humans (homunculi) in the sperm of males. Hence these entities became known as 'seed animals', or spermatozoa.

The concealment of ovulation is a feature that humans share with about 32 other species of primate, but for at least 18 other species, ovulation is advertised boldly and conspicuously. We can confidently expect concealment and advertisement to have some adaptive significance, and a number of intriguing theories have been proposed to account for both. In this section, we will examine the significance of concealed ovulation in humans as a female mating strategy.

8.4.1 *Terminology*

Oestrus, the ovarian cycle and menstruation are closely related events. Some definitions of these are provided in Box 8.1. All mammals go through an ovarian cycle, but not all mammals menstruate or show signs of oestrus. Humans, for example, show no signs of oestrus but do pass menstrual blood. Some primates show oestrus and menstrual bleeding while others show neither (Figure 8.8).

8.4.2 *Conspicuous and concealed ovulation – some hypotheses*

We should really be asking the question, 'Why is ovulation so conspicuous in some primates yet concealed in humans and a few other primates?'

Conspicuous ovulation

A number of hypotheses have been proposed to account for conspicuous ovulation. One is that sexual swellings incite competition between males. Females will then be fertilised by the 'best' males, thereby ensuring that good genes are passed to sons, who will therefore also perform well in pre- and post-copulatory sperm competition (Clutton-Brock and Harvey, 1976). Another hypothesis is that multiple mating by female primates confuses paternity and that, since many

BOX 8.1

Features of the ovarian cycle

Oestrus

Period of heightened interest in copulation in female mammals. Oestrus is associated with visible and olfactory signals such as swellings and colour changes around the anus or vulva. The word is derived from the Greek word for the gadfly, an insect that pursues cattle and drives them into a frenzy

Ovarian cycle

A hormone-regulated cycle that results in the release of an egg (ovum) from the ovary. In humans, this occurs once every 28 days

Menstruation

A discharge of the inner lining of the uterus that leaves the body through the vagina. In humans, ovulation occurs approximately 14 days following the onset of menstruation

Ovulation

The production and release of ova by the ovary

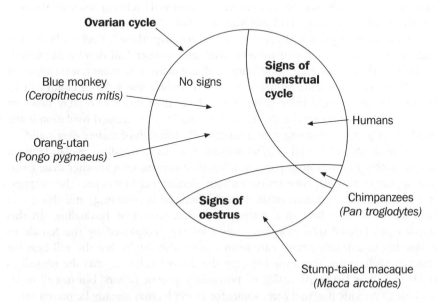

Figure 8.8   Association between oestrus and menstrual bleeding during the ovarian cycle for selected primates

All females pass through an ovarian cycle. Some species show signs of oestrus or menstruation, some both, and some none

males in a troop may suspect that they are the father, this reduces the aggression of adult males towards the offspring of the female.

Concealed ovulation

Concealed ovulation in humans and a few other primates has led to a much greater variety of hypotheses than conspicuous ovulation. The problem is not a trivial one since the concealment of ovulation only serves to make sexual activity even more inefficient. As sexual activity is costly to an organism (in terms of energy and time expended, exposure to risks of transmission of disease and vulnerability to predators), there must be some sound evolutionary reasons for masking the time when sex could be more productive.

A number of theories have been proposed. One is the 'sex for food' hypothesis. Given the ubiquity of prostitutes in history, it has been suggested by Hill (1982) that concealment of ovulation allowed early female hominids to barter sex for food. If ovulation were conspicuous, a male would know when a gift in exchange for sex would bring him some reproductive advantage, and he could withhold such gifts if the woman were not fertile. Concealing ovulation allows women to be in a position nearly constantly to exchange sex (with the prospect of paternity) for resources. Hill argues that there is ample evidence in the ethnographic literature that human males trade resources such as meat for sex. Women gain not only valuable nourishment, but also the opportunity to copulate with some of the best males in the group since males good at provisioning are also presumably able in other respects. There is some evidence, that in hunter-gatherer societies, this still happens in that women will exchange sex with the best hunter in return for meat (Hill and Kaplan, 1988).

An alternative to this is the 'anticontraceptive hypothesis'. Burley (1979) has suggested that concealed ovulation evolved after women had developed enough cognitive ability to associate conception with copulation. If women were aware of ovulation and could thus refrain from sex to reduce the risks of childbirth to themselves, they would leave fewer descendants than those who were unaware and hence could not exercise this choice. In this view, concealed ovulation is the products of genes outwitting consciousness, the triumph of matter over mind.

Benshoof and Thornhill (1979) suggest that, by concealing ovulation from her ostensibly monogamous partner, a female can mate with another male (who may appear to have superior traits) without alerting her first mate. They suggest that a woman may unconsciously 'know' when she is ovulating, and this could help her to assess when an extramarital liaison would be rewarding. In this scenario, concealed ovulation becomes a strategy employed by the female to enable her to extract paternal care from a male who thinks that she will bear his children while at the same time enjoying the ability to choose what she regards as the best genetic father. All males are potentially genetic fathers, but not all males are able to provide paternal care; some, for example, may already be paired off.

The question arises of whether males can detect when a female is ovulating and adjust their frequency of copulation. Reviewing their own data and those of others, Baker and Bellis (1995) conclude that the frequency of inpair copulation is relatively evenly spread across the menstrual cycle (with a trough during menses) and that there is no evidence that the number of such copulations increase during ovulation. In contrast to inpair copulations, Baker and Bellis also claim that data support the idea that females are more likely to have extrapair copulations during their pre-ovulatory fertile period. They also claim that a female with an existing sexual partner tends to walk a greater distance (presumably unconsciously in search of extrapair liaisons) and spends less time with her main partner during her fertile period. Baker and Bellis conclude that the data are consistent with the idea that females without a main partner try to avoid mid-cycle contacts while those with a partner seek new contacts. If we accept these data, this provides support for the notion that ovulation is probably cryptic to the male but influences the disposition of females towards polyandry.

Richard Alexander and Katherine Noonan at the University of Michigan have proposed a view of concealed ovulation as a tactic developed by women to divert men from a strategy of low-investing, competitive polygyny towards a more caring and high-investing monogamy. For a male to be assured of paternity, he needs to remain in close proximity to his partner for prolonged periods of time. It is no use a man wandering off to have sex with another woman since she may not be ovulating and his wife back at home may become the subject of the attentions of like-minded philandering males. So the male stays at home, finds his partner constantly desirable and has the reward of a high degree of confidence in his paternity. This is the so-called 'daddy at home' hypothesis (Alexander and Noonan, 1979).

It is common in primates to find males killing offspring that are not their own in order to bring the bereaved mother into oestrus again. Support for the need for females to evolve a counter-measure against this comes from studies on the 'post-conception oestrus' of grey langurs (*Presbytis entellus*) and red colobus monkeys (*Colobus badius*). In both of these species, an oestrus signal is given even when the females are pregnant. This could be interpreted as a measure to confuse males; if so, it is one of the few examples of 'dishonest signals' given by females of their reproductive status. With this in mind, Hrdy (1979) has suggested that concealed ovulation served to confuse the issue of paternity of early hominids and thus prevent infanticide by males who, if they were confident they they were not the genetic fathers of offspring, could engage in this practice. This is sometimes known as the 'nice daddy' theory.

This idea is extended by Schroder (1993), who notes that one of the few cross-cultural generalisations about sexual mores is that sex normally takes place in private, out of sight of members of the participants' own species (conspecifics). Schroder suggests that this is a vestige of a strategy employed by females and males to mate out of sight of the dominant male or the woman's normal partner. Such behaviour allowed the consolidation of multimale, multifemale groups. Instead of wandering off in the risky pursuit of females in another group, a male could stay with the natal group and help in the defence against outside males. Such a defence would serve the interests of the female in reducing infanticide while allowing reproductive opportunities for the males. Schroder sees concealed ovulation as a means of optimising clandestine copulation and believes that such copulations improved the options for female choice (the cuckoldry hypothesis) at the same time as reducing the risk of infanticide for the female's children.

8.4.3 *Testing rival hypotheses of concealed ovulation*

There is clearly no shortage of ideas to account for concealed ovulation, and the literature is vast and difficult. Even a superficial glance at primate sexual behaviour shows that concealed ovulation can be found in a variety of mating systems, such as those of monogamous night monkeys, polygynous langurs and multimale vervets. It is entirely feasible, therefore, that the various hypotheses are not entirely exclusive. One, such as the 'nice daddy' theory, could explain the origin of concealment, and another, for example the 'daddy at home', its maintenance. In disentangling these arguments, much progress has been made by two Swedish biologists, Birgitta Sillen-Tullberg and Anders Moller (1993). By looking at the probable **phylogeny** of the anthropoid primates, they claim to be able to evaluate the various hypotheses.

Sillen-Tullberg and Moller gathered data on the mating system and signs of ovulation in 68 extant taxa (the unit being either a species or a genus) of primates. Their results are shown in Table 8.2. The conclusion from an initial inspection of these data is that there is a correlation between monogamy and concealed ovulation. Ovulation is not advertised conspicuously in any monogamous system. This could be thought to lend support to the 'daddy at home' theory, but the converse predictions are not upheld. Thus, concealed ovulation is also the majority state in unimale systems and not uncommon in multimale groups, in which, at least for the latter, paternity confidence would be low. If concealed ovulation is adaptive, it must serve different functions in different groups.

Table 8.2 The distribution of the three states of visual signs of ovulation with respect to mating systems among extant anthropoid primate taxa (data from Sillen-Tullberg and Moller, 1993)

	Visible signs of ovulation		
Mating system	**Absent**	**Slight signs**	**Conspicuous**
Monogamy	10	1	0
Unimale (harem)	13	6	4
Multimale (promiscuous)	9	11	14

The crucial test is of course when concealed ovulation first appeared. Using procedures to disentangle primitive, derived and convergent characters, Sillen-Tullberg and Moller constructed a phylogenetic tree of changes in the visual signs of ovulation. They concluded that the primitive state for all primates was probably slight signs of ovulation, and that concealment had evolved independently 8–11 times (the range indicating the effect of slightly different assumptions in the modelling). We can now compare this with the phylogeny of mating behaviour. As may be expected, this is more difficult to establish. The results obtained by Sillen-Tullberg and Moller are shown in Figure 8.9.

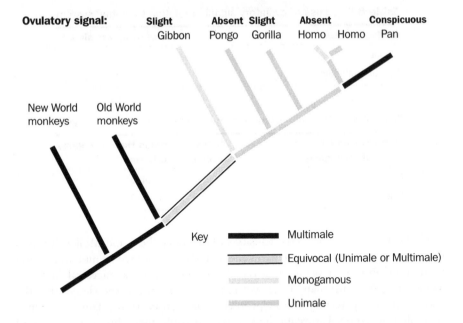

Figure 8.9   Distribution of three types of mating
system among the anthropoid primates

   The ancestors of the hominoid lineage are shown as either uni- or multimale. For the hominids, fossil evidence showing a high degree of sexual dimorphism among early hominids, males being larger than females, tends to rule out monogamy among our ancestors. The probable system is unimale for early hominids, later *Homo* species adopting monogamy or opportunistic polygyny. Figure 8.9 also shows that humans and gibbons invented monogamy while chimps reinvented promiscuity. Some humans practise polygyny, and this is shown as a variant. By comparing the lineages of ovulatory signals and mating systems, it is possible to establish the prevailing mating systems, when ovulatory signals disappear or conversely, avoiding assumptions about which variable is independent, what type of mating system evolves in the various states of ovulatory signal. The results for the 68 taxa are shown in Tables 8.3 and 8.4.
   Sillen-Tullberg and Moller draw three sets of conclusion from their work:

1. Ovulatory signals disappear more often in a non-monogamous context. This tends to support Hrdy's infanticide hypothesis

2. The fact that, once monogamy has been established, ovulatory signals do not usually disappear throws doubt on the 'daddy at home' hypothesis of Alexander and Noonan as the origin of concealed ovulation

3. Monogamy evolves more often in lineages that lack ovulatory signals. The lack of ovulatory signals could be an important condition for the emergence of monogamy.

Table 8.3   Loss of ovulatory signal in relation to mating system

Mating system	Loss of ovulatory signal
Monogamy	0–1
Uni- or multimale	8–11

Table 8.4   Evolution of monogamy in relation to ovulatory signal

Sign of ovulation at the origin of monogamy	Number of times that monogamy evolved at this sign
Absent	4–6
Slight	1–3

In short, if we accept the conclusions of Sillen-Tullberg and Moller, it seems that concealed ovulation may have changed and even reversed its function during primate evolution. It began as a female strategy to allow ancestral hominid women to mate with many males yet remain secure in the knowledge that the confusion over paternity would protect their infants from infanticide. Once concealed ovulation was established, a woman could then choose a resourceful and caring male and use concealed ovulation to entice him to stay with her to provide care. Women became continually sexually active, and men correspondingly found women continually desirable – even without signs of ovulation. It is an interesting thought that the continual sexual receptivity of human females and the high frequency of non-reproductive sex in which humans engage began as a female strategy to outwit men. An ancestral male anxious to be secure about his paternity could no longer afford the time budget to guard a group of females and prevent sexual access by other males. Instead, he was forced into a monogamous relationship, providing care for his partner and what he thought were his own offspring, in return expecting sexual fidelity from his wife. This in turn led to a male psychology that was particularly sensitive to any evidence of unfaithfulness, with repercussions for the emotional life of men. Such is the origin of the strength of sexual jealousy that still today accounts for a good deal of human conflict (see Chapter 10).

There is now some speculation on whether or not ovulation is truly concealed. The biologist John Manning has provided some evidence suggesting that ovulation may not be totally concealed, at least not from a committed partner. By examining mammographs of women who attended a breast screening unit in Liverpool, Manning's team was able to compare any observed asymmetry in breast size with the position of the women in their ovarian cycle (Manning *et al.*, 1996). Since asymmetry in breast measurement is larger than that in non-sexually selected traits such as finger length or wrist and ankle measurements, Manning suggests that breasts may serve as honest signals to males of the reproductive fitness of a prospective partner. Manning observed a slight increase in asymmetry in breast measurement followed by a sharp increase in symmetry at around the mid-cycle point, when a female is most fertile. It follows that a male who is closely attendant upon a female and is aware of her detailed body shape could, at least in principle, consciously or subconsciously detect when she is most fertile. This form of 'partially concealed' ovulation is consistent with the idea that inconspicuous ovulation is associated with monogamy. It is known that men are sensitive to degrees of facial symmetry in assessing attractiveness. The issue now then is whether they are able to detect and respond to the very slight levels of 'cyclical asymmetry' that Manning observed.

The problem of pendulous breasts may also be related to concealment of ovulation. Humans are unique among the primates in that the female carries permanently enlarged breasts. In other words, the size of the breasts varies little during the menstrual cycle. Why should this be the case? Baker and Bellis (1995), agreeing with the work of Smith (1984), suggest that this is part of the mechanism of sexual crypsis (hiding), which is a strategy evolved by the female to hide the times when she is most fertile. If a female keeps permanently enlarged breasts, it becomes difficult for a male to assess when she has ceased lactation and is again ovulating. According to this scenario, permanently enlarged breasts are part of the drive towards concealed ovulation. This still, however, leaves the problem of why other sexually cryptic monogamous primates such as some New World monkeys did not evolve enlarged breasts.

8.4.4 *Menstruation*

In stark contrast to the concealment of ovulation, which is probably almost if not entirely complete to males, we have the loud signal of menstrual bleeding, which in the case of humans is profuse and would have been difficult to hide in prehistoric cultures. Bleeding also places an additional nutritional demand on women.

It has been suggested that menstrual bleeding may have served the function of removing pathogens transmitted to the female reproductive tract by male sperm (Profet, 1993), but other primates presumably manage the same task either by not bleeding or by reabsorbing blood. Menstruation could be an honest signal to the male that the female is fertile and worth some mating effort. But why should the female give such a loud signal if she has already concealed her ovulation? The fact that menstruation does not coincide with ovulation does at least imply that the male must spend some time with the female sharing food and offering protection. Once the female is pregnant, however, what is to stop the male pursuing other menstruating females?

In a remarkable synthesis of ideas, Knight *et al.* (1995) propose that early hominid females of the Pliocene era used menstruation as a means of co-ordinating their ovarian cycles with each other and with the lunar cycle. The suggestion is that women began menstruating together at the start of a new moon. This signalled a sex strike to men, who were excluded from sexual activity until they returned from the hunt. It is argued that menstruation in synchrony with the lunar cycle was used by females to exclude 'lazy' but mobile males from sexual cheating. The moon served as an external natural clock to synchronise female behaviour in a wide area. Males then had no choice but to hunt and return with food to exchange for marital sex. Knight *et al.* also suggest that the sudden use of iron oxides (red ochre) around 110 000–75 000 years ago can be interpreted as the appearance of the first cosmetics designed to imitate menstruation.

Knight's work has met with a mixture of approval and scepticism. One serious problem is whether menstrual synchrony among women really does take place. Early reports that women living communally (McClintock, 1971) synchronise ovulation have been criticised on methodological grounds (Wilson, 1992b). Work by Baker and Bellis (1995) failed to show any menstrual synchrony for menstruating women, but more recent work by Stern and McClintock (1998) suggests that women do release pheromones that could cause synchrony of menstruation in a social group. Despite various theories, menstrual bleeding remains one of the most enigmatic features of female sexuality.

Thus, the physical evidence on human mating and ancestral sexual practices is revealing and illuminating in some cases, such as the size of male testes and body dimorphism, and complex but suggestive in others, for example the concealment of ovulation. Many enigmas remain, but one message from physical anthropology is perhaps that humans are nearly but not quite monogamous. If males can accumulate enough wealth, they will practise opportunistic polygyny, either serially or simultaneously. Meanwhile, the continuous sexual receptivity of women enforces a strong bond between herself and her partner, but similarly enables a spot of polyandry if appropriate – not so different from what we really knew all along.

Evidence that both males and females make considerable investment in a relationship comes from the fact that both sexes have a highly refined sense of sexual attractiveness. We do not mate at random: we choose our partners carefully. Coupled with this, there is strong agreement on what features are attractive in males and females. As we evolved away from the ancestral unimale groups of the Australopithecines, we also invented standards of male and female beauty. The evolutionary logic of such standards is the subject of the next chapter.

SUMMARY

▧ An evolutionary approach to human sexuality helps us to understand the mating strategies pursued by ancestral and contemporary males and females.

▧ Humans are like many mammals in the sense that females are the sex that limits the reproductive success of males. The large harems created for the exclusive use of ancient despots and emperors show that, in some conditions, males can behave opportunistically to achieve an extreme degree of polygyny.

▧ Such harems, however, probably represent an aberration of the 'normal' mating behaviour characteristic of the human species. The extended period of postnatal care needed for human infants, and the need for males to be confident of paternity, probably ensured that early *Homo sapiens* were monogamous or only mildly polygynous.

▧ Such a conclusion is reinforced by interpretations of data on sexual dimorphism and male testis size. The fact that males are slightly larger than females tends to suggest some degree of intrasexual competition among males for mates. Human testes are too large, however, to point to a unimale mating system, such as found among gorillas, and too small to be consistent with multimale and multife-male 'promiscuous' mating groups.

There are some enigmatic features of female sexuality, such as the concealment of ovulation and the visible signs of menstruation, that have received much theoretical attention. The balance of current theories suggests that sexual crypsis (concealed ovulation) began among early hominid females in a unimale setting, enabling them to choose desirable males for mating without the risk of infanticide from suspicious males. In effect, concealed ovulation served as a strategy whereby women resisted signalling to men the period of maximum fertility and thereby confused paternity estimations. Once established, it is suggested that women could then use crypsis to extract more care from a male.

KEY WORDS

Monogamy ■ Oestrus ■ Ovarian cycle ■ Ovulation ■ Phylogeny

Polyandry ■ Polygyny ■ Sexual dimorphism ■ Testis

FURTHER READING

Baker, R. R. and Bellis, M. A. (1995) *Human Sperm Competition.* London, Chapman & Hall.
A book that makes some controversial claims about human sexuality based on unusual and original research.

Betzig, L. (ed.) (1997) *Human Nature: A Critical Reader.* Oxford, Oxford University Press.
A useful book that contains numerous original articles on human sexuality together with a critique in retrospect by the original authors.

Ridley, M. (1993) *The Red Queen.* London, Viking.
A delightful book that explores the nature of sexual selection and its application to humans.

Short, R. and Potts, M. (1999) *Ever Since Adam and Eve: The Evolution of Human Sexuality.* Cambridge, Cambridge University Press.
Superbly illustrated and authoritative work. A humane account of the evolution and significance of human sexuality.

9

Human Mate Choice:
The Evolutionary Logic of
Sexual Desire

Love-thirty, love-forty, Oh! Weakness of joy,
The speed of a swallow, the grace of a boy,
With carefullest carelessness, gaily you won,
I am weak from your loveliness, Joan Hunter Dunn.

('A Subaltern's Love-Song', Sir John Betjeman)

The onset of sexual desire exerts a powerful force over human lives. One of its most obvious features, and of human mating in general, is that it is highly discriminating – we have strong preferences about who will and who will not suffice as a potential partner in both the short and longer terms. However idealised our view of romantic love, there is abundant evidence that humans employ a variety of hard-headed criteria in assessing the desirability of a mate, including income, occupation, intelligence, age and, perhaps above all, physical appearance. The appearance of the human body is an important reservoir of information and has exerted a potent influence over the whole of our culture. Western art since the time of the Greeks and certainly since the Renaissance has, periodically, celebrated the beauty, grace and symbolic significance of the human body. We are enthralled and fascinated by the appearance of members of our own species. Whether there are universal and cross-cultural standards in the aesthetics of the body is still a debatable point, but certainly in Western culture there exists a multimillion dollar industry to help us better to shape our bodies towards our ideals or disguise the effects of age. Corporate advertising figured out long ago that one of the best ways to market a product is to associate it with a handsome specimen of one or both sexes.

So where does this aesthetic sense come from? To a committed Darwinian, the answer would be clear: our perceptual apparatus will be designed to respond positively to features that are honest indicators of fitness – for aesthetics read reproductive potential. Similarly, we are descended from ancestors who made wise choices in selecting their partners and using the same logic, to the extent that we have inherited their desires and inclinations, we can expect any criteria we use in choosing a mate to be fitness-enhancing. This chapter is focused around both these issues: the aesthetics and the decision-making criteria exercised when males and females choose a sexual partner.

9.1 Evolution and sexual desire: some expectations and approaches

Darwinians view attractiveness in terms of reproductive fitness rather than as the relationship between an object, an observer and some abstract Platonic form. Features that are positive indicators of reproductive fitness in a potential mate should be viewed as attractive by males and females. In this sense, beauty is more than skin deep – it is to be found in the 'eye' of the genes. Despite the mild degree of polygyny indicated by the evidence presented in Chapter 8 and the few cases of opportunistic extreme polygyny, it is clear that, in most relationships, men and women make an appreciable investment of time and energy. Consequently, both sexes should be choosy about future partners, but in different ways.

Of all the features used in appraising a potential mate, two in particular have produced robust empirical findings that reveal inherent differences between male and female taste. These are physical attractiveness and the status of males. In the case of male status, the application of the principles established in Chapters 4, 5 and 6 predicts that, since females make a heavy investment in raising young, and since biparental care is needed following birth, females will be attracted to males who show signs of being able to bring resources to the relationship. This ability could be expected to be indicated by the dominance and status of a male within the group. Dominance could be selected for by intrasexual selection if males compete with each other, and by intersexual selection if females exert a preference for dominant males. A crude indication of dominance would be size, and we have already noted that humans are mildly dimorphic. In the complex social groups of early humans, there was, however, bound to be a whole set of parameters, such as strength, intelligence, alliances, and resource-holding and provisioning capabilities, that indicated the social status of the male. Some of these would be subtle and context-dependent, and a female would be best served by a perceptual apparatus that enabled her to assess rank and status using context-specific cues and signals.

If females respond to indicators of potential provisioning and status, males should be attracted to females that appear fecund and physically capable of caring for children. Since the period of female fertility (roughly 13–45 years of age) occupies a narrower age band than that of the male (13–65 years), we would also expect the age of prospective partners to be evaluated differently by each sex. Men should be fussier about age than women and hence rate physical features that correlate with youth and fertility higher on a scale of importance than should women.

To test these expectations, we can examine human preferences using data from a number of sources:

- What people say about their desires in response to questionnaires
- What people look for when they advertise for a partner
- Statistical evidence on the mating behaviour of people.

In the sections that follow, each of these approaches will be employed.

9.2 Questionnaire approaches

9.2.1 Cross-cultural comparisons

The use of a questionnaire on sexual desire in one culture lays itself open to the objection that responses reflect cultural practices and the norms of socialisation rather than universal constants of human nature. In an effort to circumvent this problem, David Buss (1989) conducted a questionnaire survey of men and women in 37 different cultures across Africa, Europe, North American Oceania and South America, and hence across a wide diversity of religious, ethnic, racial and economic groups. As might be expected, numerous problems were encountered with collecting such data, but Buss' work remains one of the most comprehensive attempts so far to examine the sensitivity of expressed mating preferences to cultural variation. From the general considerations noted above, Buss tested several hypotheses (Table 9.1).

Table 9.1   Predictions on mate choice preferences
tested cross-culturally by Buss (1989)

Prediction	Functional (adaptive) significance
Women should rate earning potential in a mate more highly than should men	The fitness of a woman's offspring can be increased by the allocation of resources
Men should rate physical attractiveness higher than women do	The fitness and reproductive potential of a female is more heavily influenced by age than is that of a man
Men will, on the whole, prefer women younger than themselves	Men reach sexual maturity later than women. Also as above
Men will value chastity more than women will	'Mom's babies, daddy's maybes'. For a male to have raised a child not his own would have been, and still is, highly damaging to his reproductive fitness
Women should regard ambition and drive more positively than men do	Ambition and drive are linked to the ability to secure resources and offer protection, both of which are fitness-enhancing for a woman

The results in terms of the number of cultures in which there was a significant ($P < 0.05$) difference between the qualities addressed in each hypothesis above are shown in Table 9.2.

Table 9.2 Number of cultures supporting or otherwise hypotheses on gender differences in mate preference (data from Buss, 1989)

Hypothesis	Number of cultures supporting hypothesis	% of total	Number of cultures contrary (con) to hypothesis or with a non-significant (ns) result	% of total
Women value earning potential more than men do	36	97	1 ns	3
Men value physical attributes more than women do	34	92	3 ns	8
Women value ambition and industriousness more than men do	29	78	3 con 5 ns	8 13
Men value chastity more than women do	23	62	14 ns	38
Men prefer women younger than themselves	37	100	0	0

The results show moderate to strong support for all the hypotheses. Data on age difference also allow a calculation of mean age preferences for mating. On average, men prefer to marry women who are 24.83 years old when they are 27.49 years old, that is 2.66 years younger than themselves. Women, on the other hand, prefer to marry men who are 3.42 years older. Interestingly, there are data available for 27 of the 33 countries sampled for actual ages of marrying; the actual difference is that men marry women 2.99 years younger than themselves. The fact that these figures agree so closely suggests that, at least in terms of age, preferences and practice are reassuringly similar.

The reasons for these consistent and near universal differences in mating age have been the subject of much speculation, no consensus as yet emerging. Some ideas concerning the evolution of this trait that have been proposed are that:

- Age serves as an advertisement by the male of his relative resistance to disease and general fitness in order to have reached that age
- Age is an indicator of status since status tends to increase with age
- Males mature sexually later than women. A female desires a male who will be efficient at copulation.

Against all these pressures must be balanced the female's desire for a male who is still young enough to provide support before dying.

The consistent cross-cultural sexual differences in the importance placed on attractiveness in a partner do not of course mean that the absolute value of attractiveness is forever fixed by our genes. Studies on the mate preferences of people in the United States have now covered a 50-year period (1939–89). Over this time, the importance of good looks in a marriage partner on a scale of 0–3 increased for men from a mean of 1.50 in 1939 to 2.11 in 1989, the corresponding increase for the judgement of women being 0.94 to 1.67 (Buss, 1994). The rise in the rating could relate to the use of fashion models in the media and advertising industry. What is interesting is that, despite this cultural change, differences between the sexes remain very similar.

9.2.2 *Urgency in copulation*

If we accept the reliability of questionnaire studies such as those carried out by Buss and others, it seems that the human mind is sexually dimorphic in the psychology of sexual attraction. With this in mind, Symons and Ellis (1989) hypothesised that males and females would respond differently to the following question:

> If you had the opportunity to copulate with an anonymous member of the opposite sex who was as physically attractive as your spouse but no more so and as competent a lover as your spouse but no more so, and there was no risk of discovery, disease or pregnancy and no chance of forming a durable liaison, and the copulation was a substitute for an act of marital intercourse, not an addition, would you do it? (Symons and Ellis, 1989, p. 133)

The prediction was that since sexual novelty benefits a male's reproductive interests more than a female's, males would be more inclined to answer yes to this question. The answers were in categories of 'Certainly would', 'Probably would', 'Probably not' and 'Certainly not'. The results are shown in Figure 9.1.

9.2.3 *A qualified parental investment model: the effect of levels of involvement*

Such questionnaire studies have consistently shown clear differences in the psychology of sexual attraction, but they have also revealed some degree of convergence. Of the top ten criteria for a good long-term partner listed by subjects in a study by Buss and Barnes (1986), seven out of the ten were the same for men and women, although they were of course given different priorities. These were kindness and understanding, intelligence, personality, health, adaptability, creativity and graduate status. We could explain this by arguing that the dimorphism in taste is only slight because of the effect of near-monogamous mating and the large investment made by both males and females in living together and raising children. A subtle way of probing what gender differences might exist beneath the consensus was applied in a study by Kenrick *et al.* (1996). They argued that humans are different from most other mammals in that there is a large range in the possible amount of investment made by couples in any relationship. It could range, in the case of a male, from a few drops of sperm in a one-night

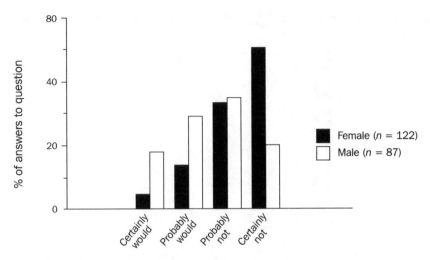

Figure 9.1 Responses to questions on casual sex expressed as a percentage of the respondents answering (data from Symons and Ellis, 1989)

stand to a lifelong commitment to a spouse and the raising of children. Kenrick *et al.* thus investigated whether the criteria for choosing a mate varied with the level of involvement in a relationship. The investment implications of different levels of involvement can be expected to vary according to gender (Table 9.3).

Table 9.3 Degree of involvement and expected implications for investment according to gender

Level of involvement	Date	Sexual relations	Exclusive dating	Marriage
Investment by female	Low	High	High	High
Investment by male	Low	Low	High	High

It is obvious that progressing from dating to sexual activity is associated with a sharper rise in potential investment implications for women than for men. With this in mind, Kenrick *et al.* asked 93 undergraduate students to consider the importance of 24 criteria (physical attractiveness, status and so on) in accepting a mate at the four different levels of involvement. Figure 9.2 shows the findings for earning capacity.

In terms of all the criteria assessed, a number of patterns emerged. As expected, significant differences began to appear at the sexual level of involvement for characteristics defining family values and health, females finding these more important than males once the level of sexual commitment is reached. Women rated status, however, more highly than did men at all levels of involvement.

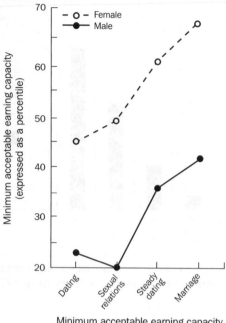

**Minimum acceptable earning capacity
at each level of involvement**

Figure 9.2   Minimum level of acceptable earning capacity
at different levels of involvement (from Kenrick *et al.*, 1996)

A level of, for example, 67 per cent for females at marriage indicates that, to be
acceptable, a male must earn more than at least 67 per cent of other males but
could earn less than 37 per cent of all males

This study was of course that of a rather narrow profile of the population in one culture, and it is curious that, for some characteristics, the significance of the gender difference declined from exclusive dating to marriage. The robustness of these findings in the face of different social conventions and across different population profiles remains to be established.

There are clearly problems with many studies based on questionnaires, particularly when unselective samples are used. One sometimes gets the impression that American undergraduates are constantly being plagued by interviewees asking about their sex lives. Nevertheless, the findings tend to be in agreement with evolutionary expectations. If, as social science critics would say, responses are conditioned by social norms, we still have the problem of explaining why so many social norms correspond with evolutionary predictions.

9.3 The use of published advertisements

An intriguing way to gather information on mating preferences is to inspect the content of 'lonely hearts' advertisements in the personal column of newspapers and magazines. A typical advertisement reads:

> Single Prof. male, 38, graduate, non-smoker, seeks
> younger slim woman for friendship and romance.

Notice that the advertisement offers information about the advertiser as well as his preferences for a mate. Such information carries some advantages over questionnaire response surveys in that it is less intrusive and less subject to the well-known phenomenon that interviewees will tend to comply with what they take to be the expectations of the questioner. Moreover, the data are 'serious' in that they represent the attempts of real people to secure real partners. Against this must be placed the fact that the data are selective and probably do not represent a survey across the entire population profile.

Greenless and McGrew (1994) examined 1599 such advertisements in the columns of *Private Eye* magazine. The results for physical appearance and financial security are shown in Table 9.4.

Table 9.4 Percentage of advertisers seeking and offering physical appearance and financial security according to gender (data from Greenless and McGrew, 1994)

	% women (n = 297)	% men (n = 703)	Significance of difference (P<)
Physical appearance			
Offered	71	50	0.0001
Sought	33	48	0.0001
Financial security			
Offered	43	69	0.0001
Sought	33	9	0.0001

The results are consistent with the questionnaire surveys of Buss and others, and support the following hypotheses:

1. Women more than men seek cues to financial security
2. Men more than women offer financial security
3. Women more than men advertise traits of physical appearance
4. Men more than women seek indications of physical appearance.

The work of Greenless and McGrew has been repeated many times, with similar findings. Figure 9.3 gives details of the findings of a survey by Dunbar (1995).

It also appears that men of high status seek, and are able to attract, women of higher reproductive value. At the anecdotal level, most people know of ageing male rock stars or celebrities who marry women many years their junior. In a study of a computer dating service in Germany, Grammer (1992) found a positive correlation between the income of men using the service and the number of years separating their age from the (younger) women they sought. The reversal of the trend shown in Figure 9.3b for the age group 60–69 may indicate the fact

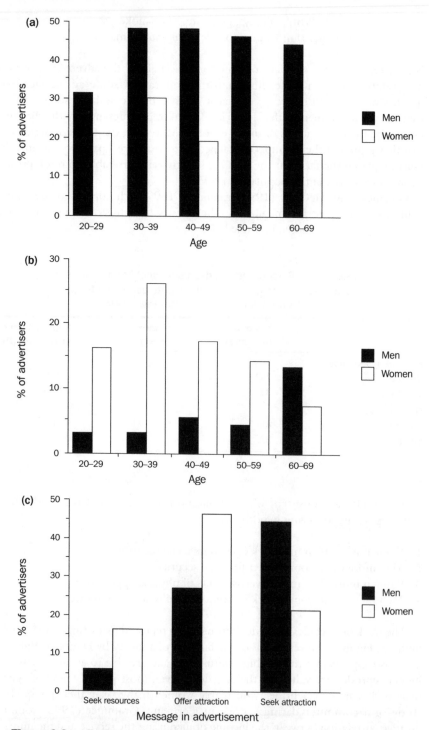

Figure 9.3   Features of 'lonely hearts' advertisements. (a) Those seeking attractiveness. (b) Those seeking resources. (c) Features of 'lonely hearts' advertisements averaged over all age groups (adapted from Dunbar, 1995)

that more women are alive than men in this category. Reproduction is not really an issue now, but men may be able to choose women with resources.

9.3.1 *Origin of mate choice preferences: evolutionary psychology or structural powerlessness*

Studies on expressed preferences reveal nothing about the ontogeny of those preferences. Some social scientists have proposed an alternative to adaptive explanations, which Buss and Barnes (1986) call the structural powerlessness and sex role socialisation hypothesis. This suggests that since, in patriarchal societies, women have less access to power and wealth than men, the chief way in which a woman can attain status and acquire resources is to marry up the social ladder (hypergamy) and trade looks for status. The hypothesis fails on the first count, however, of explaining why preferences are remarkably similar across a wide range of cultures. If it is argued that all these cultures share the same features of patriarchy, we then need an even grander theory to explain this. A more serious problem is that the structural powerlessness model makes a prediction at variance with the facts. If women seek high-status males to advance their own standing, it follows that women should be less selective with regard to status and wealth as their own premarital wealth and power increases. The evidence, however, suggests otherwise: high-status women still value high-status men. Buss (1994) found that women with a high income tend to value the financial status of men even more than women on a lower income do.

9.4 **The use of stimulus pictures**

9.4.1 *Male assessment of females*

It is clear that good looks are important to men seeking a partner, but what type of looks should be preferred? The belief that there is a vast variability in notions of beauty between cultures in time and place has tended to thwart the scientific search for universal and adaptive norms for beauty. Darwin, reviewing standards in a variety of cultures, concluded there was no universal standard for beauty in the human body. Reference is often made to the rather fleshy nude women appearing in paintings by Titian and Rubens, which are compared with modern-day models to emphasise the changing ideals of attractiveness (although whether or not the artists were attempting to depict an ideal is questionable). In 1993, however, Devendra Singh, a psychologist working at the University of Texas, published some important work suggesting that there may be some universals in what the sexes find attractive. Singh argued that two conditions must be met by any universal ideal. First, there must be some plausible linkage of features designated as attractive to physiological mechanisms regulating some component of reproductive fitness, and hence a positive correlation between variation in attractiveness and variation in reproductive potential; in other words, attraction must equate with fitness. Second, males should possess mechanisms to judge such features, and these should be assigned a high degree of importance in the estimation of attractiveness (Singh, 1993).

Singh argues that the distribution of body fat on the waist and hips meets the conditions above. More specifically, he suggests that the **waist to hip ratio (WHR)** is an important indicator of fitness and attractiveness. The WHR (that is, waist measurement divided by hip circumference) for healthy premenopausal women usually lies between 0.67 and 0.80, whereas for men it usually lies between 0.85 and 0.95. It is well known that obesity is associated with a higher than average health risk, but what is more surprising is that the distribution of fat in obese women, and hence the WHR, is also a crucial factor in predicting their health status. Singh collected a body of evidence to show that women with a WHR below 0.85 tended to be in a lower-risk category compared with women with a WHR above 0.85 for a range of disorders such as heart disease, diabetes, gallbladder disease and selected carcinomas.

If the WHR does have an adaptive significance, it should not be subject to the vagaries of fashion. To test this, Singh examined the statistics of Miss America winners and *Playboy* centrefold models. The results for Miss America are shown in Figure 9.4.

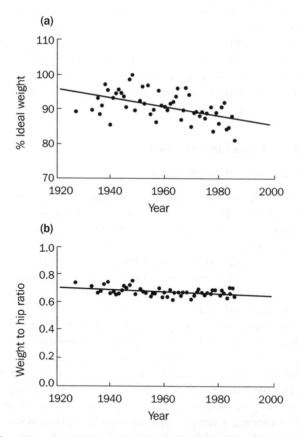

Figure 9.4 Miss America statistics. (a) Weight as percentage of 'ideal' over a 60-year period. (b) Waist to hip ratio over a 60-year period (after Singh, 1993)

Ideal weight was established from life insurance tables

The results confirm the widespread suspicion that the weight of fashionable models has fallen over the past 60 years. What is equally significant is the fact that the WHR has remained relatively constant. This suggests that while weight as an indicator of attraction is, to some degree, subject to fashionable change, the WHR is far more resilient. It follows that the WHR is a possible candidate for a universal norm of female beauty. It is stable among women deemed to be attractive, and it correlates with physical health and fertility. The next step is to ascertain whether it is a factor in the assessment of beauty.

To test perceptions of attractiveness in relation to weight and WHR, Singh presented subjects with a series of line drawings (Figure 9.5), with the instruction that they rank the figures in order of attractiveness. The results were unambiguous: for each category of weight, the lowest WHR was found to be most attractive. In addition, an overweight woman with a low WHR was found to be more attractive than a thin woman with a high WHR. This again suggests that attractiveness is more strongly correlated with the distribution of body fat, as it affects the WHR, than with overall weight *per se*.

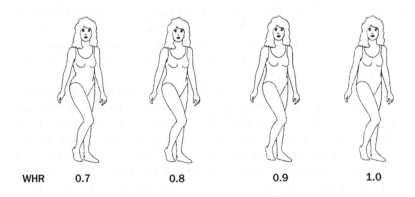

| WHR | 0.7 | 0.8 | 0.9 | 1.0 |

Figure 9.5   Stimulus figures given to subjects in
Singh's study (after Singh, 1993)

A series of four such figures was presented in three categories: underweight,
normal and overweight. Only the normal-weight category is shown here

Singh's initial work was carried out on Caucasian males. To test whether different cultures shared the same WHR preference, Singh applied the same procedures to a group of Indonesian men recently arrived in the United States (94 per cent of whom were of Chinese descent) and a group of Afro-American men. The findings were virtually identical to those of the Caucasian study. The normal-weight group was found to be the more attractive overall and again, within all groups, a WHR of 0.7 was found to be most attractive. Women reached conclusions similar to those of men when evaluating the line drawings. It appeared to Singh that neither ethnicity nor gender significantly affected the WHR dimension of attractiveness (Singh, 1995).

Singh's work has aroused a great deal of interest and some controversy. One possible objection is that men are exposed to a barrage of images of women from the media, so preferences may be culturally determined. It is probable that many men learn their cues for what counts as attractive from the behaviours of other men and respond positively to the features of media-generated stars and icons.

Dugatin has reported the action of social cues on mate choice even in guppies. Female guppies are known to prefer males with orange colouration, but they have been observed to choose a male with less colouration if other females have already chosen him. The exact function of this is unclear, but such copying may save a female guppy time and thus reduce the energy costs and risks involved in finding a mate. The logic, although not exactly foolproof, may be that if several females have already chosen a male, he must have desirable qualities (Dugatin, 1996).

To test the possible influence of cultural cues, Yu and Shepard presented the same images used in Singh's study on male preferences to one of the few remaining cultures that exists in isolation from Western influences, the Matsigenka indigenous people in South-East Peru. The results were strikingly different from those of Singh's study of American subjects. The most attractive figure emerged as the one that was labelled by Westerners as overweight, with a WHR of 0.9. The figure found to be most attractive by Singh's subjects (of normal weight, with a WHR of 0.7) was labelled by a typical Matsigenka male as having had fever and having lost weight around the waist. In the same study, the authors also examined the preferences of South American indigenous people who had been exposed to Western influence and found male preferences to be more in keeping with United States standards.

The authors conclude that many so-called cross-cultural tests in evolutionary psychology may 'have only reflected the pervasiveness of western media' (Yu and Shepard, 1998, p. 32). The authors also point out that an adaptationist explanation is still possible. In traditional societies, physical appearance may be less important in mate choice since individuals are well known to each other, and couples have direct information, such as age and health status, about potential mates. In Westernised countries, a daily exposure to strangers may have sharpened our need to make judgements using visual cues.

Whereas Singh looked for a correlation between WHR and health indicators, another group of researchers in America, led by Bobbi Low, took a different line, but one which is still consistent with the view that WHR is a prime indicator of attractiveness. Low argued that broad hips and large breasts in a woman are indicators of the ability safely to bear and nurture children. Men should, therefore, be attracted to women with wide hips and large breasts. The development of broad hips and functional mammary tissue does, however, carry its own disadvantages for the woman, such as the impairment of locomotion. Low thus suggests that women have evolved physiological mechanisms to deposit fat on the hips and breasts as 'dishonest signals' to the male whose perceptual apparatus is geared up to respond to wide hips and large breasts. It is easier to fool a male into thinking that you have large hips than actually to develop them (Low *et al.*, 1987).

If ample breasts and hips were simply indicators of overall body fat and hence the diet of the female prior to pair bonding (as a well-fed female should be desirable to a male), the male should not be too fussy about the distribution of body fat. The work of Singh shows, however, that the distribution of the fat is crucial. Low suggests that, to counter this dishonest signal, men have demanded a display that wide hips are not simply the result of overall body fat by looking for signs of fat elsewhere, such as on the waist and arms. Women have responded by depositing fat on the hips but not the waist.

9.4.2 *Female assessment of men*

The prevalence of magazines depicting women in various states of undress would seem to imply that men are more easily aroused by visual stimuli than are women. Magazines devoted to pictures of men exist but cater largely for the homosexual market rather than for serious viewing by women. This does not, however, mean that women are insensitive to male physique. Singh (1995) also applied the concept of WHR to the female assessment of male attractiveness. It is well established that fat distribution in humans is sexually dimorphic and that, among the hominoids, this is a feature unique to humans. After puberty, women deposit fat preferentially around the buttocks and thighs while men deposit fat in the upper body regions, such as the abdomen, shoulders and nape of the neck. Such gynoid (female) and android (male) body shapes vary surprisingly little with climate and race. Singh reviewed evidence to show that the WHR is correlated with other aspects of human physiology, such as health and hormone levels (Figure 9.6).

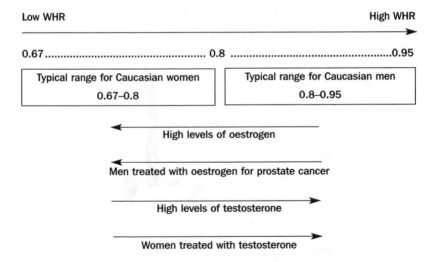

Figure 9.6   Distribution of waist to hip ratio (WHR) according to gender and hormone levels (data from Singh, 1995)

A low WHR is indicative of a high oestrogen level; a high WHR indicates a high level of testosterone

Singh presented women with line drawings (Figure 9.7) of men in various weight categories and with varying WHRs asking them to rate the men's attractiveness. In all body weight categories, figures with a female-like, low WHR were rated as being least attractive. The most attractive figure was found to be in the normal-weight category with a WHR of 0.9 (Figures 9.7 and 9.8). Interestingly, when females were asked to assign attributes such as health, ambition and intelligence, they tended to assign high values to the high WHR categories. Men with WHRs of 0.9 not only look better, but are also assumed to be brighter and healthier.

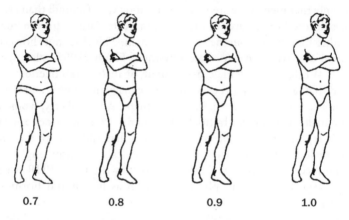

Figure 9.7   Stimulus pictures of men with various waist to hip ratios (WHRs) in the normal-weight category (from Singh, 1995)

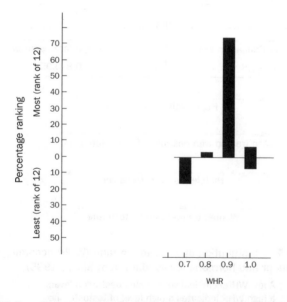

Figure 9.8   Percentage of female subjects rating the picture as the most attractive (rank of 1) or least attractive (rank of 12) according to the waist to hip ratio (WHR) for the normal-weight group (after Singh, 1995)

9.5 Fluctuating asymmetry

Evidence is now emerging that many human traits, such as fat distribution, hair colour, hair distribution and so on, are the products of **sexual selection**. As indicators of fertility, youthfulness, clear eyes, narrow waists and smooth skin are of great value and, not surprisingly, males find these attractive. However, as Ridley (1993, p. 287) observes, 'beauty is a trinity of youth, figure and face'. Facial features are more difficult to analyse in an evolutionary manner. There are countless variations of what is regarded as beautiful. Some features, for example large pupils and red lips, can be interpreted as indicators of sexual receptivity, and others as signs of health and vitality, but what about overall shape and proportion?

In 1883, Galton noticed that, by merging photographs of the faces of many women, he could produce an 'average face' that tended to be more attractive than most of the individual faces. This preference for the average could have a number of explanations. The average could represent an optimal adaptation in some way to local conditions, males and females thus being programmed to value the optimum. The average face is likely to be a highly symmetrical one, and it could be that symmetry is used as an indicator of beauty and hence reproductive fitness. Consistent with this notion, Gangestad and Thornhill (1994) have found a positive correlation between the symmetry of men, measured on such features as foot length, ear length, hand breadth and so on, and estimations by women of their attractiveness.

If we follow the symmetry argument, the obvious question that arises is why symmetry should be attractive. The most probable explanation is that it requires a sound metabolism and a good deal of physiological precision to grow perfectly symmetrical features. The development of bilateral characters such as feet, wings, fins and so on is open to a variety of stressful influences, such as **parasite** infection and poor nutrition, that result in an asymmetrical finished product. On this basis, the extent of symmetry observed in such characters may serve as an honest signal of phenotypic and genotypic quality.

Research on the importance of symmetry uses the concept of **fluctuating asymmetry (FA)**. FA refers to bilateral characters (that is, one feature on each side of the body) for which the population mean of **asymmetry** (right measure minus left measure) is zero, variability about the mean is nearly normal, and the degree of asymmetry in an organism is not under direct genetic control but may fluctuate from one generation to the next. There is now abundant evidence that FA is increased by mutations, parasitic infections and environmental stress, so FA consequently becomes a negative indicator of phenotypic quality (Manning *et al.*, 1996). The measurement of absolute and relative fluctuating asymmetry is shown in Figure 9.9.

Manning has suggested that if body size dimorphism in humans is a result of sexual selection, with females preferring taller males, it could be expected that body size is also an indication of phenotypic quality. Now, phenotypic quality, from the evidence presented above, can also be expected to be negatively correlated with the degree of asymmetry in an organism: the larger the degree of asymmetry (that is, the higher the FA value), the poorer the quality of the organism. On this basis, we should find a weaker relationship for female body size since body size in females has not been the subject of a sexual selection pressure. The preliminary findings from a study on just 70 adult males offer some support for this (Manning, 1995). When the average relative FA of four traits was compared with body size, a positive correlation was observed for females and a negative one for males.

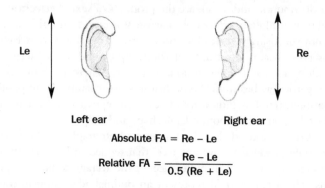

Absolute FA = Re − Le

$$\text{Relative FA} = \frac{Re - Le}{0.5\,(Re + Le)}$$

Figure 9.9   Measurement of relative and absolute fluctuating asymmetry

Manning *et al.* have also suggested that since the maintenance of symmetry requires metabolic energy, those males who have 'energy-thrifty genotypes' are better placed to maintain a low FA (Manning *et al.*, 1997). Those males with high resting metabolic rates should show greater signs of asymmetry since less energy is available to divert to symmetry maintenance. Figure 9.10 shows data in support of this based on a preliminary study of 30 males and 30 females. The relationship for females was not significant.

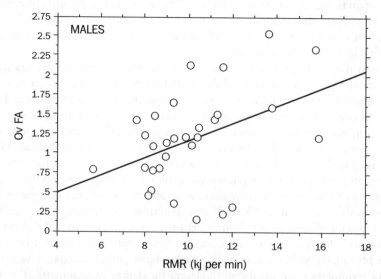

Figure 9.10   Simple linear regression of overall fluctuating asymmetry (ov FA) (that is, the sum of FAs for individual traits) plotted against resting metabolic rate (RMR) for 30 male subjects (from Manning *et al.*, 1997)

Note positive correlation $P = 0.019$

In both these studies, Manning is the first to acknowledge that more research and a larger data set are needed to substantiate these findings. FA promises to be a fruitful area of research.

9.6 **Male preference for youth, the female menopause and longevity**

I will greatly multiply thy sorrow and thy conception; in sorrow thou shalt bring forth children. (Genesis 3, 16)

In the mythology of the biblical book of Genesis, the agony of childbirth was the price that Eve paid for tasting the tree of knowledge. In evolutionary terms, childbirth also carries a considerable cost to the child and mother but one that is more than offset by the benefits of increased brain size. It turns out that the risks of childbirth provide the key to understanding the **menopause** and hence the preference for youth consistently displayed by males. In fact, the menopause provides a useful illustration of how evolutionary psychology and life history factors interact.

We have already noted evidence suggesting that men are attracted to women in a fairly narrow and youthful age band (narrower at least than the corresponding age range of men that women desire) and that this has a simple evolutionary explanation in terms of the shorter period of fertility in women compared with men. Studies have consistently shown that youthful female faces are rated as more attractive than older faces (Jackson, 1992). This could of course have been strongly influenced by cultural factors, but Judith Langlois and colleagues have found that even infants as young as 2–12 months agree with adult judgements with regard to attractiveness (Langlois, 1987; Langlois *et al.*, 1990).

At around the age of 50, the menopause sets in for women (the average for developed countries being 50.5 years) whereas male fertility continues beyond this and gradually declines with age rather than experiencing the abrupt cessation seen with female fertility. Now, although this fact can help to explain mate choice preferences, it is a feature that calls for an explanation in itself. Most mammals of both sexes, including chimpanzees and gorillas, experience a gradual decline in fertility with age, totally unlike the menopause of human females. So why the sudden shutdown experienced by human females? An explanation may be found by examining the mortality risks of human childbirth.

At birth, the human infant is enormous compared with the offspring of chimps and gorillas; 7 pound baby humans emerge from 100 pound mothers compared with 4 pound gorilla babies emerging from 200 pound mothers. Consequently, on a relative scale, the risk of death to the mother during childbirth is huge for humans and minute for chimps and gorillas. Because human infants are effectively born premature (see Chapter 6), they remain dependent on parental (especially maternal) care for a long period of time. For a woman who already has children, every extra child, while increasing her reproductive success, involves a gamble with the risk that she may not survive to look after the children. Now, the risks of death in childbirth increase with age, and there comes a point at which the extra unit of reproductive success of another child is exactly balanced by the extra risk to her existing reproductive achievement. Beyond this point (at which an economist might say the marginal benefit equals the marginal cost), it is not worth proceeding. To protect the mother's prior investment in her children, natural selection probably instigated the menopausal shutdown in human fertility. Since childbirth carries no risk to the father, and fathers can always increase their reproductive success with other partners, men did not evolve the menopause (Diamond, 1991). Given the difference in **parental investment** in the early years of a child's life, we can see that children born to old men but young women have a better chance of survival than those born to old women.

If reproductive shutdown at the menopause occurs to protect a mother's investment, it could be supposed that females should live just long enough after the menopause to protect and raise their children to independence. In fact, women tend to live longer than is strictly necessary. Among other mammals, a mother will spend only about 10 per cent of her life after her last birth, whereas human females can live nearly a third of their lives after their last child. This has led to much speculation about the evolutionary function of grandmothers. But grandmothers aside, it is a reasonable hypothesis to suppose that women who bear children late in life and survive childbirth subsequently live longer. Voland and Engel (1989) claim to have found support for this hypothesis. Modern medical care will probably iron out any effect in modern cultures, so they examined the records of 811 women born between 1700 and 1750 in a rural district of Germany and found that life expectancy did increase with the age of the last child. In the same study, these authors confirmed that childbirth on the whole decreases life expectancy since married childless women tended to live longer than married mothers. However, if women became mothers, their life expectancy was increased significantly by a late age of the last child (Figure 9.11).

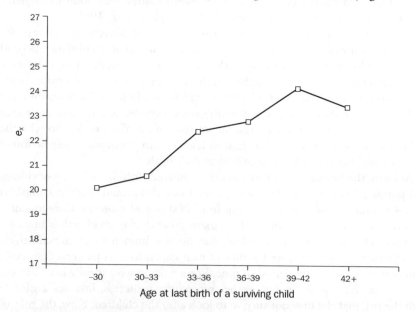

Figure 9.11   Future life expectancy of 47-year-old women (e_x) by their age at the birth of the last surviving child (from Voland and Engel, 1989)

Pearson's correlation coefficient = 0.06980; $P = 0.02787$

Ageing itself is a result of natural selection tending to favour investment in early life reproduction rather than constant maintenance and repair. Genes that divert resources away from combating the damage caused by ageing towards reproduction will (since they promote reproduction) tend to be preserved. This is Kirkwood's 'disposable soma theory' proposed over 20 years ago (Kirkwood, 1977). If this theory is right, female longevity should increase the fewer the number of children that a woman has. By examining data on the British aristocracy, Westendorp and Kirkwood (1998) find evidence in support of this.

Although men remain fertile to an older age than women, there is not much consolation to be had. The male mortality risk is higher at virtually every age of their lives, their lifespan is shorter, and they are more prone to accidents. Adam's punishment for tasting of the fruit of the tree of knowledge was that:

In the sweat of thy face shalt thou eat bread
Till thou return to the ground;
For out of it wast thou taken:
For dust thou art,
And unto dust shalt thou return. (Genesis 3, 19)

Life is hard and then you die. From natural selection, we shouldn't really expect much else.

9.7 Facial attractiveness: averageness, symmetry and honest signals

9.7.1 Honest signals of genetic worth

In Chapter 5, we saw how the lek paradox could be avoided by viewing organisms as hosts locked into a constantly evolving battle with parasites. In this co-evolutionary struggle, there is no optimum solution to be found: each adaptation is only temporary, and each solution sets up a selective pressure for an anti-solution (Thornhill and Gangestad, 1993). Parasites generally use the resources of the host – nutrients, enzymes, proteins and so on – for their own reproductive purposes. It is probable that, in any population, the majority of pathogens will be best adapted to the most common biochemical pathways of their hosts. One defence against this is sex. By outbreeding, parents produce offspring that are different from themselves; any parasites that were successful with parents may be less so with their offspring since the offspring will have a whole new array of proteins and a different defence system.

It follows that the greater the genetic polymorphism at the population level – that is, the more alleles at any given locus on the genome of the species – the more likely it is that at least some individuals will stay ahead of parasites. If genetic variation confers fitness on a population, the same thing may be true at the level of an individual: the more heterozygous an individual is, the more variability to be found in its genome, the more varied the proteins that are produced and thus the harder it will be for a parasite to exploit its host efficiently. But organisms face a problem: genetic variation that is good from the point of excluding parasites could be disruptive if variation departed from what was adaptive for any given set of conditions. Just as too much **inbreeding** may cause a depression of fitness by generating homozygosity for recessive alleles, and homozygosity may depress fitness by allowing parasites to exploit the lack of variation in proteins, so excessive outbreeding may introduce less well-adapted variations. There is clearly a balance to be struck.

On one side of the equation, sexual selection leads us to believe that mates will choose each other taking into account the resistance to parasites that a potential partner posseses and hence could potentially confer on its offspring. Following this line of reasoning, Thornhill and Gangestad suggest that four sorts of preference might be expected to evolve through sexual selection in an environment where parasites are prevalent:

1. A preference for heterozygosity
2. A preference for parasite-resistance alleles
3. A preference for indicators of development stability are hence signs of a well-functioning genome
4. Preferences for handicaps that can only be afforded by parasite-resistant individuals. Such handicaps may include hormones that suppress the **immune system**, and hence the bearers of such handicaps must have sound immune systems to be able to survive. (Thornhill and Gangestad, 1993)

The question arises of how a choosy partner will be able to estimate the degree of heterozygosity and the state of the immune system in a potential mate. We noted in Chapter 3 that there is evidence of both mice and humans being attracted to smells that indicate differences in the major histocompatibility complex between mates, but there may be more obvious ways. If males send out honest signals to females in the form of secondary sexual characteristics, well-adapted males can advertise their worth. Alternatively, they can reveal a handicap that they are able to bear as a result of an efficient immune system. It transpires that testosterone could serve as just such a handicap: it suppresses the immune system such that only well-adapted males can tolerate a high level of it, additionally revealing its presence by affecting the growth of secondary sexual display characteristics. Figure 9.12 shows how the system is conjectured to work.

The system shown in Figure 9.12 results in the outcome that only fit males can afford to suppress their immune system with a high level of testosterone. Consequently, they display characteristics that indicate this fact. Evidence in support of this scheme for non-human animals comes from the work of Saino *et al.* on barn swallows. The length of tail of a barn swallow seems to be an honest advertisement of testosterone. Males with short tails were not only less attractive to females, but also, when injected with testosterone, suffered a higher mortality than their long-tailed counterparts (Saino *et al.*, 1997).

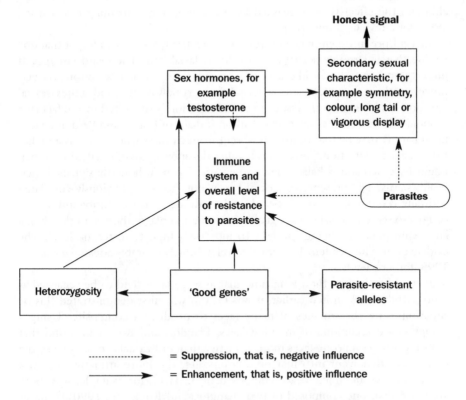

Figure 9.12   Conjectured relationship between resistance to parasites, secondary display characteristics and honest signals

9.7.2 *Faces as purveyors of information*

The obvious question is 'To what degree do humans display sexually selected secondary characteristics?' One promising area to search is that of aesthetic judgements of faces. Faces carry a mass of information and, unsurprisingly, humans are extremely responsive to each other's faces.

Eibl-Eibesfeldt proposed that the qualities that males find attractive in human females, such as a small, upturned nose, large eyes and a small chin, correspond to 'infantile' features. Such features are found in the young of other creatures that both sexes find attractive or 'cute'. It may be that women have developed these features to evoke the caring response that males feel towards their young. In effect, women's faces have targeted in on the perceptual bias in male brains (Eibl-Eibesfeldt, 1989). Other features in women's faces that men find attractive are also associated with signs of youth. As women age, increasing levels of testosterone stimulate hair growth. Significantly, men tend to find hairless faces in women more attractive – as evidenced by the efforts of women to remove facial hair. Hair colour also tends to darken with age, and the attractiveness of blonde hair could be that it is a reliable sign of youth. Skin darkens with age and with pregnancy. There is some evidence that men find paler skin more attractive, although the effect is often masked by the fashion for sunbathing (see Barber, 1995, for a review of this area).

In a culture in which shaving is routinely practised, it is easy to forget that one feature that is strongly sexually dimorphic is facial hair. The beard emerges at puberty and thus looks like a prime candidate for sexual selection. As yet, however, there is no consensus on whether runaway or good genes sexual selection has been at work. Fisherian sexual selection is suggested by the fact that in some cultures, such as the native Andean Indians of Bolivia and Peru, men lack facial hair. It may be that beards are absent from cultures where there is a higher risk of parasite infection. Since parasites alter the appearance of facial skin, women might have demanded hairless faces to ensure that men honestly signalled their health status. The presence or lack of hair is not, however, obviously correlated with any ecological factors. A beard could indicate sexual maturity and age. It could also serve as a dishonest signal by seeming to enlarge the size of the chin, a large chin generally being thought to indicate a high testosterone level. The evidence on whether females actually prefer beards is ambiguous (see Barber, 1995, for a review).

It has been known for a long time that average faces tend to be regarded as more attractive. As noted earlier, it was Galton who first demonstrated this by superimposing the images of many faces to produce a composite. Using a computerised technique of merging faces, Langlois and Roggman found that not only were composite faces more attractive than individuals ones, but also the more faces that went into making a composite, the more attractive, up to a point, the face became. A face composed of 32 faces, for example, was more attractive than one composed of two (Langlois and Roggman, 1990). Symons suggested in 1979 that the average was rated as attractive because the average of any trait would tend to be optimally adapted for any trait, since the mean of a distribution presumably represents the best solution to the adaptive problem (Symons, 1979). Thornhill and Gangestad add to this the fact that protein heterozygosity tends to be highest in individuals who exhibit the average expression of heritable traits that are continuously distributed. It could be, then, that **facial averageness** correlates with attractiveness because averageness is an indication of resistance to parasites (Thornhill and Gangestad, 1993).

The experimental problem to overcome here is that there are at least two variables: averageness and symmetry. Composite faces also tend to be more symmetrical since asymmetries are ironed out. Symmetry, as already noted, is a reliable cue to physiological fitness that takes the form of resistance to disease and a lack of harmful mutations that compromise development stability. Yet there must also be other factors at work for whereas symmetrical faces are regarded as being more attractive than faces with gross asymmetry, computer-generated, perfectly symmetrical faces are not perceived as being attractive as natural faces with a slight asymmetry (see Perrett *et al.*, 1994). Some of these conjectures are summarised in Table 9.5.

Average Caucasian female

Average Caucasian male

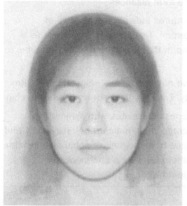

Average Japanese female

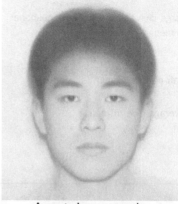

Average Japanese male

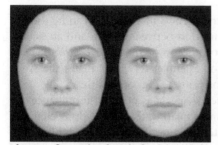

Average Caucasian female feminized (left)
and masculinized (right)

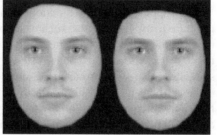

Average Caucasian male feminized (left)
and masculinized (right)

Figure 9.13   Facial attractiveness: averaging and sexual dimorphism

The first 4 images in this picture show the effect of merging photographs of 30 Caucasian females (mean age 20.6 years), 25 Caucasian males (mean age 21.0 years), 28 Japanese females (mean age 21.4 years) and 28 Japanese males (mean age 21.6 years) to create four 'average faces'. Most would agree that the faces appear attractive. But Dave Perrett *et al.* (1998) at the University of St Andrews found that by feminising faces both males and females were rated as even more attractive. This finding would appear to place a limit on the degree of sexual dimorphism that can emerge in human faces and itself refutes the hypothesis that average faces are optimally attractive.

Table 9.5 Facial features and conjectures on possible function in natural or sexual selection

Facial feature in male or female	Possible role in selection
Lack of facial hair in women and attempts to remove facial hair	Indicates youth
Pale skin in women	Indicates youth since skin darkens with age
Neotenous (childlike) features regarded as attractive in women	Could elicit nurturant response in men. Also small chin and nose indicate a low level of testosterone
Symmetry in male and female faces regarded as attractive	Symmetry is an indicator of physiological precision, protein heterozygosity and hence resistance to or freedom from pathogens
Large smile	Indicates sociability, health and well-being. Ability to form alliances
Large chin and prominent cheekbones in men	Such features indicate a high level of testosterone. Testosterone may serve as a handicap in that it suppresses the immune system
Hair on face of men (attractive?)	Could be runaway sexual selection. Indicates maturity. Enlarges the appearance of the chin
Average face attractive	Averageness correlates with symmetry. The average could indicate heterozygosity and resistance to pathogens. It could also indicate an optimum level for a continuously distributed trait

The supposed adaptive functions of attractiveness in Table 9.5 can be used to generate many hypotheses, such as a connection between health and attractiveness, fertility and attractiveness and so on. Some of these have already been tested. Shackelford and Larsen (1999), for example, report a small-scale study suggesting that facial attractiveness may correlate with good health.

Since there is sexual dimorphism in face shape, it is possible to enhance the femininity or masculinity of any face by manipulating features such as chin size and cheekbone prominence using computer imaging. Perrett *et al.* did just this, looking for the effects of masculinisation and feminisation on perceptions of attractiveness. They found that feminising a female face increased its attractiveness, a preference they saw as contrary to the averageness hypothesis: female faces became more rather than less attractive if they departed from the average in a feminine direction. The surprising result was that when an average male shape was feminised, it too became more attractive (Figure 9.13). There arises the possibility that women looking for a long-term partner prefer a less masculine face since a high testosterone level could indicate social dominance but an unwillingness to invest in a relationship. Women may employ multiple strategies:

choosing a feminised face for a long-term partner but finding masculine faces attractive for short-term **extrapair copulations** (Perrett *et al.*, 1998).

Figure 9.14   Hugh Grant and Andie McDowell

Facial beauty may be the result of sexual selection. Attractive faces tend to be symmetrical and show signs of good health such as good skin tone and whiteness of the eyes. Females with neotenous (childlike) faces, including large eyes, small noses and full lips, are rated as attractive. Male features that are judged to be attractive, such as high cheekbones and a strong jaw and chin, correlate with a high level of testosterone

SUMMARY

Although males and females share many mate choice criteria, evolutionary considerations would predict a difference between the attributes sought by males and by females in their prospective partners. Females are predicted to look for high-status males who are good providers, whereas males are predicted to look for young, healthy and fertile females who are good child-bearers. Men are also much more likely to seek sex without commitment. There is considerable empirical support for these and related predictions.

Physical beauty in both males and females is expected to be correlated with signs of reproductive fitness. Evidence is now accumulating that body shapes found to be attractive are those which carry fitness indicators. Facial beauty in both sexes is likely to be related to overall fitness, health and immunocompetence. Faces may convey honest signals about fitness and reproductive value.

KEY WORDS

Asymmetry ■ Extrapair copulation ■ Facial averageness
Fluctuating asymmetry (FA) ■ Immune system ■ Menopause
Parasites ■ Parental investment ■ Sexual selection
Waist to hip ratio (WHR)

FURTHER READING

Buss, D. M. (1994) *The Evolution of Desire*. New York, HarperCollins.
A book based on Buss' own research and that of others. It clearly explains expected sex differences in desire.

Buss, D. M. (1999) *Evolutionary Psychology*. Needham Heights, MA, Allyn & Bacon.
An excellent discussion of male and female mating strategies.

Greary, D. C. (1998) *Male, Female: The Evolution of Human Sex Differences*.
  Washington DC, American Psychological Association.
A well-referenced work that contains much material on sex differences, including mate choice criteria.

10

Conflict within Families
and Other Groups

I know indeed what evil I intend to do
But stronger than all my afterthoughts is my fury,
Fury that brings upon mortals the greatest evils.

(Euripides, *The Medea*)

The Greeks knew a thing or two about dysfunctional families, and it is not surprising that, on several occasions, Freud turned to their myths and legends to find labels for what he supposed were problems of the human psyche. He posited, for example, an Oedipus complex, whereby males subconsciously desire the death of their father in order to sleep with their mother, and an Electra complex involving the secret desires of daughters for their fathers. From an evolutionary perspective, both are misleading notions and equally unlikely. Much human behaviour, especially that involving conflict, is, however, maladaptive. In *The Medea* story above, Jason abandons his wife Medea for a more desirable bride. Medea, in her anger, kills the bride's father, the bride, and even her own children by Jason. Medea can be taken to signify both a wronged woman's fury and the dark and inexplicable irrational forces in human nature. *The Medea* is a work of fiction of course, but people do such things. We hardly need, however, to invoke a Medea complex; Darwinism (as you might expect) also has something to say about murder and infanticide.

This chapter, then, is concerned with the application of evolutionary theory to understanding interactions involving conflict between offspring and their parents, between siblings and between spouses. To tackle these problems, theoretical perspectives established in earlier chapters will be deployed. In Chapter 3, it was noted that the existence of altruism posed a special problem for Darwinian theory and that the breakthrough came with Hamilton's notion of inclusive fitness. The concept of inclusive fitness also helps us to understand conflict between related individuals. It seems obvious that, since offspring contain the genes of their parents, parents will be bound to look after their genetic investment. But parents have loyalties that are divided between care for their current offspring and the need to maintain their own health in order to produce future offspring. It follows that offspring may demand more care than parents are willing to give, and in this situation we should expect to observe a mixture of altruism and conflict. Chapters 8 and 9 established that partners in a sexual relationship may have different interests and strategies, and it is these conflicting interests that help to explain violence and strife within marriages.

One extreme manifestation of human conflict is homicide. From an analytical point of view, homicide has the advantage that considerable statistical information is available. Two American psychologists, Margo Wilson and Martin Daly, have pioneered the use of homicide statistics to test evolutionary hypotheses, and this chapter considers their work.

10.1 Parent–offspring interactions: some basic theory

10.1.1 *Parental altruism*

Everyone is aware that the animal world abounds with examples of parents (often mothers) making great sacrifices in their efforts to protect and nurture their young. Perhaps the most extreme form of motherly care can be seen in some spider species (of which there are about ten) where the young eat their mother at the end of the brood care period. In the case of *Stegodyphus mimosarum*, the spiderlings do not eat each other or spiders of other species, yet they devour their mother with relish; it seems that this 'gerontophagy' serves as the mother's final act of parental care. From a Darwinian perspective, parents will cherish their progeny because their inclusive fitness is thereby increased. In turn, parents will be loved by their offspring because they can provide help to increase the fitness of the offspring. Parents will also be loved, although with less fervour, because they can increase the inclusive fitness of current offspring by producing more siblings. Altruism between parents and offspring is thus covered by Hamilton's theory of inclusive fitness. Parents will donate help b at a cost c to themselves as long as:

$$\frac{b}{c} > \frac{1}{r}$$

where $r =$ **coefficient of relatedness** between parents and offspring. For diploid outbred offspring (that is, parents who are not related), $r = 0.5$ (Figure 10.1).

10.1.2 *Parent–offspring conflict*

The widespread occurrence of parental care, and the clear biological function it served, probably diverted biologists for many years from the fact that conflict between parents and offspring may also have a biological basis. It awaited the work of Trivers to point out that **parent–offspring conflict** is also predicted from evolutionary theory (Trivers, 1974). The theory of such conflicts is worth examining for the light it throws on human behaviour, especially maternal–fetal conflict.

To follow Trivers' original argument, suppose one parent (the mother) gives birth and care to one infant each breeding season and that the infant needs and benefits from the care. The problem faced by the parent is when to cease

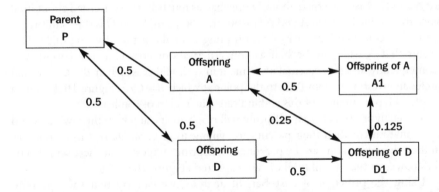

Figure 10.1 Coefficients of genetic relatedness (*r* values) between parents and offspring, between siblings and between nieces or nephews

providing care. A theoretical solution requires that we examine the costs and benefits of parental care to the parent (P) and the offspring (A). Table 10.1 shows how the benefits and costs are distributed between parent and offspring.

Table 10.1 Distribution of benefits and costs (to parent P and offspring A) of parental care between parent and offspring

	Benefits (*B*)*	Costs (*C*)*
Parent (P) gives care to offspring (A)	Increase in inclusive fitness since there is a 50 per cent chance that the caring gene is also present in offspring A	Investment in offspring A reduces the ability to invest in offspring D
Offspring (A)	Increase in fitness in terms of own reproductive success	Detracts from parent's ability to invest in sibling D and hence the offspring of D (D1), which are related to A by *r* = 0.5 and 0.25 respectively

Note that *B* and *C* have already been weighted by *r*, and so are not the same terms as in Hamilton's equation of 'help if: *b/c* > 1/*r*'.

Suppose we define the benefit to the parent in terms of units of probability that A will survive and reproduce, and the cost to the parent as units of reduced probability that it can produce another offspring D that will survive and reproduce. Now, parent P will care for A and D equally: they both have an *r* value of 0.5. As long as *B* > *C*, care to offspring A will be favoured. Note that, in this case, *B* has already been '*r* adjusted'. As A matures, the law of diminishing returns dictates that further care to A will bring fewer benefits than investing in a new offspring D. When *B* < *C*, the interests of parent P (or equivalently the helper genes carried by P) are best served by withdrawing from A and investing in D. Although the parent ultimately cares equally for A and D, A is concerned about D only half as much as it is concerned about itself since the *r* value between A and A is 1 but that between A and D is 0.5. Similarly, A is twice as concerned about its own offspring than about the offspring of D. This means that A will only be half as concerned about cost *C* as the parent will. Consequently, whereas parental investment in A should cease when *B* = *C*, A would prefer an investment from P up to the point at which *B* = *C*/2. Figure 10.2 shows a graphical representation of this and indicates the region of conflict.

The analysis predicts that offspring will reach an age at which they will demand more investment than their parents are willing to give. Similarly, this asymmetry in the value placed on care by parent and offspring suggests that there will also be differences in the overall level of care expected (Figure 10.3).

Using the principle of cost–benefit analysis, we can say that the optimum situation for the parent and the offspring is when their own benefits minus their costs are at a maximum. This is at a different point for the parent and the offspring. For the parent it rests at Q, while for the offspring it resides at R. The net effect is that the offspring desires more care than is optimum for the parent, and conflict is expected to ensue.

It has proved difficult to test Trivers' theory quantitatively because of the difficulties of quantifying costs and benefits. There is qualitative evidence, however, that is in broad agreement with theoretical expectations. One prediction from conflict theory is that offspring will prefer parents to direct resources at themselves (especially when they are young) rather than expending effort in producing more offspring. Young human parents often jokingly complain that 'kids are the best contraceptives'. Among chimpanzees, this may literally be true. Tutin (1979) noted that immature chimps often attempt to interrupt the copulations of their parents.

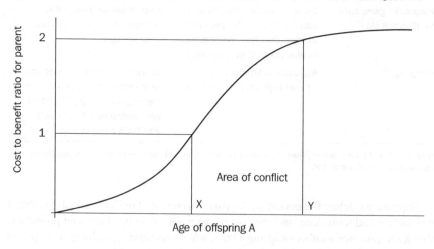

Figure 10.2   Relationship between the cost to benefit ratio for an investing parent and the age of the offspring (adapted from Trivers, 1974)

X is the position when it is optimum for parent P to withdraw care from A.
Y is the position when it is optimum for offspring A to allow
parent to withdraw care

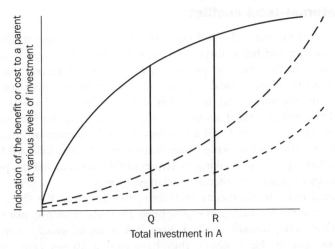

Q = Optimum care from the perspective of the parent, that is, when $(B - C)$ is maximum
R = Optimum care from the perspective of the offspring, that is, when $(B - C/2)$ is maximum

——— Benefit (B) to parent
— — Cost (C) to parent
– – – Cost to A = $C/2$

Figure 10.3   Curves of benefits and costs to parent
and offspring against total investment in offspring

As the age of a parent increases, and it nears the end of its reproductive life, so its chances of producing any more offspring reduce and the costs to itself of giving care to existing offspring decrease. It should follow that older parents are more willing to invest in the young than are younger ones. This seems to be supported by observations on the Californian gull (*Larus californicus*), although other interpretations are possible, such as the fact that older gulls will be more experienced in caring for their young (Trivers, 1985).

The logic of genetic relatedness dictates that individuals will care for themselves and their offspring more than they will for their siblings and the offspring of their sibs. At its extreme, this asymmetry could result in **siblicide**, which is in fact widespread in nature. In some species of eagle, the mother normally lays two eggs even though, in nearly all cases, only one offspring survives. The mother perhaps lays the second egg as an insurance policy against infertility in a single egg. On hatching, the elder chick kills the younger (see Mock and Parker, 1997).

The work of Trivers has stimulated a fresh look at behaviour that was once thought of as being unproblematic. In pregnancy, for example, it is easy to assume that the interests of the mother and fetus are virtually identical and that the considerable investment by the mother in the fetus is a clear case of kin-directed altruism. Recent work by Haig (1993), examined in the next section, shows that the situation is more complicated and provides a useful testing ground for Trivers' theory of conflict.

10.2 Maternal–fetal conflict

One of the most intimate relationships in the natural world must surely be that between a mother and her embryonic developing child. The mother is the life support system for the fetus: she provides it with oxygen from every breath she takes and food from every meal she eats. It is tempting to think that, in this precarious state, the interests of the mother and fetus must be identical, and that conflicts of the sort examined above can only arise after the birth of the child, when the mother will soon be in a position to produce more offspring. Even here, however, gene-level thinking brings some surprises. The Harvard biologist David Haig has applied the thinking of Trivers to this situation and has produced his own theory of genetic conflicts in human pregnancy.

The crucial point to appreciate in Haig's analysis is that the fetus and the mother do not carry identical genes: genes that are in the fetus may not be in the mother if they were paternally derived. Even if they are maternally derived (and therefore present in the mother), they have only a 50 per cent chance of appearing in future offspring of the same mother. It is quite feasible, then, argues Haig, for fetal genes to be selected to draw more resources from the mother than is optimal for the mother's health or optimal from the mother's point of view of distributing resources among her current and future offspring. We will consider four examples in which Haig's theory has met with empirical support.

10.2.1 Conflicts over glucose supplied to the fetus

When a non-pregnant woman eats a meal high in carbohydrates, her blood sugar level rises rapidly, but then falls as the insulin secreted in response to the raised glucose level causes the liver to store the excess glucose as glycogen for later use. In contrast to this, a similar meal taken during late pregnancy will cause the maternal blood glucose level to rise to a higher peak than before and moreover stay high for a greater length of time, despite the fact that the insulin level is also higher. In effect, the mother is less sensitive to insulin and compensates, but not entirely, by raising her insulin level. This is of course puzzling. Why should a woman develop a reduced sensitivity to insulin and then bear the cost of having to produce more?

Haig's theory of genetic conflict suggests an answer to this problem. It is in the interests of the fetus to extract more blood sugar for itself than it is optimal for the mother to give. The mother is concerned for her own survival after giving birth and, in addition, is more concerned for existing and future offspring than is the fetus. The fetus can send signals to the mother via the placenta just as the mother can send signals to the fetus. Haig suggests that one result of this is that, in late pregnancy, the placenta produces allocrine hormones that decrease the sensitivity of the mother to her own insulin, thereby allowing the glucose level to rise to benefit the growth of the fetus. The mother responds in this tug-of-war by increasing her insulin level. Further evidence in favour of this theory is that the placenta possesses insulin receptors and that, in response to a high insulin level, it produces enzymes that act to degrade insulin and thus disable the mother's counterattack. We can picture the escalation of measure and countermeasure as resulting in a pair of forces acting upon the level of some parameter such as glucose, each attempting to move it in an opposite direction towards two opposed optimum levels (Figure 10.4).

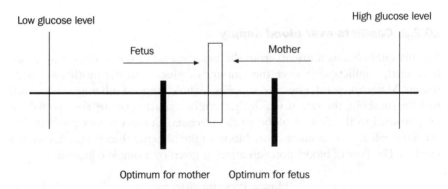

Low glucose level

High glucose level

Fetus

Mother

Optimum for mother   Optimum for fetus

Figure 10.4   Schematic representation of the effort by the mother and fetus to drive the blood glucose level to different optima

The central cursor can be thought of as moving on a sliding scale subject to pressure from the mother and fetus

10.2.2 *Conflicts over decision to miscarry*

It has been estimated (Roberts and Lowe, 1975) that about 78 per cent of all human conceptions never make it to full term. Most conceptuses miscarry before the 12th week and many before the first missed menstruation. When spontaneous abortions are karyotyped (that is, the structure of their chromosomes examined under an optical microscope), they are seen to have genetic abnormalities. Early miscarriages seem to be the result of some sort of quality control mechanism employed by the mother to terminate a pregnancy before she has committed significant resources. If this is the case, there must be a threshold of quality below which a mother will attempt to terminate the pregnancy and save resources for future offspring and above which she will accept the pregnancy to full term. Fetal genes will also have an interest in the decision of whether or not to terminate. If the quality is extremely low, it may be better for genes to 'abandon hope' of survival in the current fetus and 'hope' for survival (with a 50 per cent chance of being represented) in a future offspring. In this case, it would be preferable from the point of view of the fetus not to jeopardise the chances of future viable offspring by dragging from the mother resources that will be wasted.

But here's the rub: the cut-off point for quality will be different in the mother and the fetus. In essence, the fetus is predicted to respond to a lower quality cut-off point than the mother does. It is significant, then, that the maintenance of pregnancy in the first few weeks of gestation is dependent on maternal progesterone. Later on, the fetus gains control of its fate by releasing human chorionic gonadotrophin into the maternal bloodstream. This hormone effectively stimulates the release of progesterone and serves to maintain the pregnancy. One would expect that the older the mother, the lower the quality that she will still accept as she nears the end of her reproductive life. It may be no accident that births with genetic abnormalities increase in frequency with the age of the mother.

10.2.3 *Conflicts over blood supply*

The placenta obtains nutrients from the bloodstream of the mother, and, as we have seen, conflicts arise over the amount of glucose in the mother's bloodstream. Maternal blood, however, supplies a whole range of other nutrients, and, roughly speaking, the rate at which they can be extracted by the embryo will be proportional to the flow of blood to the placenta. Conflict theory predicts that the fetus will act to stimulate more blood to the placenta than is optimal for the mother. The flow of blood along an artery is given by a simple equation:

$$\text{Flow} = \frac{\text{Pressure difference}}{\text{Vascular resistance}}$$

It follows that the fetus could increase its supply of nutrients by either (a) dilating the arteries of the mother and thus reducing the vascular resistance, or (b) increasing the blood pressure in the maternal supply. Maternal genes are predicted to act to counter both of these manoeuvres. Haig reviews a range of evidence and concludes that hypertension (raised blood pressure) in pregnancy is one result of **maternal–fetal conflict**. Significantly, for example, high birthweight is positively correlated with maternal blood pressure whereas low birthweight is associated with hypotension (low blood pressure).

10.2.4 *Conflict after parturition*

As does the fetus, so a young infant may also behave to extract more resources from a mother than is optimal for the mother to provide. The presence of benzodiazepines (substances that inhibit neurotransmission and thus serve as sedatives) in human breast milk may also be a response of the mother to excessive infant demands. At the level of the genes, we could expect maternally derived genes to be less demanding than those that are paternally derived since paternal genes are probably not present in the mother and will be selected to exploit the mother for their own advantage. Evidence in favour of this prediction comes from conditions in which genetic abnormalities are observed. In Prader–Willi syndrome, infants have a maternal copy of a region of chromosome known as 15q11–13 (to be read as 'chromosome 15, long arm, bands 11 to 13') but lack a paternal copy. These unfortunate infants have a weak cry and a poor sucking response, as well as being generally inactive and sleepy. In contrast to this, children with Angelman syndrome have paternal copies of 15q11–13 but no maternal ones. In these cases, sucking is prolonged but poorly co-ordinated. Children are highly active when awake and suffer sleeping difficulties. It looks here as if we have a system in which paternally derived genes demand excessive resources from the mother and are restrained by maternally derived genes.

These two conditions illustrate the general phenomenon of **genomic imprinting**, a situation in which a gene behaves differently depending on whether it was contributed by the father or the mother. A good example concerns the *Igf-2* gene, which produces an insulin-like growth factor that promotes embryonic development. In mice and humans, copies of the *Igf-2* gene from the mother are switched off and copies from the father left on. This is consistent with evolutionary conflict theory. It is in the interests of the paternally derived *Igf-2* gene to demand more resources than the mother is prepared to give. The reason for this stems from the theory considered earlier: the mother wishes to reserve her resources for future offspring, but the current paternal *Igf-2* gene has no guarantee, unless the mother is 100 per cent monogamous, which is unlikely, that it will appear in future offspring. In this light, as Pagel neatly summarises, 'genomic conflict is one cost of infidelity' (Pagel, 1998, p. 19). Given the excessive demands of the paternal *Igf-2* gene, the best counter-tactic of the mother is to switch off her *Igf-2* copy and cease its contribution to the production of the growth factor. Although genomic imprinting provides evidence for selection at the level of the gene, all such cases are unlikely to be explained by conflict theory. Iwasa, for example, has recently advanced a 'dose compensation' explanation (see Pagel, 1998).

Haig himself points out that the presence of conflict does not in itself imply system instability or failure. A useful analogy is that of evenly matched tug of war teams: as long as both sides keep pulling, the system is stable. However, the system collapses when one side lets go. In this light, both Prader–Willi syndrome and Angelman syndrome can be seen as resulting from one side of the conflict not operating, with disastrous effects. This whole scenario reminds us of course that evolution is all about compromise.

10.3 Human violence and homicide

The application of adaptationist thinking to human conflict within families was pioneered by Martin Daly and Margo Wilson in the 1980s. Daly and Wilson completed their PhDs in animal behaviour in the early 1970s and were inspired by Wilson's (1975) *Sociobiology* to apply evolutionary theorising to step-families. Since then, they have written numerous articles and books dealing with human conflict in the form of **homicide**.

Before we examine their work, it is important to understand the way in which adaptive reasoning can be applied in this area. In modern nations, homicide is, for the most part, damaging to the fitness of an individual. The perpetrators are likely to be found and either incarcerated or executed. Many homicides also have a negative effect on inclusive fitness since it is often one relative who kills another. To cap it all, many homicides are followed by suicide, which is hardly fitness-maximising. It would at first sight seem bleak territory on which to erect adaptationist arguments, and, unsurprisingly, homicide has usually been regarded as a result of inherent human wickedness, a failure of social upbringing or the result of some sort of pathological condition. The originality of the work of Daly and Wilson has been to realise that, amid all these causes, there may also be the effect of psychological mechanisms that can be understood in selectionist terms.

Here, evolutionary psychology differs from the traditional approaches of behavioural ecology and sociobiology. Sociobiologists, when studying the behaviour of animals other than humans, looked for the adaptive significance of contemporary behaviour. With some qualifications, it was assumed that behaviour should be optimal with respect to fitness. The evolutionary psychologist, however, looks for motivation mechanisms and 'domain-specific' mind modules that were shaped during the environment of evolutionary adaptation (the Pleistocene era when the genus *Homo* appeared; see Chapter 6). The output of these modules is adaptive only on average and largely in the environment in which they were shaped. Like all mechanisms, they can suffer problems of calibration. A jealous rage that once served the interests of an early male in driving away would-be suitors from his wife may translate in the 20th century to a man murdering his wife's lover and receiving a life sentence.

Homicide is a gruesome business, but the serious nature of the crime makes statistics on killings more reliable than those on probably any other act of violence. It is against patterns of homicidal statistics that predictions from evolutionary psychology can be tested.

10.3.1 *Kinship and violence: a paradox?*

Criminal statistics show that most violence occurs within the family. This presents a challenge to evolutionary psychology since kinship is supposed, on balance, to promote altruism. What can be the advantage of killing a relative and thus reducing inclusive fitness? Here, we need to tease out the statistics carefully and be aware of the need for standardising the data. If we examine the mortality statistics for most nations, it would be easy to show that more people die in their homes from all causes than are killed by, for example, riding a motorbike. Does it follow that riding a motorbike is safer than staying at home? The answer is of course no: to draw any conclusions about relative risk to life we need to standardise for, among other things, age, sex and time of exposure to the hazard. Many people, including the elderly, spend a lot of time at home; fewer people spend even less time riding motorbikes. To understand properly the effect of kinship on violence, we need to take into account the other variables. The majority of reported violence takes place within families because families spend a lot of time together and the opportunities for misjudgement and loss of control are correspondingly greater.

In an effort to standardise for time of exposure, Daly and Wilson (1988a) examined statistics for homicide within groups of cohabitants in Detroit for the year 1972. Detroit combined the advantages of proximity to the home of Daly and Wilson with a sympathetic deputy chief of police, Dr Bannon. The term 'cohabitant' included both kin and non-kin. The results are shown in Table 10.2.

The conclusion is that, in group living, a homicide victim is much more likely to be unrelated to the murderer than to be his or her kin. It could be then, as Daly and Wilson themselves suggest, that kinship has an overall mitigating effect on violence as would be expected from considerations about inclusive fitness. There are other possibilities, however, not controlled for in the data above. Relatives are likely to have been known for a longer period by the assailant than

Table 10.2   Risk of homicide by relationship to cohabitant
(adapted from Daly and Wilson, 1988a)

Relationship category	% of adults living with member of category	Observed number of victims	Expected vumber of victims	Relative risk: Observed / Expected
Spouses	60	65	20	3.32
Non-relatives	10`	11	3	3.33
Offspring	90	8	29	0.27
Parents	40	9	13	0.69

The expected number of victims is the number of victims predicted if homicide were evenly distributed according to the pattern of cohabitation in the first column.

non-relatives, allowing stronger bonds to form and, perhaps more crucially, allowing some understanding to form about the personalities of the parties concerned. The mitigating effect may, then, be a product of mutual under-standing gained by prolonged contact rather than any kinship effect as such. It is difficult to see how all the confounding variables could be controlled, but the data so far are at least suggestive.

In the West, fratricide is often regarded as one of the primal sins described in the Genesis story of Cain killing Abel. This story may overstate its significance in human history; after all, if Cain were going to kill, he had no choice but to choose a relative. But given that relatives do kill each other, evolutionary thinking should be able to make some predictions about the patterns that are observed, and it is to this that we turn next.

10.3.2 Infanticide

We have already noted that **infanticide** is not uncommon in the animal kingdom. The males of such animals as lions and lemurs will often kill the offspring of unrelated males to bring a female back into oestrus. Among the vertebrates, it is mostly males who kill infants, but where males are the limiting factor, the roles are reversed. Among the marsh birds called jacanas, polyandry is often found, and sex roles are often inverted. A large territorial female may have several nests containing her eggs in her territory, each nest presided over by a dutiful male. If one female displaces another and takes over her harem of males, she sets about methodically breaking the eggs of her predecessor. In these examples, infanticide can be understood as a means of increasing the fitness of the killer.

We may find such instances distasteful, but we rightly regard examples of human infanticide as being even more horrific. In most countries, infanticide is of course illegal, and debates continue about the legitimacy of feticide. It could, however, be that in the environment in which early humans evolved, infanticide in some circumstances represented an adaptive strategy. If we can judge the reproductive experiences of early humans by those of modern-day hunter-gatherers, we begin to appreciate the tremendous strain that raising children probably entailed. Infant mortality would be high, fertility would be low, partly as a result of the prolonged feeding of infants, and the best that most women could hope for would be 2–3 children after a lifetime of hard work. Under these conditions, raising a child that was defective and had little chance of reaching sexual maturity, or a child that was for some reason denied the support of a father or close family, would be an enormous burden and contrary to a woman's reproductive interests. At a purely pragmatic level, infanticide may have sometimes been the best option to maximise the lifetime reproductive value of a woman. One might term the withdrawal of support by the father the 'Medea effect'. Given the risks of misunderstanding, it should be noted that this reasoning is not to elevate reproductive value as some supreme moral principle or touchstone applicable to contemporary culture; by the same token, however, it would be futile to be moralistic about the behaviour of *Homo sapiens* in the Old Stone Age.

If infanticide did once (as it still does in some cultures) represent a strategy for preserving future reproductive value, we might expect the frequency of infanticide to decrease as the mother's age increases. This follows from the fact that, as a mother ages, her residual reproductive value declines: when there is less at stake, less drastic behaviour may be expected. Daly and Wilson present evidence that this effect is at work both among the Ayoreo Indians and modern-day Canadians. Figure 10.5 shows infanticide in relation to the age of the biological mother.

The results from Figure 10.5 are in agreement with predictions, but it is difficult to rule out other effects. Women may become better mothers as they age, learning from experience; younger mothers may suffer more social stress. The effect could thus be one of socially learnt skills and culturally specific stress factors rather than an adaptive response.

Reproductive value of offspring and infanticide

In the analysis above, Daly and Wilson derived testable predictions from the way in which the reproductive value of the parent varied as a function of age. The child too has reproductive value as the carrier of the parent's genes and as a potential source of grandchildren. From these considerations, predictions can also be derived.

From a gene-centred perspective, children are of value to their parents because they have the potential to breed and continue to project copies of genes into future generations. This dispassionate approach suggests that the value of a child would increase up to puberty and decrease as the child approached the end of its reproductive life. This follows from the fact that the chances of, for example, a 10-year-old girl reaching sexual maturity at 16 (a typical age for first menstrua-tion in a child in a hunter-gather society) are greater than those of a 2-year-old. The 10-year-old has benefited from 10 years of parental investment and has only 6 more years to survive the hazards of life to reach 16, whereas the 2-year-old has 14 years to go. The variation in valuation will be more marked and steep in cultures with a high infant mortality rate, but even in industrialised countries, where infant mortality has dropped dramatically over the past 100 years, the effect will be present. The reproductive value of an offspring will vary according to the shape of the curves shown in Figure 10.6.

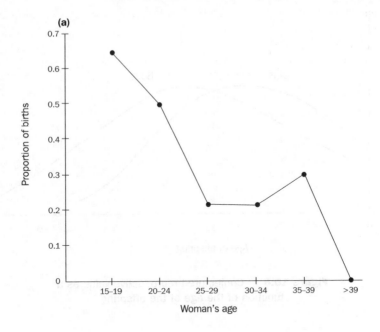

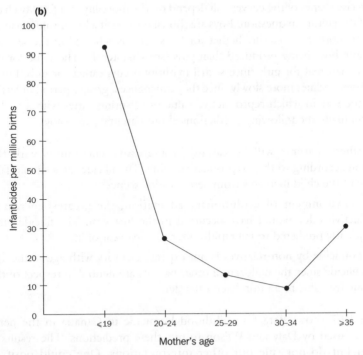

Figure 10.5   Rate of infanticide versus age of biological mother.
(a) The proportion of births that led to infanticide among a sample of Ayoreo
women who were known to have lost at least one baby. (b) The risk of
infanticide at the hands of the natural mother within the first year
of life, Canada 1974–83 (adapted from Daly and Wilson, 1988a)

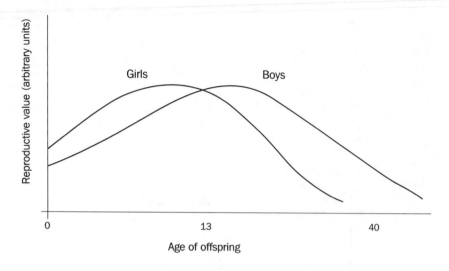

Figure 10.6   Conjectured reproductive value as a function of the age of the offspring

The precise shapes of the curves will depend on the mortality and fertility characteristics of the culture in question. Boys are shown starting at a lower value than girls since infant mortality is usually higher for boys. This is reflected in the sex ratio, slightly more boys being produced than girls (see Chapter 4). The value for boys peaks later than that for girls since sexual maturity occurs earlier for girls, but the value for boys declines more slowly since they experience a greater period of fertility.

From the way in which reproductive value of offspring varies with age, Daly and Wilson made the following predictions about parental psychology:

- In conflicts, parents will be careful in aggressive encounters with their offspring according to their reproductive value. The filicide rates should fall as the age of the child increases from zero to adolescence

- In the environment of evolutionary adaptation, the greatest increase in reproductive value would have occurred in the first year. The filicide rate is consequently predicted to fall rapidly after the first year of life

- Child homicide by non-relatives is not expected to vary with age in the same way as filicide since the children of other parents are neutral in respect of their reproductive value to an unrelated offender.

Figure 10.7 shows data for childhood homicide in Canada in the period 1974–83 as used by Daly and Wilson to test these predictions. The results are consistent but do not rule out other interpretations. One could posit, for example, an 'irritation factor' that varies with the child's age. The lower risk for fathers may then represent a lower time of exposure, and the fall in risk with age may be a product of coming to terms with the difficulties of child-rearing. This theory would predict a higher level of homicide by parents against adolescents considering the conflicts between teenagers and parents – but then again teenagers are harder to kill. The decline in child homicide with the age of the child also applies to homicides initiated by step-parents, who, presumably, have no interest anyway in the reproductive value of their step-children (see Figure 10.8 below). This too indicates that other factors may be at work.

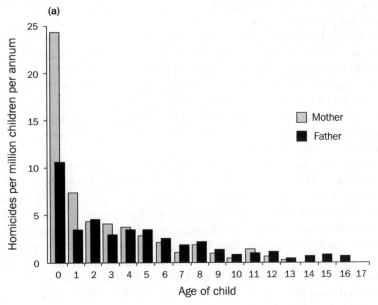

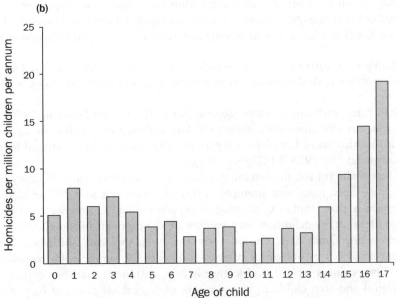

Figure 10.7   Child's risk of homicide. (a) Risk of homicide at the hands of father or mother. (b) Risk of homicide at the hands of a non-relative (from Daly and Wilson, 1988a)

Infanticide and step-parents: the Cinderella syndrome

Popular traditional culture abounds with tales of wicked step-parents who fail to provide proper care for their children. The story familiar to many in the West is that of Cinderella. Cinderella's biological mother has died and her father has remarried. Cinderella is raised in the house of her step-mother, who has two children of her own by a former marriage. Despite the obvious virtues of Cinderella compared with her 'ugly sisters', she is treated harshly. Fortunately, a way out of her plight is provided by the *deus ex machina* of a charming prince. It would be wrong to try to construct a scientific argument on the basis of a fairy story, but it is worth noting that cultures all around the globe have their own variant of the Cinderella myth. The essence of such stories is that step-fathers and step-mothers are wicked and not to be trusted. Again, this proves little by itself since all the stories could have descended from a common archetype. Such is the bad image of step-parents in traditional folklore that, faced with a high divorce and remarriage rate in the Western countries, children's books now go out of their way to show step-parents in a more positive light.

The stories may, however, reflect a fundamental component of the human experience: the raising of children by non-genetic parents. Throughout human history, step-parents must have played a role in the raising of young when, for a variety of reasons, one parent was killed or left the family unit. A prediction that follows from selectionist thinking about parenting is that parental solicitude should be discriminating with respect to the offspring's contribution to the reproductive interests of the parent. In step-families, it would follow that parents should be more protective of their biological children than their step-children. We are not reliant on folktales to test these predictions. Daly and Wilson (1988a, 1988b) supply a range of cross-cultural statistical evidence showing that children in step-families are injured or killed in disproportionate numbers compared with those in their biological families. The effect seems to be extremely robust. As Daly and Wilson remark:

> Having a step-parent has turned out to be the most powerful epidemiological risk factor for severe child maltreatment yet discovered. (Daly and Wilson, 1988a, p. 7)

A child living with one or more step-parents in the United States in 1976, for example, was 100 times more likely to be fatally abused than a child of the same age living with his or her biological parents. This effect is also illustrated by the Canadian data for 1973–84 (Figure 10.8).

These findings have, understandably, been met with shock and incredulity, and there have been numerous attempts to show they are not sound. One obvious objection is that families containing step-parents may be unrepresentative of family life since, by definition, step-parents are the products of some breakdown of an existing family unit. The previous family unit may have failed because of violent behaviour by one of the parents, and hence the sample is not representative. The problem with this objection is that, within step-families containing biological and step-children, it is the non-biological offspring of one of the partners who tend to be hurt. The effect has also been noted in other cultures. Flinn (1988) documented the interactions between men and their children in a village in Trinidad. He found that significantly more interactions involving

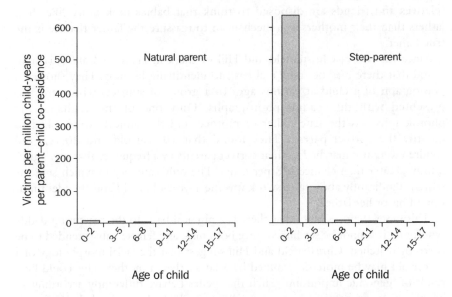

Figure 10.8   Risk of a child being killed by a step-parent compared with a natural parent in relation to the child's age (from Daly and Wilson, 1988a)

conflict between fathers and their step-children than between fathers and their genetic offspring. There are still other interpretations of course: step-children may resent the intrusion of an outsider and thus promote conflict. Again, however, we must face the prospect that Daly and Wilson may have touched upon an evolutionary-derived component of the human psyche.

It hardly needs saying that this does not cast a shadow on all step-families or that violence in these circumstances is somehow excusable. Fortunately, homicide is still an extreme rarity; it is probably more of a puzzle to explain, in selectionist terms, why so many step-families do work and so many children receive love and affection from non-biological parents. Crime statistics can be depressing, so it is good to realise that counter-Darwinian (or at least not obviously selected) behaviour is also part of being human.

Reassuring daddies: face recognition

One way to reduce the risk of infanticide is to ensure that the parents recognise the child as their own and thus care for it. Mothers have a pretty good idea whether or not the child is theirs, but fathers need to be reassured. When we encounter a couple with a new baby or perhaps a couple with their young child whom we have not met before, it seems a natural part of social conversation to comment on the resemblance between the child and the mother and/or father. We can in fact easily convince ourselves that we see a definite likeness. Scientifically of course, this is not a very exact procedure. We know that the child belongs to the parents, so it is easy for us to think we can identify a resemblance, either imagined or real. Daly and Wilson suggested in 1982 that mothers, relatives and friends are disposed to think that babies look more like their fathers than their mothers as a mechanism to reassure the father that he is the true father.

Since this work, Christenfeld and Hill of the University of California have found that there may be some real basis to identifying likeness. They showed a photograph of a child at various ages to a group of subjects who were also presented with three adult photographs. Only one of the adults in the photographs was the true father or mother, and the subjects were asked to identify the correct parent. They found that 10-year-old and 20-year-old children were not matched to their correct parents by a frequency that was significantly greater than chance (33 per cent). The only category in which subjects scored significantly above guesswork was the 1-year-old child and the identification of his or her father.

This raises the possibility that there is a physical basis to the reassuring daddy syndrome. It cannot be sex linked since both male and female babies tended to be correctly matched. Christenfeld and Hill suggest that the facial morphology of a 1-year-old may be more determined by a father than a mother. This could be a result of 'genomic imprinting', such that genes behave differently according to whether they are paternally or maternally derived. We have noted this before in maternal–fetal conflict. In mice embryos, for example, genes for extracting resources from the mother make greater demands if they are paternally derived than if they are obtained from the mother. This is because the paternal genes have been selected to obtain resources at the expense of other offspring from the mother. The mother, on the other hand, knows that all her offspring are her own, so distributes her resources more judiciously. In the case of children, genes for facial features at 1 year of age inherited from the father may be preferentially expressed. It would be in the mother's interests to limit her expression of resemblance genes in order to ensure that the father recognises his child, and so cares for it (Christenfeld and Hill, 1995). This idea of Christenfeld and Hill involves a rather sophisticated mechanism to be expected of genes. It may be that genes are not that 'clever', but this whole area is ripe for further exploration.

10.4 Human sexual conflicts

> The cuckoo then, on every tree,
> Mocks married men; for this sings he
> Cuckoo,
> Cuckoo, cuckoo; O word of fear,
> Unpleasing to a married ear.
>
> (Shakespeare, *Love's Labours Lost*, V ii)

Shakespeare was a good psychologist. The risk of cuckoldry – the 'word of fear' – has damaging repercussions for the male, far more so than for the female. In this final section, we examine how this asymmetry in sexual interests has conditioned and moulded conflict between the sexes.

10.4.1 *Marriage as a reproductive contract: control of female sexuality*

Marriage is a cross-cultural phenomenon and, although ceremonies differ in their details and marriage law varies across cultures, marriage has a set of predictable features across virtually all societies. Marriage entails or confers:

- Mutual obligations between husband and wife
- Rights of sexual access that are usually but not always exclusive of others
- The legitimisation of children
- An expectation that the marriage will last.

Daly and Wilson (1988a), in line with the work of Levi-Strauss, have pointed out that a feature additional to the list above is that marriage often involves an exchange of wealth between the kin of the bride and groom. Such transactions are usually controlled by men. In the United Kingdom, the last vestige of this is the fact that fathers 'give away' the bride. In most cultures, however, a bride price is paid: men purchase wives. In their survey of the ethnographic data for 860 cultures, Daly and Wilson showed that, in about 58 per cent of cases, the groom purchases a bride by transferring wealth to or labouring for the bride's kin. In fewer than 3 per cent of the cases, a dowry was paid by the bride's kin to the groom's family or to the newly weds. These figures suggest that men are viewing women as property to be traded. It could be that this is another dimension of 'patriarchy', with no Darwinian undertones. One extra feature, however, suggests otherwise. In some cultures, such as the Kipsigis of Kenya, the bride price is linked to the reproductive value of the wife (Borgerhoff-Mulder, 1988).

The small percentage of cultures in which the bride pays a dowry may be amenable to an adaptationist explanation. Gaulin and Boster have proposed that dowry usage may be a form of female–female competition. In highly stratified societies where there is a large disparity in wealth but where monogamy is socially imposed, there will exist a body of resourceful and high-status men who, despite their good fortune, are able to take only one bride. Gaulin and Boster argue that females will compete with each other for access to such high-quality males, and, to increase their marketability, they offer a dowry (Gaulin and Boster, 1990).

Viewed through Darwinian eyes, marriage begins to look like a reproductive contract. This is seen at its clearest when fears arise that the contract has been breached, such as when one male is cuckolded by another. In reproductive terms, the consequences of cuckoldry are more serious for the male than the female. The male risks donating parental investment to offspring who are not his own. This male predicament is reflected in laws dealing with the response of the male to incidents where he finds his partner has been unfaithful. In the United States, a man who kills another caught in the act of adultery with his wife is often given the lesser sentence of manslaughter rather than murder. This is a pattern found in numerous countries: sentences tend to be more lenient for acts of violence committed by a man who finds his wife *in delicio flagrante*. The law carries the assumption that, in these circumstances, a reasonable man cannot be held totally responsible for his actions.

Violence after the event may serve as a threat to deter would-be philanderers or to rein in the affections of a wife whose gaze may be wandering. Before the event, however, males of many species guard the sexuality of their partners. Research on animals has shown that, in numerous species where parental investment is common, males have evolved anti-cuckoldry techniques. Male swallows, for example, follow their mates closely while they are fertile, but when incubation begins, mate-guarding ceases and males pursue neighbouring fertile females. When the same males perceive that the threat of cuckoldry is high, such as when they are experimentally temporarily removed, they seem to compensate by increasing the frequency of copulation with their partner. Indeed, male swallows do have something to fear: Moller (1987) estimated that about 25 per cent of the nestlings of communal swallows may be the result of extrapair copulation.

If anti-cuckoldry tactics have evolved in avian brains where members live in colonies, both birds are ostensibly monogamous and both sexes contribute to parental care, might we not expect similar concerns to have evolved in humans? The answer, according to Wilson and Daly (1992), is a resounding yes. Numerous cultural practices can be interpreted as reflecting anxieties about paternity. A few of the more obvious cases will be described, as documented by Wilson and Daly and other authors.

Veiling, chaperoning, purdah and incarceration

Obscuring a woman's body and facial features as well as ensuring that she is always accompanied when she travels are common practices in patriarchal societies and can be seen as ways of restricting the sexual access of women to other men. It is usually only practised on women of reproductive age, children and postmenopausal women being excluded.

Foot-binding

This was once used in China, partly as a display by a man that he was wealthy enough to be able to dispense with the labour of his wife. It also serves as a way to restrict her mobility and hence her sexual freedom.

Genital mutilation

Unlike male circumcision, female genital mutilation is specifically designed to reduce the sexual activity of the victim. Practices range from partial or complete clitoridectomy to infibulation. Girls aged 13–18 of the Sabiny tribe in Uganda are taken to a village clearing where their clitorises are sliced off. It is estimated that about 6000 girls are mutilated in this way each day in Africa and that more than 65 million women and girls now alive in Africa have been 'circumcised' (Hosken, 1979). The practice is condemned by some governments, such as that of Uganda, but the Sabiny women resent attempts to ban the practice as cultural interference. In some African countries, infibulation involves the suturing of the opening of the labia majora, opening them up again on marriage. Infibulation makes sexual intercourse impossible, and so is a guarantee that a bride is intact. Women are cut

Figure 10.9   A Bengali Muslim woman covered in a veil

Although an injunction to modesty in female dress appears in the Koran, the
wearing of a veil in public (hijab) is not an essential requirement of Islam, and its
cultural origins are obscure. Hijab, practised by girls after puberty, functions to
discourage amorous male advances. From an evolutionary perspective, hijab is a
means by which men in patriarchal societies maintain paternity confidence

open on marriage, and more cuts are needed for the delivery of a child. In all,
female genital mutilation is to be found in 23 countries worldwide. If there is any
practice that sinks the pretensions of cultural relativism, it must be this.

Women as men's legal property

Studies on the history of European adultery laws reveal the double standards that
have until recently operated. It seems that, before recent reforms, the reproductive capacity of a female was treated as a commodity that men could own and
exchange. Some of the salient features of these laws are as follows:

1. Laws tended to define adultery in terms of the marital status of the woman,
   the marital status of the man usually being ignored
2. Adultery often took the form of a property violation, the victim being the
   husband, who might be entitled to damages or some other recompense
3. If a wife was adulterous, this represented clear grounds for divorce; if a
   husband was adulterous, divorce was a rarer consequence
4. As late as 1973, Englishmen could legally restrain wives who were intent on
   leaving them.

Menstrual taboos

In some cultures, menstruation is treated as a taboo, and women are ostracised
while they menstruate. Various hypotheses have been proposed to account for this

custom, such as the idea that the taboo serves as a protection against bacterial 'monotoxins' or that it forms part of the general domination of women by men. In her work on the Dogon, an agricultural community living in Mali, West Africa, Strassman (1992, 1996) advances the idea that menstrual taboos enforce reproductive signalling. The taboo takes the form that, during menstruation, women must retire to a menstrual hut. A mixture of supernatural threats and fines in the case of non-compliance enforces conformity with this procedure. By measuring the levels of hormone metabolites in urine samples, Stassman was able to establish that the vast majority who do visit the hut are in fact menstruating. Her overall conclusion is that Dogon menstrual taboos force females to signal their position in their fertility cycle. Honest signalling is ensured by threats of reprisals. Signalling menstruation is one of the few ways in which men can assess paternity confidence.

Sperm competition

Sperm competition could be interpreted in terms of anti-cuckoldry tactics. If sperm from another male are present in the reproductive tracts of a female, other tactics have presumably failed and post-insemination measures need to be employed. Baker and Bellis (1995) have provided evidence to support what they have called the 'kamikaze sperm hypothesis'. The basic idea is that some of the sperm from one man will attack foreign sperm if encountered in the reproductive tract of the female (see Chapter 5). The work of Baker and Bellis in relation to kamikazi sperm has been heavily criticised on methodological and experimental grounds, and it remains uncertain whether or not such a mechanism really does exist (Birkhead *et al.*, 1997).

10.4.2 *Jealousy and violence*

> Not poppy, nor mandragora,
> Nor all the drousy syrups of the world
> Shall ever medicine thee to that sweet sleep
> Which thou owedst yesterday.

> (Iago, *Othello*, III iii 333)

Iago is one of the most evil of Shakespeare's characters, and he understood well the power of jealousy acting on the human mind. Tormented by his suspicions, Othello first kills his wife Desdemona and then himself. In the light of selectionist thinking, the emotion of jealousy in males is an adaptive response to the risk that past and future parental investment may be 'wasted' on offspring that are not their biological progeny. In females, given the certainty of maternity, jealousy should be related to the fact that a male partner may be expending resources elsewhere when they could be devoted to herself and her offspring. Men also lose the maternal investment that they would otherwise gain for their offspring since this is directed at a child who is not the true offspring of the male. It is to be expected that jealousy and its consequences would be asymmetrically distributed between the sexes. We should also expect it to be a particularly strong emotion in men since humans show a higher level of paternal investment than any other of the 200 species of primates. In a near-monogamous mating system, men have more to lose than women do.

To test for sexual differences in the experience of jealousy, Buss *et al.* (1992) issued questionnaires to undergraduates at the University of Michigan asking them to rank the level of distress caused by either the sexual or the emotional infidelity of a partner. The result suggested that men tend to be more concerned about sexual infidelity and women more about emotional infidelity.

The same effect was observed when subjects were 'wired up' and tested for physiological responses to the suggestion that they imagine their partner behaving unfaithfully either sexually or emotionally. The difference was less marked for women, but men consistently and significantly showed heightened distress to thoughts of sexual, compared with emotional, infidelity.

Such effects are what would be predicted from an evolutionary model of the emotions. Men would tend to be more concerned about the sexual activity of their partners since it is through extrapair sex that a male's investment is threatened. Women, on the other hand, should be less concerned about the physical act of sex *per se* than about any emotional ties that might lead her partner and his investment away from herself.

Wilson and Daly claim that such sexual differences in the intensity of jealousy can be seen at work in violent crimes between spouses. Table 10.3 shows data on spousal homicide in Canada over a 9-year period. It should be noted that jealousy is a major cause of homicide as well as figuring as a motive in male violence against females to a much larger degree than the reverse. In fact, jealousy as a motive is probably underreported in this data set since the terms 'argument' and 'anger/hatred' as recorded by the police may contain a background of jealous emotion.

Table 10.3   Police attribution of motive in 1060 spousal homicides in Canada 1974–83 (adapted from Daly and Wilson, 1988a)

Motive	Killer is husband	% of total	Killer is wife	% of total
Argument	353		160	
Jealousy	195	24	19	7.7
Anger/hatred	84		22	
Other	180		47	
Totals	812		248	

10.4.3   *Divorce and remarriage*

A failed marriage, thankfully, rarely ends up in the homicide statistics, but patterns of divorce and remarriage also provide a convenient testing ground for evolutionary hypotheses. Divorce statistics are in close agreement with findings on published advertisements in which men seek looks and youthfulness while women are more concerned with status and resources (see Chapter 8). Buckle *et al.* (1996) examined statistics for divorce and remarriage for populations in Canada and England and Wales. They found that, in stating grounds for divorce, men were more concerned about adultery while women were more concerned about cruelty. This is interpreted in the language of evolutionary psychology as paternity certainty and care for offspring respectively (Figure 10.10).

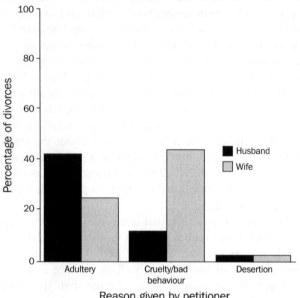

Figure 10.10 Percentage of divorces as a function of the grounds given by male and female petitioners for England and Wales (mostly 1974–89) (after Buckle *et al.*, 1996)

Women were also much more likely to terminate a marriage at an early stage than were men; in fact, over 90 per cent of all divorces in England and Wales (mostly from 1974 to 1989) were initiated by females under 25. This is just what would be expected in the light of the shorter reproductive span of females compared with that of males. If a woman suspects that she has made a bad decision and she wants children, it is imperative to dissolve the marriage in the early stages. Waiting reduces her reproductive potential and her future marriagability. The opposite effect is observed for males: as they get older, they are more likely to seek a divorce (Figure 10.11). Men remain fertile for longer and could raise a second family with a younger wife. Predictably, on remarrying, men seek wives about 6 years younger than themselves whereas for first marriage the figure is only 2 years.

The consistency of such data with evolutionary expectations reinforces the idea of marriage as a 'reproductive contract' in which men and women pursue different strategies to optimise their reproductive fitness. Humans are not the only animals to dissolve the pair bond in the case of childlessness. Ring-doves, for example, are generally monogamous during the breeding season but experience a 'divorce rate' of about 25 per cent. The major reason seems to be infertility: if the couple fail to produce chicks, they separate and look for other mates (Erickson and Zenone, 1976).

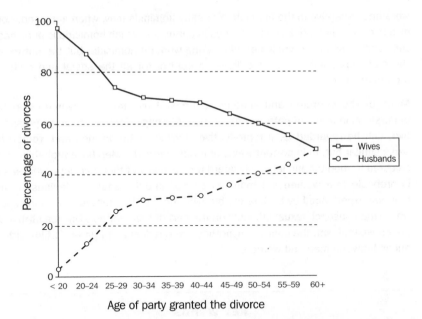

Figure 10.11   Percentage of divorces ($n = 1\,766\,424$) granted in England and Wales (mostly 1974–89) as a function of the age and sex of the petitioning party (after Buckle et al., 1996)

SUMMARY

A knowledge of genetic relatedness (r) is useful in helping to understand both altruistic and antagonistic behaviour. Conflict will arise when the reproductive interests of two individuals differ. This is predicted between sexual partners and even between parents and offspring since offspring may demand resources at the expense of current and future siblings.

Maternal–fetal conflict provides a good example of the dynamic balance that is reached when the optima for resource allocation differ between two genomes.

Human violence resulting in death provides a set of data that allows predictions from evolutionary psychology to be tested. There is evidence, for example, that the risk of a child being killed by its mother declines as the age of the child and the mother increase. These facts are in keeping with the concept of reproductive value, but there may be other factors at work.

Whereas kinship is usually predicted to reduce levels of violent behaviour, there were probably circumstances in the environment of evolutionary adaptation of early hominids in which, for example, infanticide served as an adaptive strategy to maximise the lifetime reproductive value of a parent. One of the most robust findings of the work of Daly and Wilson is that step-children experience a much higher risk of infanticide that genetic children. Psychological mechanisms that were once adaptive in the life history of early hominids may, when activated, now give rise to maladaptive consequences in cultures in which homicide is punished. Daly and Wilson have published pioneering work on homicide, but the nature of the statistics used makes it difficult to control for all the social and cultural factors at work.

Much of the coercion and violence exercised by men over women can be understood in terms of paternity assurance. Marriage, from a Darwinian perspective, can be regarded as a reproductive contract between men and women to serve similar but, in important ways, different interests. Men have probably been selected to monitor and jealously guard the sexuality of their partners, whereas it is probable that women are more concerned about resources. Jealousy is an emotion experienced by both sexes but with more violent repercussions in cases where men suspect sexual infidelity on the part of their partner. Divorce statistics are consistent with the contract hypothesis and reflect age-related fertility differences between men and women.

KEY WORDS

Coefficient of relatedness ■ Genomic imprinting ■ Homicide
Infanticide ■ Jealousy ■ Kin ■ Maternal–fetal conflict
Parent–offspring conflict ■ Siblicide

FURTHER READING

Daly, M. and Wilson, M. (1988a) *Homicide.* New York, Aldine de Gruyter.
Daly, M. and Wilson, M. (1998) *The Truth About Cinderella.* London, Orion.
Both these books contain the ground-breaking interpretations of crime statistics made famous by Daly and Wilson.

Mock, D. W. and Parker, G. A. (1997) *The Evolution of Sibling Rivalry.* Oxford, Oxford University Press.
A thorough work that develops some sophisticated theory, but one which does not deal with humans.

11

Altruism, Co-operation and the Foundations of Culture

Nature, Mr Allnut, is what we were put in this world to rise above.

(Katherine Hepburn, as Rosie Thayer, in *The African Queen*)

Ethics as we understand it is an illusion fobbed off on us by our genes to get us to co-operate.

(Wilson and Ruse, 1985, p. 52)

Whether we are entirely bound by the natural world or are capable of rising 'above' it is still a dilemma in evolutionary theory. Is culture biology writ large? Or is culture something that humans have created in spite of biology? The popular lay view is probably the latter: culture is something that elevates us above the beasts of the field and the rest of the animal kingdom; children are, after all, socialised to behave in a civilised manner, and this means not behaving like animals. This perspective is usually allied to a view of morality and human nature suggesting that what is natural or innate, such as selfishness, greed, avarice and aggression, tends to be bad, while what is good, such as co-operation, moral restraint and respect for others under the rule of law, is that which we have assiduously grafted onto nature by a few thousand years of civilised existence. From this perspective, Utopias are always thwarted by 'human nature' breaking through our naïve or idealistic hopes. This is of course a misconception and one that is highly damaging to the naturalistic paradigm. If all that evolutionary thinking can do is show us that the brutal side of our nature is shared with the other beasts, then its scope is rather limited.

So what we are 'really like'? Behind the smokescreen of culture, with its obfuscating religious systems, secular laws, customs and habits, what do we find? A heart of darkness or a state of primitive innocence? The search is, in fact, profoundly misleading. For many Darwinians, there is no antinomy between nature and culture: we are what we see; our culture is our nature.

The evolutionary theorist, therefore, has plenty of work to do. Thus, in this chapter, we will explore the inroads made by selectionist thinking into areas traditionally associated with the cultural rather than biological aspects of human nature. The chapter starts by developing the ideas on **reciprocal altruism** first introduced in Chapter 2 and moves to consider attempts within game theory to show how co-operative behaviour could start and flourish. It will be suggested that the manner in which altruistic exchange can benefit both donor and recipients could offer an example of a biological basis to morality. The chapter ends by considering, necessarily briefly, some general models constructed to tackle the broader problem of the relationship between genes and culture.

11.1 Game theory and the origins of human altruism

It is a safe enough bet that the early replicators were entirely selfish. From this primal state, the step towards kin-directed **altruism** is a short one, and it is easy to envisage how this could have occurred: the care given to direct offspring could move to indirect offspring, and all the benefits of inclusive fitness would follow. It is also easy to see how reciprocal altruism benefits both parties once it is established. A favour can be given to another individual at small cost to oneself on the understanding that the favour will be returned. There are conditions that need to be set for this, such as the ability to recognise who donated the favour, a good chance of meeting that individual again and so on, but these are not improbable ones. The fundamental logic of this type of altruism was explored in Chapter 2.

The big problem, however, is to account for the origin of co-operative behaviour, given that, in the very first interaction, it would pay to act selfishly. The problem of the evolutionary emergence of co-operation is highly pertinent to humans. Humans spend a great deal of time co-operating and delaying their immediate gratification for future rewards. We could concede, along with Rosie Thayer in the quotation above, that this is something that transcends the natural and distinguishes us from mere brutes, but Darwinians are unlikely to give up this ground without a fight. In this next section, we will explore some models showing that co-operation may be a fitness-maximising strategy. If this is so, the simplistic linkage of nature–bad and culture–good is exploded. We may be caring and morally sensitive creatures by virtue of our biology.

11.1.1 The prisoners' dilemma

In traditional moral thinking, the focus is on how individuals treat each other. In many real-life situations, our behaviour is contingent upon the behaviour of others. Situations such as this have been modelled using **game theory** (Axelrod and Hamilton, 1981). A good starting point to investigate the moral basis of behaviour is a game known as the **prisoners' dilemma**.

The word 'prisoner' relates to one context to which the logic of this game could apply. If two suspects are apprehended at the scene of one of a series of crimes, one tactic that the police could adopt is to separate the individuals and question them independently. If the overall evidence that the police have gathered is flimsy and a successful prosecution will rely upon a confession, each suspect may be offered the promise of a lighter sentence if they turn 'king's evidence' and inform (defect) on the other. If they both defect, they implicate each other and both receive a full jail sentence. If both co-operate and refuse to be tempted to defect, then, given the lack of evidence, each receives a smaller sentence for a minor part of the crime. Figure 11.1 shows a pay-off matrix for this type of scenario.

		Player B	
		Co-operate 'It was neither of us'	Defect 'It was him'
Player A	Co-operate	R = Reward = 1 year	S = Sucker's payoff = 5 years
	Defect	T = Temptation to defect = 0 year	P = Punishment = 4 years

Figure 11.1   Pay-off matrix for two prisoners caught in a dilemma

Note that the values are the pay-offs to Player A

The game of course is entirely hypothetical, but it illustrates that 'rational behaviour' (in the sense of maximising the returns to oneself) can result in the least favourable outcome; if both parties defect or inform, they are each worse off. They are, in effect, punished for failing to co-operate with each other. The dilemma then is whether to co-operate or defect. Defection will bring the best rewards if the other co-operates, but each prisoner does not know what the other will do. It is difficult to see how co-operation could evolve in this system. One might imagine that suspects would confer before the crime and both agree to co-operate, but this still begs the question of why co-operate rather than defect.

The game of prisoners' dilemma was first formalised in 1950 by Flood, Dresher and Tucker (see Ridley, 1993). It is a situation that applies to many interactions in life. As Ridley points out, the gigantic trees in tropical rain forests are the products of prisoners' dilemmas: if only they would co-operate and agree to, for example, not grow over 20 feet tall, all would be able to put more energy into reproduction and less into growing gigantic trunks to tower over their neighbours. But they cannot.

Human folly is also often the result of prisoners' dilemma situations. The villages of San Giminano in Italy and Vathia in Greece are remarkable for their tall towers built by the villagers, at great personal expense, trying to outdo their neighbours and rivals. The arms race between the superpowers in the post-war period left both America and the former Soviet Union worse off. In all these cases, the protagonists were locked into one form or another of the prisoners' dilemma.

The problem has relevance to human behaviour in that many human interactions in the past must have taken this form. It might be expected then that the long evolution of humans and other animals should have given natural selection time to solve the problem. In essence, we are looking for 'an evolutionary stable strategy' – a strategy that, if pursued in a population, is resistant to displacement by an alternative strategy. Headway was made in relation to this problem when it was realised that social life is akin not to one chance encounter between individuals of the prisoners' dilemma form but to many repeated interactions. The problem thus became to explain what strategy each should pursue if the game is played over and over again. It then transpires that defection is not always the best policy.

Prisoners' dilemma with repeated contact – it's natural to be nice

To investigate this, it is best to elevate the problem to a more abstract level. Figure 11.2 shows a representation of the dilemma using values for reward and punishment used by the political economist Robert Axelrod. When we are faced with such situations in life, we have a wide range of options. We could play meek and mild, and always co-operate. In the face of defection, we would then 'turn the other cheek' and continue to co-operate, to our ultimate detriment. Another option is to always play rough ('hawkish') and constantly defect. Imagine a population of individuals that interact only once; then the strategy 'always co-operate' is easily displaced by a mutant 'always defect'. A single defector introduced into a population of co-operators will thrive, the population of co-operators slumping and becoming extinct because a co-operator will always lose on meeting a defector. Eventually, the population will be composed of all defectors. They will, as a group, not do as well as the co-operators, but selection does not act for the good of the group. It is even possible that this defecting strategy will lead the species to extinction. When all the population is composed of defectors, this is an evolutionary stable strategy: it is resistant to invasion by a mutant co-operator.

One of the closest matches between biological systems and game theory so far comes from recent work on bacteriophages. Phages are viruses that infect bacteria, subverting the metabolic systems of their host into making copies of the phages, to the detriment of the health of the bacterial cell. In such cases, the bacterial cells eventually burst, releasing more phages into the cellular environment. Turner and Chao studied the phage Φ6 and a mutant variant called ΦH2. These are mere chunks of RNA that depend entirely on bacterial cells for replication and, in terms of their simplicity, are among the lowest forms of life. Compared with Φ6, ΦH2 can be regarded as a defector in that it manufactures fewer of the intracellular products needed for its own replication. In cells infected with Φ6, ΦH2 has a free ride in that it exploits the products manufactured by Φ6. By ingenious experiments, Turner and Chao compared the fitness of Φ6 and ΦH2 in cells where either was dominant; in other words, they measured the fitness of the co-operator Φ6 in a cell full of defectors ΦH2, the fitness of ΦH2 in a cell full of Φ6 co-operators and so on (Turner and Chao, 1999). The fitness values are shown in Figure 11.3.

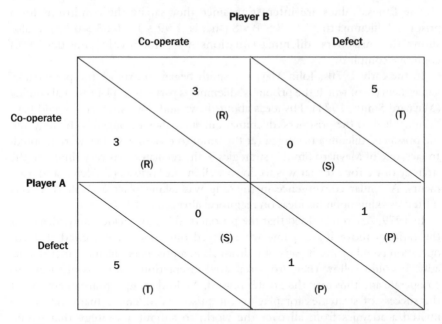

Figure 11.2   Prisoners' dilemma values for co-operation and defection

Values are now shown for each player either side of the diagonal line

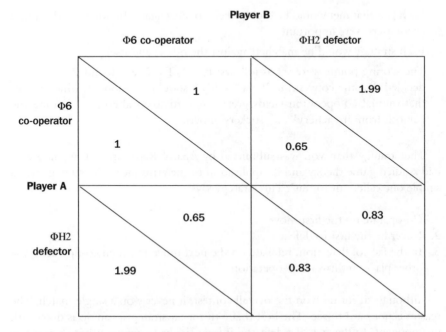

Figure 11.3   Fitness values for bacteriophages Φ6 and ΦH2, the fitness of Φ6 in encounters with itself being set as 1 for reference purposes (data from Turner and Chao, 1999)

The fitness values are interesting since they satisfy the conditions for a prisoners' dilemma that $T > R > P > S$ (that is, $1.99 > 1 > 0.83 > 0.65$). It also shows that prisoners' dilemma situations need not involve any degree of cognitive complexity.

In the early 1970s, John Maynard Smith began to explore the potential of game theory – of which the prisoners' dilemma is part – to explain animal conflict (Maynard Smith, 1974). His ideas about 'hawk and dove strategies' could have been applied to the prisoners' dilemma, but since they belonged to biology and the prisoners' dilemma belonged (at the time) to economics, they were ignored. In fact, one of Maynard Smith's strategies – 'the retaliator' – is very similar to the strategy of **tit for tat** that was to do so well in later prisoners' dilemma tournaments. A similar convergence in thinking was taking place as Robert Trivers started working upon his ideas on reciprocal altruism in 1971.

In 1979, Axelrod realised that the prisoners' dilemma yielded only 'defect' as the rational move if the game were played only once. He argued that co-operation could evolve if pairs of individuals in a prisoners' dilemma met repeatedly. It could follow that, from repeated interactions, both would learn to co-operate and thus reap the greater reward. Axelrod set up a tournament to test the success of strategies for playing the prisoners' dilemma many times. He invited academics from all over the world to submit a strategy that would compete with the others in many repeated rounds. Sixty-two programs were submitted and let loose on Axelrod's computer. The important features of this first tournament were as follows:

- Each pair that met would be highly likely to meet again. In other words, future encounters were important
- Each strategy would be matched against the others and itself
- The scoring points were set as follows: $R = 3$, $T = 5$, $S = 0$ and $P = 1$. It was decided that the convention $R > (T + S)/2$ should be adopted, which means that mutual co-operation yields greater rewards than alternately sharing the pay-offs from 'treachery' and 'suckers' moves.

The strategy that won was submitted by Anatol Rapaport, a Canadian who had studied game theory and its application to the arms race. The strategy was a simple one called 'tit for tat'. This strategy says:

1. Co-operate on the first move
2. Never be the first to defect
3. In the face of defection, retaliate on the next move but then co-operate if the other player returns to co-operation.

Although tit for tat won the overall contest, it never won a single match. The reason is not hard to see. The best that a tit for tat strategist can do is draw with its opponent. In the face of a defector, it loses the first encounter but then scores as many points afterwards. By definition, it can never strike out to take the lead because it is never the first to defect. Tit for tat wins overall because, even if it loses, it is never far behind. The problem with 'nastier' strategies is that they have to play each other. If players constantly defect on each other, their total reward is low. For a time, it looked like tit for tat could serve as a model for moral behaviour: it does pay to co-operate after all. Tit for tat might win a general contest, but could it ever invade an aboriginal population of selfish defectors? The answer is yes if a few tit for tat strategists appeared and met each other (Box 11.1). For a while, it seemed as if tit for tat could demonstrate the possibility of the evolution of human morality, that niceness could prosper.

Problems with tit for tat

This strategy seemed to hold high hopes for modelling the behaviour of humans and other animals. After the publication of these results in Axelrod's book, *The Evolution of Co-operation* (1984), criticism began to mount. It became clear that tit for tat was not evolutionary stable, and its success was sensitive to the precise details of the rules that Axelrod had set. In other words, it was possible to devise a strategy that would beat tit for tat.

The problem is that tit for tat is very sensitive to errors. Whether or not an opponent has defected or co-operated on the previous move requires the transfer of a message. In the real world, as opposed to the cyberworld of computer tournaments, messages become corrupted, and occasional mistakes are made. At an error rate in signalling of 1 per cent, tit for tat still came out on top. At 10 per cent, tit for tat is no longer the champion. In these circumstances, tit for two tats or generous tit for tat (GTT) can prevail. Here, the strategy is to forgive up to two defections. This prevents the effect of noise or mistakes leading to mutual and constant recrimination.

However, this strategy also has its problems since GTT allows co-operators to flourish. Co-operators will do as well as GTT, and it is conceivable that their number might drift to a position at which they formed the majority. It is now that defectors could strike. They could gain from co-operators, who would then dwindle in number. If there were not enough GTTs left, the population would become all-defecting. So playing tit for tat may drive away defectors if communication is clear, but if errors are made GTT begins to flourish. GTT allows co-operators to grow in number, defectors can then thrive upon co-operators, and we could be left with a population of all defectors.

The central problem with all these games is that it is difficult to design a strategy that would always win whatever the others in the pool. If only the others were known, we could design a winner, but this is not the case. The theoretical promise of using prisoners' dilemma-type games to model the evolution of altruism out of an initial pool of selfish behaviour has been set back by such problems.

Pavlov

Further work on game theory by Novah and Sigmund (see Ridley, 1993) found another robust strategy, which they called 'Pavlov'. Pavlov changes its behaviour after a poor pay-off. It has an initial bias to co-operate, but if a mistake is made and it receives an S pay-off, it changes from co-operate to defect. When the other

BOX 11.1

The stability of tit for tat

The rewards demonstrated below show that 'always defect' (AD) cannot by itself invade tit for tat. Conversely, although tit for tat cannot invade AD, pockets of tit for tat could, if N were very large, prosper by kin selection. Once a critical mass is reached, the whole population could become tit for tat.

		Player B	
		Co-operate	Defect
Player A	Co-operate	R	S
	Defect	T	P

Note that the pay-offs are shown to player A and by convention R > T+S/2

Now consider tit for tat playing N rounds with another tit for tat and also AD.

Player	Type		Total pay-off
1.	Tit for tat	Move C C C C C C	
		Pay-off R R R R R R	NR
2.	Tit for tat	Move C C C C C C	
		Pay-off R R R R R R	NR

Player	Type		Total pay-off
1.	Tit for tat	Move C D D D D D	
		Pay-off S P P P P P	S + (N − 1) P
2.	Always defect	Move D D D D D D	
		Pay-off T S S S S S	T + (N − 1) P

The values used by Axelrod were: R = 3, T = 5, S = 0 and P = 1

The pay-off matrix becomes:

	Tit for tat	AD
Tit for tat	RN	S + P(N − 1)
AD	T + P(N − 1)	PN

It follows (inserting the values for R, T, S and P may help) that:
1. AD cannot invade alone if RN > T + P (N − 1) which it becomes when N is large
2. Tit for tat cannot invade AD since S + P (N − 1) is < PN
3. But as N becomes very large, S + P (N − 1) approximates to PN, so if a few tit for tats gain hold, they may possibly then prosper by kin selection.

Pavlovian player receives a P, it changes its move back to C. Novah admitted that a better name would be 'Win stay, lose shift'. In fact, Rapaport had considered this strategy but dismissed it as simplistic because it does poorly against 'always defect' (Figure 11.4). The values for the matrix can be readily confirmed by setting out the scores in the same manner as shown in Box 11.1.

	Pavlov	'Always defect'
Pavlov	RN	N(S + P)/2
Always defect	N(T + P)/2	PN

Figure 11.4 Pay-off matrix to show Pavlov cannot invade a world of 'always defect'

This at first sight seems unpromising since it must be assumed that initial conditions in evolutionary history, such as 'always defect', must have been entirely selfish. The critical point, however, is that Pavlov is only a simpleton in a world of always defect. Once tit for tat has cleared away always defect in a world of minor mistakes, Pavlov could take over. Perhaps Pavlov enables us to see how the selfish gene learned to co-operate.

11.1.2 *Applications of game theory*

It is probably a fair judgement that the theory of games has outstripped its applications. There are some areas, however, where it has been profitably applied and these are considered below.

Mutualism

Game theory now allows us, theoretically at least, to distinguish between mutualism and reciprocal altruism. It can be imagined that, in mutualism, the rewards of co-operation exceed those of treachery (Figure 11.5).

		Player B	
		Co-operate	Defect
Player A	Co-operate	R = 6	S = 2
	Defect	T = 2	P = 1

Figure 11.5 Hypothetical pay-off matrix for mutualism, rewards being shown to player A (for key, see Box 11.1)

In this situation, it does not pay to defect. For example, if one lioness does not help another in the hunt, there may be little food for either of them. A variety of studies on co-operative hunting behaviour, such as that practised by lions and African wild dogs, endorse the conclusion that such behaviour is mutualistic (Creel and Creel, 1995). It is a fairly safe assumption that early hominids engaged in a variety of mutualistic interactions. The co-operation required to bring back large game to feed dependant relatives (especially children) would have encouraged mutualistic behaviour. It is possible in fact that the greatest threat to our ancestors was the existence of other hominid groups tightly bound in a system of mutualistic co-operation and willing to instigate violent conflict. This threat may have driven up group size to the level discussed in Chapter 6 and left its mark on the modular mind in the form of in-group and out-group reasoning.

It might also be expected that the communal rearing of cubs found among female lions would lead to reciprocity, since if a mother shared her milk with unrelated cubs, she could expect her cubs to benefit from the milk of other mothers. It was found, however, that although female lions share their milk fairly indiscriminately with close kin, there is little evidence of reciprocal altruism among non-kin (Pusey and Packer, 1994). Studies on birds and fish have yielded results open to a variety of interpretations. One of the best-documented illustrations of reciprocal altruism is the behaviour of vampire bats discussed in Chapter 3. Despite this, unequivocal evidence for the existence of reciprocal altruism in non-human populations is quite rare.

Application of reciprocal altruism to trench warfare in World War I

Axelrod (1984) drew upon the work of the sociologist Ashworth (1980) to apply the concept of prisoners' dilemma contests to trench warfare on the Western Front in World War I. Both Axelrod and Ashworth argue that the diaries of the infantrymen on both the British and German sides show that an unofficial policy of 'live and let live' emerged between troops who faced each other for prolonged periods.

As is well known, the war was one essentially of attrition. If one side restrained, that is, chose to co-operate, the other would gain an advantage by killing more troops ($T > S$). On the other hand, if both sides defected and attacked, although both suffered losses, troops on both sides would regard this as preferable to their side alone losing troops. In other words, $T > P$ and $P > S$. If both sides co-operated and restrained fire, a stalemate resulted, so $R < T$. From the viewpoint of troops on both sides, however, mutual restraint or co-operation is preferable to both sides defecting (attacking) since, if both sides lose roughly equal numbers of troops, there is no relative gain – something less desirable than no relative gain without the risk of loss of life, that is $P < R$. It also follows that $R > (T + S)/2$ since both sides would prefer to remain alive and in stalemate than to take equally a half share of $(T + S)$. The conditions for an iterated prisoners' dilemma were met because $T > R > P > S$ and $R > (T + S)/2$.

Axelrod claims that eyewitness reports from the infantry show clear evidence of tit for tat behaviour. Riflemen, for instance, would aim to miss each other but hit some other non-human target accurately to show that their aim was good. If casualties were caused, an equal number were exacted on the other side. The artillery often refrained from shelling the supply roads behind the enemy lines since it would be straightforward for the other side to retaliate, and both would then be deprived of fresh rations (R > P). High command suppressed such truces whenever it could; several soldiers were court-martialled, and whole battalions were punished. The action that finally ended the reciprocal co-operation took the form of random raids ordered by the officers. The raiders were ordered to kill or capture the enemy and, unlike the aim of a rifle or shell, such raids were difficult to fake.

The value of evidence from personal testimonies is that it illustrates that humans in some situations consciously appreciate the value of reciprocal altruism.

Figure 11.6   Troops in World War I caught in a desperate dilemma

According to the British sociologist Tony Ashworth, troops facing each other along the Western Front in World War I would sometimes implement an unofficial policy of live and let live. The American political economist Robert Axelrod sees this as in keeping with a tit for tat co-operative strategy

This is consistent with a view that such a behaviour pattern represents a fundamental component of the human psyche or the modular mind. We should, however, be wary of inferring more than consistency from this. The ability to articulate the reasons for co-operation in the trenches could also be taken as evidence of a conscious application of culturally derived rationality.

Environmental problems and the prisoners' dilemma

It would be foolish to imagine that all human interactions resolve into prisoners' dilemmas, but the diversity of situations to which the model can be applied is quite surprising. The abuse of environmental assets for individual gain is one such situation. Consider a crowded city where people spend hours each morning in sole occupancy of a car crawling to work at a snail's pace. Most of these people would be better off if they travelled by bus: there would be less traffic, and the buses would move faster. But in a world of all buses, the temptation to defect and drive a solitary car would be enormous, since the convenience and speed of car travel (as you would not have to stop to pick up co-operators) when the traffic is light would be rated very highly. The motorists caught in morning traffic jams are, in effect, reaping the rewards of all participants defecting. Moreover, unlike repeated encounters between pairs of people, there are too many players for a co-operative strategy to evolve.

In 1968, the biologist Garrett Hardin coined the phrase 'tragedy of the commons' as a metaphor for the nature of many environmental problems (Hardin, 1968). Hardin used the idea of the free grazing of common land such as was found in medieval Europe to convey the result of individuals each maximising their own self-interest. Suppose that three herdsmen each graze three cattle on a piece of land over which they have grazing rights. Equilibrium will be established between the rate of growth of grass and soil formation, and the loss of grass and soil by grazing. At a certain level of grazing, the equilibrium is stable, and cattle can be fattened without irreversible damage to the environment. But then it occurs to one of the herdsmen that he could place an extra beast out to graze. All the herdsmen would suffer to some degree, in the sense that their animals would not reach their former size, but the loss to the individual herdsman would be more than compensated by the possession of an extra animal. The tragedy arises of course because all the herdsmen think like this. The result is overgrazing and a collapse of that ecosystem.

Thirty years after Hardin's original conception, we can recognise that this is another form of the prisoners' dilemma. By maximising his own utility, each herdsman is in effect defecting. When they all defect, they are all worse off. Hardin's idea is actually more useful as a metaphor than as a realistic description of how the commons were managed in the Middle Ages, but, looking around, we can observe the tragedy of the commons in action in relation to many environmental problems of the 20th century. The seas have for years been treated as belonging to nobody and have therefore been regarded as a suitable dumping ground for pollutants. In some places, over-fishing has devastated fish stocks. Similarly, when we burn fossil fuels, we extract considerable benefit from the energy released and pass the cost on to the global commons, to be shared by others. To use the terminology of economists, the costs are externalised and the polluter only pays a fraction of the true damage cost. To overcome the tragedy of the commons, and make the polluter pay the full cost of the damage caused, is the problem of problems in green economics. According to Pearce (1987), we need to internalise the externalities. Interpreted in terms of game theory, we need to adjust the rewards and penalties to encourage people and institutions to co-operate rather than defect. If, for example, carbon dioxide emissions or car usage were heavily taxed, the temptation to defect would not be as great.

Ridley applies game theory to environmental problems and draws out the political lesson (by his own admission 'suddenly and rashly') that one solution to the prisoners' dilemma is to establish ownership rights over the commons and confer them on individuals or groups. According to Ridley (1996), through ownership, and effective communication within the owning group, comes the incentive to co-operate out of self-interest. The conclusion arising from Ridley's extrapolation from biological game theory to free market political economy will not be to everyone's taste, but it represents one serious contribution to the problem of how to channel self-interest to the greater good.

Indirect reciprocity

Ridley's analysis highlights the importance of effective communication to encourage co-operation. Recent work on reciprocal altruism confirms this point in a novel way. We have seen that reciprocal altruism can be expected to operate when animals have a good chance of meeting again to return favours. Such situations arise where animals gather in small colonies. Early human groups were almost certainly like this. The most common criticism of this model of reciprocal altruism is, however, that real-life encounters are often between individuals who have little chance of meeting again. So can the idea of reciprocal altruism be extended to cover non-repeated exchanges between individuals?

Nowak and Sigmund (1998) think that it can. They found, using computer simulation, that altruism could spread through a population of players who had little chance of meeting again as long as they were able to observe instances of altruism in others. The simulation involved creating donors who would help the recipient if the recipient were observed to have helped others in previous exchanges. The logic of this manoeuvre is that a donor will be motivated to donate help to someone with a track record of altruism since help is likely to be fed back indirectly in the future. More importantly, however, each act of altruism increases the 'image score' of the donor in the minds of others and thus increases the probability that he or she will be the recipient of help in the future from other observers. The problem with large groups is of course that any player will know the image score of only a fraction of the population. By making some simplifying assumptions, Nowak and Sigmund derived a remarkable relationship. They set initial conditions such that a player who was observed to have helped in the last encounter was awarded a score of +1 and a player who had defected was awarded 0. In the next encounter, the new player defected on those with a score of 0 but co-operated with those on a score of +1. If we call the fraction of the population for which any player knows the score q, the cost to the donor c and the benefit to the recipient b, Nowak and Sigmund found that co-operative behaviour can become an evolutionary stable strategy when $q > c/b$. That is:

The probability of one player knowing the $>$ $\dfrac{\text{Cost to donor}}{\text{Benefit to recipient}}$
image score of the other

The interesting feature of this relationship is its similarity to Hamilton's equation whereby altruism spreads if $r > c/b$ (see Chapter 3). The similarity between the equations may be nothing more than a coincidence, and certainly in humans our proclivity to co-operate or otherwise is based on something more subtle than the last observed encounter. Nevertheless, this work is important in that it shows that reciprocal altruism can operate without repeated exchanges between the same pairs as long as sufficient information flow takes place.

The act of blood donation is a fascinating case study of human altruism. People regularly give blood for the welfare of others without any expectation of payment. However, the reputation of blood donors is improved by the fact that they have acted altruistically. It is not surprising that the American Red Cross Society issues pins to be worn by blood donors when they have given blood. A pin is a mark of positive image in the analysis above. Similarly, society rewards those who have acted in the common good with medals and other honours. Such a system enhances the image of the bearer and allows him or her to benefit later.

The importance of image formation and reputation in the models of Nowak and Sigmund is consistent with the evolutionary role ascribed to gossip by Dunbar (1996a) and others. The basic function of gossip is that it provides us with information about members of groups to which we belong that has some bearing on our reproductive interests. Typically, gossip is concerned with the reliability of others (their image score), their sexual behaviour, their control over resources, political alliances and so on. We are intrigued by gossip about people we know and bored by gossip about people whom we do not know and are never likely to meet. Even among a gathering of intellectuals, conversation, unless it is regulated, breaks down to the level of gossip. Watch what happens to a formal meeting when the rules of procedure are temporarily lifted by, for example, a coffee break: gossip will break out.

A problem arises when we try to explain why we gossip about celebrities whom we are never likely to meet and who will probably not have much effect on our future welfare. The public at large, for example, seem to have a huge appetite for salacious tittle-tattle about presidents and royalty. There may be a number of factors at work here, for example the relief we receive from knowing that even the super-rich have problems. Barkow (1992) advances the notion that the media activates psychological mechanisms that evolved to ensure that we gained information about prominent members of our group. In effect, the constant exposure to images of the rich and famous leads us, in his view, to mistake them for members of our group. Witness the grief felt by so many when Diana, Princess of Wales, was killed in 1997. If this evolutionary perspective on gossip is correct, we would expect the content of gossip to differ in little other than fine detail across different cultures. Here then are some testable predictions awaiting investigation.

Problems

The problems in applying game theory fall into two categories. First, it is empirically extremely difficult to measure T, R, S and P, so even though we can in theory distinguish mutualistic from prisoners' dilemma situations, it is difficult in practice to know what is going on when we witness co-operative behaviour. Second, the game theory approach makes numerous restrictive assumptions such as symmetrical pay-offs and contests or interactions only between pairs. Real-life interactions are more complicated. Indirect reciprocity may be a helpful development here.

11.2 Altruism: emotion and morality

Given the conditions under which we would expect reciprocal altruism to flourish – repeated encounters and sufficient cognitive ability to recognise helpers and cheaters – we could expect to find reciprocal altruism and indirect reciprocity practised frequently in human populations. It is also probable that interactions that favoured altruists, and hence strategies such as tit for tat, were common in the life of early hominids. It is possible then that such interactions have left their mark on the mental life of humans, perhaps in the form of distinct modules dealing with emotions. If so, we should find that we are keen to co-operate but equally determined to punish defectors; we should remember those who gave us favours and not forget those who cheated. Evidence for the importance of detecting cheaters in our psychological make-up was provided in Chapter 6 when the modular mind was considered. This section explores further the possibility that there is some evidence of humans indeed being 'wired' with developmental genetic algorithms that are adapted to a life of reciprocal social exchanges.

11.2.1 *The emotional life of an altruist*

Trivers (1985) has argued that the practice of reciprocal altruism formed such an important feature of human evolution that it has left its mark on our emotional system. Humans constantly exchange goods and help, over long periods of time with numerous individuals. Calculating the costs and benefits and deciding how to act require complex cognitive and psychological mechanisms. Trivers suggests that a number of features of our emotional response to social life can be related to the calculus of reciprocal exchanges, and our emotional response serves to cement the system together for the benefit of each individual. Some typical emotional responses that are amenable to this sort of analysis are guilt, moralistic aggression, gratitude and sympathy. In addition, we have a highly codified system of justice to ensure fair play.

One of the hallmarks of a civilised society is taken to be an independent and objective judicial system. A point made by moral philosophers that can be tested by introspection, among other ways, is that we judge a situation fair when individuals can endorse it without knowing what position they occupy in the outcome. In other words, the strong sense of justice that humans possess is the desire that benefits and penalties have been distributed optimally.

We experience moral outrage and are motivated to seek retribution when an altruistic act is not reciprocated. Selection may thus have favoured a show of aggression when cheating is discovered in order to coerce the cheater back into line. If we are the cheater in a reciprocal exchange, by for example not returning a favour or not discharging an obligation, we may, if detected, be cut off from future exchanges that would have been to our benefit. So perhaps guilt is a reaction designed to motivate the cheater to compensate for misdeeds and behave reciprocally thereafter. In this way, it serves as a counterbalance against the tempting option to defect in a prisoners' dilemma. To encourage reciprocation in the first place, we have a sense of sympathy. A sense of sympathy motivates an individual to an altruistic act, and gratitude for a favour received provides a sense that a debt is owed and favours must be returned.

11.2.2 *Biological morality and induced altruism*

Humans are apt proudly to proclaim that it is their system of morality that sets them above the natural world, but evolution tells us that we should not accept the posturing of animals at face value. Some have tried to apply evolutionary thinking to provide a naturalistic account of human morality. Wilson and Ruse (1985), for example, have argued that, fundamentally, 'our belief in morality, is merely an adaptation put in place to further our reproductive ends' (p. 51) and that 'ethics as we understand it is an illusion fobbed off on us by our genes to get us to co-operate' (p. 52). These authors go on to state that the ethical codes we employ were selected for because they enhanced 'long-term group survival and harmony' (p. 52). Others, such as Alexander (1987), have also resorted to the interests of human groups as favouring the evolution of moral codes.

Badcock (1991) provides an interesting and plausible alternative account that does not rely upon group interests but shows that morality can also serve individual interests. In any human group engaging in reciprocal exchange, there exists the problem of the free rider: someone who extracts benefits without payment of their obligations. We would call this person a defector or cheat in the language of game theory. In repeated exchanges with the same individual, cheats will quickly be detected, and their rewards will drastically fall. It could pay to cheat, however, when the risk of detection is low. Imagine riding on a tube train without paying your fare. The system only operates if a majority does pay, but it is in the financial interest of any one individual to avoid payment if possible. Free riders then have an incentive not to pay themselves but strongly to advocate payment in others.

Here then, according to Badcock, could be the basis for morality in individualism: 'morality is indeed a means whereby individual human beings attempt to induce altruism... in others in their own self-interest' (Badcock, 1991, p. 121). This is in fact a classic case of the old adage used by moralisers caught defecting: 'Don't do as I do, but do as I say.' One could object that our commitment to moral codes is nothing like the calculating acts of defectors, that we subscribe to moral behaviour because we believe that it is the honourable way to act. There are a number of responses here. One is to speculate what would happen if the checks against moral lapses were loosened. Consider, for example, a transport facility such as an underground tube system that did not check whether or not you had paid for your ticket. Some people would not bother to pay, investment in the system would fall – either raising costs for those who did pay or leading to a deterioration of the service – followed by more people choosing not to pay. The end result would be the tragedy of the tube train. It is in the interest of all travellers to have those checks in place. Another response, suggested by Badcock (1991) and Trivers (1981), is that the mind deceives itself. To induce morality effectively in others by disclaiming a personal interest, part of your brain really needs to think that you have no ulterior motive. Badcock uses the analogy of a large company keen to bolster its image in the light of accusations of malpractice. The best public relations would be carried out by a separate PR department that did not know the truth. Such approaches to morality and human consciousness appear promising, particularly for the insight that they may provide into repressed states of consciousness or the unconscious.

Figure 11.7   The Good Samaritan (detail) from an etching by Hogarth (c. 1737)

In the Gospel according to Luke, Jesus illustrated the biblical injunction to 'love thy neighbour as thyself' with the parable of the Good Samaritan. Although humans often fall short of their ideals, the fact that we do offer help to the needy, irrespective of race, class or creed, poses a special problem for Darwinian psychology

11.3 The distribution of wealth: inheritance and kin investment in human culture

Wealth is, in a sense, the embodiment of indirect reciprocity. A coin or a note is a physical symbol that you are owed something for your previous efforts. You may have donated your labour to an employer, the money that was handed to you can now purchase the labour of someone else, and so it goes on. You could of course hand over your money to someone else, such as a friend or a relative, without any obvious future personal reward. Such is the situation when people make bequests in their wills. But we do not hand over our hard-earned wealth beyond the grave at random, and selectionist thinking can be applied in this context too. The next section shows how inheritance rules and practices are related to marriage systems.

This in itself provides an interesting example of how a cultural practice (rules of inheritance) can be related to genetic self-interest.

11.3.1 Inheritance rules and marriage systems

Kin selection theory tells us that parents will 'invest' biological resources in their kin because kin represent the passport that will carry their genes into future generations. To understand how biological investment should be distributed, we have already noted how the Trivers–Willard hypothesis predicts that investment in sons and daughters may not be equal. Females in good condition, for example, are predicted to favour sons. This hypothesis has so far, however, proved difficult to support with human data. In contrast, investment in the sense of providing food, protection, transport and so on is readily observed to be directed with discrimination towards kin in accordance with kin selection theory (see Chapter 3).

Now, if we extend this idea and consider investment in the legal or property sense in human cultures, we can also make some predictions that are amenable to testing. Hartung (1976) postulated that where wealth correlates with reproductive success, parents should pass on their wealth to male rather than female offspring. This follows from the fact that wealth can be used by a male where mating is polygynous to increase his number of wives and hence his number of children. In monogamous mating contexts, the bias in favour of males should be less pronounced: it is no use leaving resources preferentially to sons if the culture does not permit more than one wife (although it may help them to secure remarriage should they lose a first wife). Hartung (1982) tested these ideas in a survey of 411 cultures. He found a strong male bias in inheritance where polygyny was common (Table 11.1).

Table 11.1   Male bias in the inheritance system of a culture according to mating behaviour (data summarised from Hartung, 1982)

Mating System	% of cultures with a strong male bias
Monogamy	58
Limited Polygyny	80
General Polygyny	97

Hartung took pains to address the problem of independence of cultures and hence data points (the so-called Galton's problem), but Mace and Cowlishaw (1996) have suggested that a phylogenetic approach provides a better series of controls. They traced back 261 cultures in 11 language families to their inferred ancestral state of mating behaviour. They then looked for cases in which the marriage system, or the inheritance rules or both changed as the culture evolved. Where change did occur, the end states of inheritance bias and marriage system after the transition were recorded. The data are shown in Table 11.2.

The work of Mace and Cowlishaw supports the idea that marriage patterns and inheritance rules evolve together in ways that are adaptive. The evolution of monogamy is strongly associated with an absence of sex-biased inheritance rules. In contrast, where general polygyny evolves, it is usually accompanied by a strong male bias in inheritance. It is worth noting that whereas Hartung found that monogamy was roughly equally associated with male bias and the absence of male bias, this revised phylogenetic approach showed that an evolved association between monogamy and strong male bias is less common (25 per cent). In summary, if you live in a culture in which polygyny is common and you are rich, it would serve your biological interests to leave your wealth to your sons: with wealth, they have a chance to obtain more wives and thus produce more grandchildren. If marriage and sexual relations are largely monogamous, it is not so crucial.

Table 11.2   End states of marriage system and inheritance rules following a cultural transition (data summarised from Mace and Cowlishaw, 1996)

End state of inheritance rule	End state of marriage system		
	Monogamy (%)	Limited polygyny (%)	General polygyny (%)
Strong male bias	25	44	87
Weak or no bias	75	56	13
Number of cultures	20	18	16

11.3.2 Inheritance of wealth: practice in a contemporary Western culture

In the legal system of most Western cultures, individuals have considerable freedom of choice in deciding how to distribute their wealth in bequests to relatives and friends following their death. It is reasonable to suppose, for the same reasons outlined above, that, in our evolutionary past, the allocation of resources following death would have significantly affected an individual's inclusive fitness. So are there vestiges of any adaptive preferences in behaviour today? In attempting to answer this, Smith *et al.* (1987) analysed the bequests of a random sample of 1000 individuals recorded in the Probate Department of the Supreme Court of British Columbia. Their data were used to test four predictions:

1. Individuals will leave more of their estate to kin and spouses than to unrelated individuals. This is hardly a surprising prediction from general knowledge, but it follows from kin selection theory in the sense that resources left with kin can improve inclusive fitness, and resources left with genetically unrelated spouses may still eventually find their way to kin.

2. Individuals will leave more of their estate to close kin (with a high coefficient of relatedness, r; see Chapter 3) than to distant kin (with a low r value).

3. Individuals will leave more of their estate to their offspring than to their siblings. Although a son or daughter has the same r value (0.5) as a brother or sister of the deceased, it is likely that a bequest to, for example, a son will enhance his reproductive value more than that given to a brother. If a bequest were to be made to a brother or sister, he or she is likely to be past reproductive age or in a position where wealth has a limited effect on reproduction.

4. There will be a male bias in bequests to children of wealthy individuals and a female bias in those to children of poorer individuals. The logic of this prediction for male offspring is similar to the hypothesis of Hartung, outlined above. The female bias is predicted from the fact that in mildly polygynous mating (which may have represented our ancestral state), poor sons may not secure any wives, and thus produce no grandchildren, but poor daughters may mate with a man who already has a wife.

Predictions 1–3 were confirmed by looking at how the percentage of bequests was distributed according to closeness of kin (Table 11.3).

Table 11.3 Bequests made to relatives as a percentage of the total estate for 1000 people in British Columbia (adapted from Smith et al., 1987)

Co-efficient of relatedness (r)	Beneficiary	% of estate (mean)
0.0	Spouse	36.9
0.5	Sons	19.2
0.5	Daughters	19.4
0.5	Brothers	3.2
0.5	Sisters	4.8
0.25	Nephews and nieces	5.1
0.25	Grandchildren	3.2
0.125	Cousins	0.6
0	Non-kin	7.7

Table 11.4 Distribution of an estate to offspring according to sex (adapted from Smith et al., 1987)

Value of estate, E (Canadian $)	% bequeathed to sons	% bequeathed to daughters	Greatest % to	Significance
E < 20 350	9.9	19.4	Daughters	$P < 0.01$
E < 52 900	14.6	18.6	Daughters	NS
E < 110 850	21.9	24.6	Daughters	NS
E > 110 850	30.2	15.1	Sons	$P < 0.01$

NS = not significant

To test prediction 4, the estates of the 1000 deceased individuals were split into four wealth bands (Table 11.4). Again, the data support kin selection theory. Wealthy parents tended to bequeath a significantly larger portion of their estate to sons than to daughters, and vice versa for poorer parents.

The studies of Mace *et al.* and Smith *et al.* are worth comparing in that they illustrate how evolutionary psychology may illuminate different facets of human cultural behaviour. In Mace *et al.*'s study, it is the cultural norms and rules that reflect fitness-maximising behaviour. This is evidence that is germane to the much larger question of the relationship between genes and culture pursued in the next section.

It may seem as if male inheritance benefits only the inclusive fitness of fathers, but of course mothers benefit too from wealth passing to a son who thereby becomes reproductively successful. The study of Smith *et al.* shows how, in societies where cultural conventions are more flexible, individuals still behave through 'free choice' in order to maximise their inclusive fitness. It is interesting that the bias to male offspring from wealthy parents probably had adaptive value 100 000 years ago, and may still have in polygynous cultures, but probably has little long-term effect in Western cultures where monogamy is legally enforced.

11.4 The evolution of culture: genes and memes

It is true to observe that the major human achievements of the past 6000 years that distinguish us from the rest of the animal kingdom are the results of cultural rather than biological evolution: our genes are roughly the same as our Stone Age ancestors whereas our **culture** has changed by an incredible degree. Such is the disparity in these changes, and such is the complexity of cultural change, that is it difficult to establish with any confidence the relationship (if any) between cultural evolution and genetic evolution.

If evolutionary biology is able to say anything about human culture, we need a working definition of what culture is. The problem is that the word 'culture' has a range of connotations. To the historian of art and literature, it may signify something that is morally improving or uplifting. To an anthropologist, it may mean the fabric of beliefs and values common to a society. A biologist would identify it as something that is passed down by social learning. The definition of culture adopted here will be that culture is information (knowledge, ideas, beliefs and values) that is distinct from that stored in the human genome and which is socially transmitted. Note that this does not *a priori* rule out any evolutionary purchase on culture. The fine details of culture are obviously not represented in the genome, but the type of culture we generate and the functions that it serves could, in principle, be directed by our genes. It is of course an extravagant claim and one not to be swallowed lightly. Neither is this a definition that would be accepted by everyone. Darwinian anthropologists such as William Irons take a more phenomenological view and see culture in terms of behaviour rather than mental constructs. The logic of this position is that mental constructs such as customs and belief have biological significance only to the extent that they influence behaviour (Irons, 1979). The definition employed here, however, is more tolerant of a range of evolutionary models that have been developed to explain culture.

The crucial issue is whether there is a relationship between biological evolution and cultural change such that biology determines, in either a soft or a hard way, cultural evolution, or indeed whether culture directs or determines biological evolution. A number of models have been proposed to tackle these problems. At the risk of oversimplifying some complex work, we will reduce these to four categories and discuss each in turn. The four categories are:

1. Culture as autonomous from biology
2. Culture as a reflection of genotype
3. Genes and culture in co-evolution
4. Culture as the evolution of **memes**.

11.4.1 *Culture as autonomous*

This model starts from the observation that although very little genetic change has occurred over the past 35 000 years, culture has changed enormously. This would suggest that there is no linkage between biological and cultural evolution, as well as that culture has its own laws of development, which are at present only dimly understood and are probably best studied by the humanities and the social sciences. The evolutionary biologist retires from the field gracefully or perhaps makes the passing observation that cultural evolution seems to be Lamarckian in that the achievements of one generation can be passed to the next. In support of this view, we could point out that evolutionary thinking cannot explain every aspect of human behaviour. Our behaviour is strongly conditioned by what we learn from others, and even if this ability to learn is a product of biology, what we do learn is culture. The very definition of culture adopted earlier asserts that culture does not reside in the genome. Many cultural changes are fitness-neutral, and changes in fashion and style are probably best approached sociologically using cultural theory rather than biology. Some cultural practices are maladaptive: chastity among monks and nuns is hardly designed to promote fitness.

Considering this, it would be surprising if every component of culture had some root in biology but, equally so, it is perhaps unduly cautious to suspect that they are totally disengaged. Culture is after all the product of human minds that have been shaped by selection. Humans seem to be thriving as never before (at least in terms of population number) within complex cultures. It has already been noted that some features of what is often regarded as the province of culture, for example morality, marriage contracts and inheritance patterns, can be tackled using evolutionary concepts. One of the most telling criticisms of this approach is that it confuses the description of the phenomenon with the very thing to be explained. Noting that there is a cultural difference between two groups of genetically similar human populations does not logically imply that there is a force of culture at work: it simply means that there are similarities and differences to be explained. Finding that two cars of the same make, model and year are nearly identical except in terms of their colour does not mean that we need a new 'explanatory colour principle' to explain this fact.

11.4.2 *Culture as a consequence of genotype: culture is a programme for fitness maximisation*

There are in fact several distinct viewpoints within this category. A Darwinian anthropologist would argue that culture is a tool for genetic replication, so culture represents a rationalisation of fitness-maximising behaviour. It may not appear overtly as such, but fitness maximisation is not a conscious goal anyway. Cultural differences exist because people live in different environments and hence ways of maximising fitness through culture will differ. In this model, if aspects of culture appear maladaptive, it may be because the environment has changed faster than the culture can. Cultural change, like as genetic change, has inertia such that optimality is not always achieved.

Whereas only a fool would suggest that the precise details of culture are described by the human genome, it is possible with some plausibility to suggest that genes themselves give rise to development processes that predispose humans to develop certain cultural forms. This seems to be the position argued by Wilson in his pioneering book *Sociobiology: The New Synthesis* (1975), in which he suggested that cultural forms also confer a genetic survival value. Religious systems, for example often prescribe correct forms of behaviour in relation to food, co-operation and sex. Incest taboos may reflect an instinctive mechanism designed to avoid homozygosity for recessive alleles. The language we speak obviously depends on what we have been exposed to, but the facility to acquire a language seems very much innate. In later work, Wilson and Ruse suggested that moral standards and ethical codes are a legacy of our genetic evolution: 'ethics as we understand it is an illusion fobbed off on us by our genes to get us to co-operate' (Wilson and Ruse, 1985, p. 52).

Examples of how culture may reflect biological needs are discussed by Gary Cziko (1995) as part of a more general account of the power of selectionist thinking. One concerns the farming of rice on the small Indonesian island of Bali. The planting, irrigation and harvesting of rice are organised according to not a secular or functional timetable but a religious one. The Balinese follow a religious calendar of activity in accordance with a belief in 'Dewi Sri', the rice goddess. The outcome, however, is that this religious practice serves to maintain a high yield of rice.

Other cultural practices that seem maladaptive, wasteful or at best fitness-neutral could have the effect of ensuring group cohesion by promoting group identity and encouraging altruistic behaviour. This may be one reason why cultures and languages rapidly differentiate. One way of knowing whether your altruistic behaviour is likely to be reciprocated is to examine whether the recipient is of the same culture. The difficulty with arguments such as this is the need to show that cultural beliefs and practices further the interests of individuals as well as groups. If some aspects of culture cater only for the interests of the few at the expense of the many, as Marx implied when he stated that the ruling ideas of any period are the ideas of the ruling class, we have moved away from an analysis of culture as a reflection of genotype.

We have noted earlier the stance adopted by the discipline of evolutionary psychology to the effect that our bodies and minds should be adapted to solve the problems of the Pleistocene era rather than the problems of modern life, except insofar as they are similar. In this framework, culture is to be seen as a response of universal features of human nature to transient environments. Tooby and Cosmides (1992) have articulated this model. They argue that culture can be divided into three components: metaculture, evoked culture and adopted culture. Metaculture represents those universal components of culture that are the product of natural selection, such as language, grief over the loss of loved ones, association with kin and so on. Evoked cultures are formed when different environments act upon features of human nature. The outcome will be different types of culture according to different local environmental conditions. Adopted culture is the situation when one culture spreads to another in a manner analogous to the spread of a disease.

A good example of evoked culture in this sense is the practice of food-sharing. Anthropologists have found that the degree to which individuals share food in a culture is related to the variance in food supply. Among the Ache tribe of Paraguay, meat obtained from hunting is highly variable in its supply, the chances of a hunter returning with a kill being only about 60 per cent. Correspondingly, within the Ache, meat is shared equitably among the tribe, the kill being passed to a 'distributor' who allocates portions to families according to their size. In the same tribe, however, food obtained by gathering is of low variance – it can fairly reliably be obtained given effort and persistence. Significantly, among these people, whereas meat is distributed communally, gathered food is shared only within the kin group.

The evolutionary logic to this is straightforward. With high-variance food such as meat that is difficult to store, there is little point in gorging oneself or one's family on the products of a day's hunt. Once immediate needs have been met, there is little extra benefit to be gained. There is a strong advantage, however, to sharing surpluses among the whole group since hunting involves some degree of luck, and the sharing of meat ensures that the hunter and his family will be fed if he is unlucky. With gathered food, one reason for a lack of success would be laziness rather than luck, so the benefits of reciprocal altruism are not as compelling (see Buss, 1999).

This analysis is supported by Elizabeth Cashden in her studies on the San people in the Kalahari desert. Cashden found that some San tribes are more egalitarian than others. The Gana San tend to hoard food and are reluctant to share it outside family circles whereas the !Kung San show a much higher degree of food-sharing. The reason may be that the variability of food supply is high for the !Kung San and low for the Gana San, triggering and evoking two different but evolved psychological mechanisms (Cashden, 1989).

In Chapter 5, Miller's argument that cultural creativity is a form of sexual signalling was noted. This display hypothesis is one that can be considered here in that it shows how cultural artefacts may be rooted in our biological nature. Most surveys show that a large number of works of art are produced by young men at an age when they are most interested in sex. The essential idea of Miller is that artistic displays make men more attractive to women. In support of this, there is plenty of anecdotal but suggestive evidence of the large number of sexual contacts experienced by performers such as Jimi Hendrix, Pablo Picasso, Charlie Chaplin, Rod Stewart and (before marriage) Paul McCartney. Creativity, it seems is sexy.

Miller provides some quantitative support of his hypothesis in an analysis of jazz musicians. The distribution of output in terms of number of records against age has a fairly narrow profile: most music is produced by men aged between 20 and 40, the very age when they are investing heavily in mating effort (Miller, 1999a). The hypothesis is still rather tentative and is certainly unable to explain the form that artistic creativity takes. Moreover, the literary intelligentsia will take great delight in pointing to exceptions, such as the remarkable efflorescence of poetry by Thomas Hardy in his seventies, or the late music of an aged Vaughan Williams, but Miller's ideas remain intriguing. He regards much of human culture as 'wasteful sexual signalling' that began with language, art, music, humour and clothing but was then closely followed by religion, philosophy and literature (Miller, 1999b). If Miller is right, our most cherished products of the human intellect are cultural versions of the peacock's train, wasteful and showy extravagances functioning to attract mates.

11.4.3 Gene–culture co-evolution

Common sense tells us that culture has some autonomy from genes, so perhaps we should regard it as being constrained by an 'elastic' leash. As the metaphor suggests, cultural change can then apply a force that pulls genetic change along. This approach has spawned some highly complex mathematical models, including some by Lumsden and Wilson (1981). In early models by these authors, culture is depicted as the outcome of the behaviour of individuals in situations in which the behaviour of individuals is itself influenced by the existing culture and individual 'epigenetic' rules of development. We can imagine a growing individual as being bombarded by bits of culture, or 'culturgens'. Genetically based rules of development will influence a child to accept some culturgens (an early word for memes) and reject others. Genes and culture have, therefore, shaped the final behaviour of the adult and, moreover, he or she then contributes in the re-creation of culture. Cultural and genetic change can take place together because culture sets the environment that may alter gene frequencies for genes that describe the epigenetic rule of development. In this manner, culture and genes roll on together. There are numerous other variations in approach to this subject, but two examples – yam cultivation and lactose intolerance – should serve to illustrate the general line of thinking (Feldman and Laland, 1996).

In West Africa, people often cut down trees to cultivate yams. In these areas, heavy rainfall then results in pools of stagnant water exposed to the air, which are ideal breeding grounds for mosquito. In these areas, the frequency of the sickle-cell anaemia mutant (see Chapter 3) that confers some resistance to malaria is consequently higher than would otherwise be expected. In short, a cultural practice has changed gene frequencies.

Lactose intolerance is more complicated. The next time you pour milk on your bowl of cereal (assuming that you are not intolerant to milk), spare a thought for the genes that help you to digest it. The milk of cattle and other animals has probably been an important component of the diet of some human populations for about 6000 years, and the prevalence of dairy farming among some human groups may have led to a selection pressure for genes allowing the absorption of lactose beyond childhood weaning. It is very unlikely that early hunter-gatherers were able to continue into adulthood to synthesise the enzyme lactase, which allows the digestion of milk.

Although in the West, we tend to assume that milk is a natural part of the diet, there are large areas of the world where drinking milk would result in quite severe illness. As a broad generalisation, almost all Scandinavians and most Central and Western Europeans are lactose absorbers, but most African and East and South Asian people are not. To understand this distribution of genetic difference, we need to look at the role that milk products played in selection pressures.

One answer is that malnourishment in some environments would have helped the spread of genes assisting the absorption of lactose and thus enabling individuals to survive. This is plausible, but there are probably additional factors (Durham, 1991). It is known that lactose, like vitamin D, also helps the absorption of calcium in the gut. Vitamin D is produced in the body by the effect of ultraviolet radiation on the skin. Vitamin D deficiency and hence poor calcium absorption is a serious risk for people living in high latitudes where the intensity of sunlight is low. For these people, lactose absorption had the double advantage of providing both calories and essential calcium. This example neatly illustrates how a cultural change – the invention of agriculture and the domestication of animals – can bring about a genetic shift in human populations.

Both the prevalence of sickle-cell anaemia in Africa and lactose tolerance can be seen as varieties of the Baldwin effect (Chapter 1). Essentially, in the case of lactose tolerance, an environment that favoured animal husbandry led to an advantage to those people who where initially flexible enough to digest at least some milk products. This enabled the survival of these populations, who then experienced a steady selective pressure for those who were genetically disposed to tolerate lactose.

11.4.4   *Cultural evolution as the natural selection of memes or part of gene–meme co-evolution*

Of all the models, this is the most revolutionary. It suggests that the world of ideas (the characteristic products of culture) evolves in a Darwinian fashion but not necessarily in relation to the physical objects we call genes. To appreciate the logic of this model, consider the minimum set of conditions for some sort of Darwinian selection, and thus evolution, to operate:

1. There exist in the world entities capable of self-replication
2. The process of replication is not perfect: errors are made, and the next copy may not perfectly resemble its template

3. The number of copies of entities that can be made depends on the structure of the entities and hence their manner of interaction with the outside world
4. As a result of the finite nature of resources, operating spaces and so on, these entities experience differential reproductive success.

From these four minimal conditions, we should be able to witness Darwinian evolution. It is easy to appreciate that the entities above may not be strands of DNA. It is conceivable that, on other planets, the molecular basis of replication may be completely different. It is more of a shock to realise that the entities may not need to be physical at all; they may, in short, be ideas existing in and moving between brains.

Dawkins was not the first to note this insight, but he articulated it forcefully in selectionist terms and coined the term 'meme' to describe elements of thought or culture that replicate in human brains. As an analogy, the meme idea works surprisingly well. Memes move from brain to brain like parasites moving from host to host. We can catch them vertically from our parents, as in the case of rules of behaviour inculcated in childhood, or horizontally from each other, as in the case of peer pressure or conformity to fashion (Box 11.2).

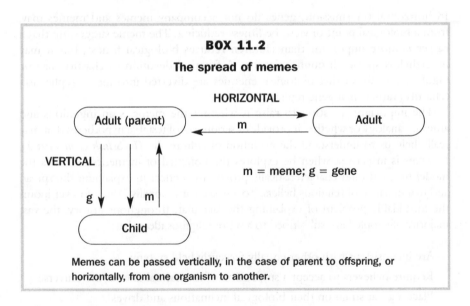

BOX 11.2

The spread of memes

Memes can be passed vertically, in the case of parent to offspring, or horizontally, from one organism to another.

Some memes are truly parasitic in the sense of damaging the survival chances of the host or the host's genes. Chastity, celibacy and self-sacrifice for noble ideals are all memes that damage their host's biological success. But this need not concern memes if memes survive: if self-sacrifice is held up (probably by a linked gene) as a laudable act, others will fall under the sway of the meme, and the meme will survive. Many memes, however, are mutualistic in that they assist their replication by ensuring the well-being of the host. Examples here include elementary rules of hygiene, methods of fashioning tools, avoiding disease and so on.

The incest taboo may be a case of genes and memes directed to the same end if, as it seems, the Westermarck effect is based on a genetic developmental programme. The taboo becomes the meme that reinforces the genetically based mechanism of avoiding incest.

Vertical

When passed vertically down the generations, memes can accompany genes. In early traditional cultures, there was probably a large degree of synergy between genes and memes. For example, the cultural meme that directs a male bias in inheritance in polygynous cultures (see above) could also enhance the genetic interests of those practising it. Similarly, the linked memes in catholicism that restrict birth control and also insist that offspring are raised in the faith have the dual effect of increasing both the spread of memes and of the genes of those professing the memes. In these cases, sociobiological and memetic explanations yield the same results.

Horizontal

In horizontal transmission, genes do not accompany memes and memes may, from a biological point of view, be fitness-reducing. The meme suggesting that a career is more important than children reduces biological fitness, but it may nevertheless spread through imitation. A fanatical devotion to chastity may be highly successful in that biological energies are diverted into meme replication (chastity) rather than gene replication.

The important question to raise is whether the meme is an ingenious and amusing analogy or whether it provides a serious set of testable hypotheses that may really help us to understand the evolution of culture. In *The Selfish Gene* (1989), Dawkins is in earnest when he explores the potential of memetics and uses the model to good (and characteristically provocative) effect in explaining the spread and maintenance of religious beliefs. To any secular rationalist, there forever looms the formidable problem of explaining the fact that, throughout history, the vast majority of people have subscribed to a set of religious ideas that:

- Are inconsistent with other equally fervently held systems
- Require believers to accept a suspension of the natural laws of the universe
- Place a great strain on their biological inclinations and drives
- And, moreover, are adhered to in spite of contradictory, little or at best ambiguous empirical support.

To Dawkins, such beliefs represent the invasion of minds by memes, and people who are victims of these memes he calls 'memeoids'. Memes, for some particularly unlikely tenet of belief, may move around linked with other memes that help their survival, such as the meme for the virtues of faith. If we define faith as 'belief despite the lack of evidence', the 'virtues of faith' meme suggests that faith is in itself a good thing to have. People of 'little faith' are castigated as being deficient in some way. Most religious systems have a 'virtue of faith' meme as part of their meme complex.

The ideas of Dawkins have been taken seriously by some theologians, their concern being illustrated by the title of Bowler's (1995) work *Is God a Virus?* As we might expect, Bowler thinks not, but others have more enthusiastically pursued the hare that Dawkins let loose. Dennett explores the notion of memes and makes the interesting point that, just as the evolution of animals on land had to wait until algae modified the earth's atmosphere, so the evolution of memes had to wait until brains large enough to contain them and transmit them came into being. He uses the notion of memes to tackle the nature of mind:

> a human mind is itself an artefact created when memes re-structure a human brain in order to make it a better habitat for memes. (Dennett, 1995, p. 365)

In support of this, we could recall the experience of having a tune in our heads that will not go away even though it may be irritating. Poets talk of poems writing themselves, novelists relate how characters seem to take on a life of their own. This is radical stuff, but eventually even Dennett is uncertain of what the new science of memetics will actually deliver.

Susan Blackmore is one psychologist who has high hopes for the new science of memetics. In her book *The Meme Machine* (1999b), she attempts to show how memes could account for the explosion in brain size that occurred about two and a half million years ago. To follow her argument, we must adopt the concept of a meme as that which is subject to imitation. Once hominids started to imitate each other, a new replicator was born: the meme. Memes are passed on by imitation. An early hominid such as *Homo habilis* might observe another fashioning a tool in a particularly effective way, or enjoying and apparently thriving on a new foodstuff. Imitation would then bring distinct rewards, at the level of both genes and memes. The genes of the imitator, and hence any genetic disposition to imitate, would be passed on if the activity conferred some survival advantage. By a parallel process, the memes would thrive since more people would now be practising whatever it was that enhanced survivability. At this stage of our evolution, genes and memes were probably closely linked in terms of the advantages they provided for the biological body.

The ability to imitate almost certainly required a basic minimum of brain machinery. Two factors in particular – Machiavellian intelligence and reciprocal altruism – may have predisposed early humans to develop the minimum brain capacity required (see Chapters 6 and 3 respectively). To reciprocate is, after all, a form of imitation, one which confers biological fitness. Machiavellian intelligence would have helped social living creatures such as early humans. The corresponding theory of mind, and hence the ability to appreciate the perspective of others, may then have facilitated imitation. Once imitation enhances genetic and memetic fitness, further selection pressures are set up. Those individuals who are good imitators and, crucially, imitate other successful imitators will do better. In addition, those who choose mates who are good imitators will also leave more viable offspring. Blackmore suggests that this combination of natural and sexual selection drives up brain size by a process of positive feedback. As brains become selected for imitating, language emerges more or less inevitably since language is one way in which memes can be propagated and obtained. As Blackmore observes:

> I suggest that the human brain is an example of memes forcing genes to build ever better and better meme-spreading devices. (Blackmore, 1999a, p. 119)

If Blackmore is right, the structure of language should show signs of complex adaptation to the task of transmitting memes (with what Dawkins observed were the essential characteristics of such replicators – high fecundity, fidelity and longevity) rather than simply conveying information about such things as social contracts or foraging tactics. The problem here is to devise a crucial experiment, as it may be that language does both. If memes and genes were linked to biological fitness before widespread culture took root, a dual function is to be expected. One interesting prediction is that if brains were shaped by sexual selection to be adept at spreading memes, we should expect females (if they did most of the original selecting) to be attracted to men who are proficient at meme spreading; men, in other words, who are intelligent, witty, possess a good memory and are culturally prolific and creative.

Memetics may be a science that can flourish in areas where sociobiology or evolutionary psychology fails to thrive. One of the long-term goals of sociobiology and evolutionary psychology is to show that culture is explicable in terms of biology. Tooby and Cosmides are explicit about this:

> Human minds, human behaviour, human artefacts and human culture are all biological phenomena. (Tooby and Cosmides, 1992, p. 21)

According to Blackmore, sociobiology is destined to fail because it is only concerned with the outcomes of the survival of one set of replicators: the genes. Once we introduce another set of replicators, memes, which have different interests, culture must be understood both genetically and memetically.

To see how this might work, consider altruism. A sociobiological account would use concepts such as kin selection, reciprocal altruism and game theory, and generally take the view that altruism is really enlightened self-interest. A traditional theological or transcendental view would account for altruism in terms of the nobility of the human spirit, our innate capacity for goodness and other Cartesian or 'ghost in the machine' explanations. Dawkins, for once, although he would be horrified, belongs in the same category as the transcendentalists when he says that 'We, alone on earth can rebel against the tyranny of the selfish replicators' (Dawkins, 1976, p. 215). The reason is that here 'we' seems to be placed outside the circle of scientific naturalism and given a life of its own. A memetic account of altruism, as suggested by Blackmore, is that an altruistic act becomes a meme that is copied. Copies of altruistic acts, and not mean-spirited acts, are copied because altruism would probably entail more contacts, and altruistic acts would be more popular and confer pleasure on their recipients. There may then be a mechanism for copying popular acts. The link in the reasoning here that needs to be strengthened is why popular acts should be preferentially copied. This could start, as suggested earlier, by hitch-hiking on the back of reciprocal altruism: a tendency to reciprocate a favour would cause the imitation of kindness to spread.

Memes are unlikely to exist alone. The meme for altruism, for example, is linked with other memes that confer a memetic advantage. A combination of memes, Blackmore calls memeplexes. Religious systems as suggested by Dawkins and others are complex memeplexes. They have strict rules of adherence and employ enhancers and threats. An enhancer encourages the spread of the meme by an appeal to its virtues such as 'blessed are the merciful for they shall obtain mercy' (Matthew 5.3). Advertising slogans are full of memes connected with enhancers: 'A Mars a day helps you work rest and play' is successful in that it promises benefits from the activity (eating a chocolate bar) as well as using rhythm and rhyme, which fit in well with our neural circuits. Some memes carry threats. One of the many Biblical memes to renounce worldly pleasures for example is sanctioned by a divine threat: 'If any man love the world the love of the Father is not in him' (John 1.2).

Blackmore makes the radical suggestion that the concept of the self is a memeplex – or selfplex as she calls it. She suggests that the 'I' acts as a sort of protective coat around other memes. By entertaining an 'I', we feel more strongly bound to the memes that go along with it, and hence we fight for their survival. Yet the self ultimately has no centre:

> That is, I suggest, why we all live our lives as a lie, and sometimes a desperately unhappy and confused lie. The memes have made us do it – because a 'self' aids their replication. (Blackmore, 1999b, p. 234)

There are of course problems with the meme theory, and there may be a circularity in some of its attempts to explain the propagation of memes. The concepts of 'self' and 'I' can only protect other memes if they carry some force in themselves. The power associated with the self needs further explication. The analogy with DNA is also not perfect: natural selection works because inheritance is discrete and not blended. If any new change in the genome simply blended with other genes, novelty would be quickly lost and evolution would grind to a halt. It is, however, not at all clear that memes act like discrete units. They seem to blend and merge in ways quite unlike the information carried on DNA. Whereas genes have a universal language, there seems little prospect of a uniform meme language in brains. Gene mutation is random and undirected whereas changes to memes are goal directed. Perhaps, however, we should not expect all replicators to behave like DNA. The most serious problem may be understanding the conditions by which some memes are copied well and others are ignored. In the case of biological features such as camouflage, it is easy to see why such genes have been selected. With memes, it is not at all obvious why, for example, the meme for flared trousers should have increased in frequency in the early 1970s, have suffered a setback in the 1980s and then (partially) revived in the late 1990s.

There is much work to do on memes and theories of knowledge. The very debate about memes shows that there are, in science, high standards to be met before an idea can become part of accepted wisdom. Were we to argue that the standards of scientific scrutiny (success in prediction, consistency, repeatability of measurements and so on) were simply themselves deeply entrenched memes (with no real purchase on the real world), this would raise the unwelcome spectre of epistemological relativism. If we argue that such standards help us in some way to track reality, we need a better theory of how memes and real objects (brains and objects in the world) interact.

Dennett (1995) speaks of Darwinism as a 'universal acid' that passes through everything, leaving behind 'sounder versions of our most important ideas'. Darwinism will probably continue to transform our notion of culture and our sense of identity, but there is more work to be done.

SUMMARY

■ Game theory shows that strategies involving co-operation, such as tit for tat, could offer a model of the evolution of human morality.

■ The undoubted importance of altruism and co-operation in the environment of evolutionary adaptation made it imperative that cheaters could be detected. The need to detect cheaters could have shaped our emotional life, such as our sense of justice and our experience of gratitude and sympathy.

■ The transmission of wealth during marriage arrangements or as a result of death contains features that, as predicted by Darwinian anthropology, increase inclusive fitness.

■ There are at least four models of how genes and culture could interact. Culture could be totally autonomous (biologists remain silent), culture could be an expression of the long reach of the human genotype, culture and genes could co-evolve together in a dynamic interaction, or culture could be the natural selection of the memes. All of these models are in their infancy.

KEY WORDS

Altruism ■ Coefficient of relatedness ■ Culture ■ Game theory
Meme ■ Prisoners' dilemma ■ Reciprocal altruism ■ Tit for tat

FURTHER READING

Barkow, J. H., Cosmides, L. and Tooby, J. (1992) *The Adapted Mind*. Oxford, Oxford University Press.
A manifesto for evolutionary psychology. See especially Chapter 1 on psychology and culture, and Chapter 18 on culture.

Blackmore, S. (1999b) *The Meme Machine*. Oxford, Oxford University Press.
Bold claims are made for the power of the memes. A provocative work with an interesting analysis of the self.

Crawford, C. and Krebs, D. L. (1998) *Handbook of Evolutionary Psychology*. Mahwah, NJ, Lawrence Erlbaum.
Some good articles reviewing all aspects of evolutionary psychology. For a review of biological theories of culture, see Chapter 5.

Ridley, M. (1996) *The Origins of Virtue*. London, Viking.
Discusses game theory and human co-operation. Provides an interesting and controversial application of game theory to politics and environmental issues.

12

Epilogue: The Use and Abuse of Evolutionary Theory

Those who forget history are condemned to repeat it.

(George Santayana, 1863–1952)

We pass lightly over sentiments like those of Santayana these days, and we do so at our peril. Familiarity may have blunted their edge, but there is one area more than any other where such thoughts should still strike us with full force, and that is in the history of attempts to define human nature. Nothing could be more important, yet in no other area has science been so betrayed. Good ideas have been neglected and bad ones pursued to tragic ends. In this field in particular, scientists have an obligation to the past and to the future to be aware of the history of their discipline and the social ramifications of their ideas.

The impact of Darwinism on other disciplines has been enormous: philosophy, theology, psychology, anthropology, literature, politics – the list could be continued – have all been radically altered by the import of evolutionary ideas. Given that the theory deals with fundamental questions about the human condition this is not altogether surprising. The results, however, have not always been welcomed and, in the sphere of politics in particular, the consequences have been messy. There have been unwarranted extensions from facts to values, and some odious political philosophies have sought their imprimatur from evolutionary thought. In this final chapter, we will examine the interaction between scientific ideas and their social contexts. In particular, we will attempt to separate readings of Darwin that fall foul of empirical or logical errors from those which are legitimate. The first part of the chapter describes the way in which evolutionary ideas have been subject to a variety of political interpretations and extrapolations over the past 150 years. The sections that follow then tackle some of the philosophical issues raised.

12.1 Evolution and politics: a chequered history

When evolutionary theories of human development and origins began to emerge in the early 19th century, they were attacked by the political and religious establishment as being radical, destabilising and a threat to the social order. The evolutionary views of Lamarck were met with contempt and dismissed as atheistic, revolutionary and subversive. Nothing better illustrates the hostility to evolutionary ideas among conservatives than the reaction to a book written by the Edinburgh publisher and amateur naturalist Robert Chambers. The book was called *The Vestiges of Natural Creation* and was published anonymously in 1844. In fluent prose, mixing religious speculation and scientific facts, Chambers gave expression to the idea that life had evolved and that species were mutable. The Anglican establishment came down hard on Chambers. Adam Sedgewick, a Cambridge don and former tutor of Darwin, called it a 'filthy abortion' that would sink man into a condition of depravity and poison the well-springs of morality. One of the reasons why Darwin delayed publication until 1859, despite the fact that he had the essential mechanism of natural selection to hand by about 1838, was that, as a respectable and prosperous middle-class Whig, he feared the use to which it would be put by radical agitators such as the atheists and Chartists who were clamouring for reform. As Desmond and Moore (1991) remark in their masterly study of Darwin's life:

> Anglican dons believed that God actively sustained the natural and social hier-
> archies from on high. Destroy this overruling Providence, deny this supernatural
> sanction of the status quo, introduce a levelling evolution, and civilisation would
> collapse. (p. 321)

When Darwin's *On the Origin of Species* finally came out in 1859, there had been a sea change in British life. Despite Darwin's own anxieties on the eve of publication, wealthy entrepreneurial Britain received his ideas gladly. After the publication of the *Origin*, there grew up a movement known as **social Darwinism**. In fact, much of the thinking contained in this movement can be found in the writings of Herbert Spencer before 1859, and the movement could more deservedly be called social Spencerism; the association with Darwin has, however, stuck. It was in fact an assortment of ideas rather than a fully co-ordinated political philosophy, but the basic premise was that evolutionary biology could teach a political lesson.

Since, as biology has shown, struggle, competition and survival of the fittest are natural phenomena that have operated to shape well-adapted and complex organisms such as ourselves, this clearly is how the social world should be organised. The natural world had operated to weed out the weak and feeble, there had been no support from any central authority, yet naked competition between individuals pursuing their own ends had indubitably led to progress. To the social Darwinists, the political message was clear: colonialism, imperialism, *laissez-faire* capitalism, disparities of wealth and social inequalities were all to be justified and encouraged. One of the leading social Darwinists in America was William Graham Sumner (1840–1910), a professor at Yale University. For Sumner, any redistribution of wealth from rich to poor favoured the survival of the unfittest and destroyed liberty:

> Let it be understood that we cannot go outside this alternative: liberty, inequality,
> survival of the fittest; not liberty, equality, survival of the unfittest. The former

carries society forward and favours all its best members; the latter carries society downwards and favours all its worst members. (quoted in Oldroyd, 1980, p. 215)

Darwin himself was not immune to the ever-present temptation to mix social and biological concepts when he observed in a letter that 'the more civilised so-called Caucasian races have beaten the Turkish hollow in the struggle for existence' (Darwin, 1881). But if the capitalists and their apologists drew succour from Darwin, so did the Communists. In a letter of 1861, Marx wrote that 'Darwin's book is very important and it suits me well that it supports the class struggle in history from the point of view of natural science' (quoted in Oldroyd, 1980, p. 233).

It is easy to see why social Darwinism appealed to industrialists, entrepreneurs and those who had gained or stood to gain from the operation of the free market. Its additional appeal for Marx was that it eliminated teleology and design from nature. Marx saw that evolution could be used to undermine his ideological enemy – organised religion. Ironically for contemporary Marxists, Darwinism has proved to be a double-edged sword. Marx's own views on human nature were ambiguous, but most Marxists have adopted the view that human nature is plastic in the sense that 'being determines consciousness'. Modern Darwinism shows that there is a universal human essence. It was the expression of this essence that brought about the downfall of the Soviet bloc – Soviet man was never quite plastic enough.

The political affiliations of another group that drew inspiration from Darwinian ideas, the **eugenics** movement, are harder to define. Eugenics is often treated as a subset of social Darwinism but is in fact dissimilar in both motivation and policies. The eugenics movement in Britain began with the work of Francis Galton (1822–1911). Galton, who was Darwin's cousin, adopted a strong hereditarian position and argued that there was a correlation between a person's social standing and his or her genetic constitution. As early as 1865, Galton had tried to sway public opinion to his view that upper classes should breed more and the lower classes less, but with little effect. However, the eugenics movement flourished in Edwardian Britain, where the social strains between rich and poor and the effects of international competition were beginning to tell (MacKenzie, 1976). The general worry was that if the lower classes were breeding faster than the upper, a general lowering of the genetic stock of Britain would result.

On the eve of World War I, the eugenics movement flourished on both sides of the Atlantic. The first International Congress of Eugenics, held in London in 1912, had Winston Churchill as the English vice-president, with Charles W. Eliot, the president of Harvard University, as the American vice-president. Eugenics societies included some distinguished geneticists as well as the likes of socialists such as Beatrice and Sidney Webb. In Britain, eugenics ideas appealed particularly to the professional middle classes. They preyed upon middle-class fears of a rising working-class population and a concern among the establishment with the poor medical condition of working-class recruits for the Boer War. It was particularly attractive to the professional middle classes and intellectuals since it suggested that experts and meritocrats like themselves should play a role in an efficient, state-organised society (MacKenzie, 1976). In America, Galton's recommendations on selective breeding were not taken seriously until

Lamarckism was discredited among biologists and social scientists (Degler, 1991). In a Lamarckian framework, if the environment worked upon individuals and the modifications thereby induced could be inherited, the main hope for social progress lay in improvements to social conditions. Once the inheritance of acquired characters is removed as a scientific possibility, selective breeding becomes a serious option for improving the race.

One of the most prominent eugenicists in America was Charles Davenport. Davenport held positions at the universities of Harvard and Chicago before becoming director of the Eugenics Record Office at Cold Spring Harbor. Davenport and his workers initially adopted the Mendelian assumption that each human trait was the work of one gene. They then traced the genealogical path of such traits as criminality, artistic skill and intellectual ability. Their warning to the nation about the effects of uncontrolled breeding is exemplified by their analysis of the Jukes family. Davenport examined the burden on society brought about by the offspring of one Margaret Jukes, a harlot and mother of criminals. He concluded that as a result of her protoplasm multiplying and spreading through the generations, the State treasury was worse off to the tune of $1.25m in the 75 years up to 1877 (Richards, 1987; Degler, 1991). As the eugenicists saw it, one way to stem the march of degenerate protoplasm was to restrict immigration into the United States of those racial types who were expected to belong to inferior stock.

In the United Kingdom, the movement was attractive to Fabian socialists who believed in state intervention to cure the inefficiencies of an unplanned economy. For this reason, it could be described better as socialist Darwinism rather than social Darwinism. The eugenicists made, by today's standards, some outrageous proposals. There were suggestions, for example, that the long-term unemployed should be discouraged from breeding since they obviously carried inferior genes. Major Leonard Darwin, fourth son of Charles Darwin, in his book *Eugenic Reform*, was strongly opposed to the advancement of scholarships to bright children from the lower classes. His reasoning was that once such children were promoted by their educational attainments to the class above, their fertility would decrease, whereas if they were left as they were, they would probably have more children, so their gifted genes would be more likely to propagate. In addition, argued Major Darwin, the existence of scholarships would worry the parents of children already in the higher social classes since they would now face more competition, and this would further reduce their already low fertility. Looking back, these ideas appear comic, but in other countries they led to extreme and tragic consequences. In the 1920s, 24 American states passed sterilisation laws and, by the mid-1930s, about 20 000 Americans had been sterilised against their will in an effort to stamp out inferior genes.

By the 1930s, natural scientists in Britain and America were realising that the early deliberations of the eugenicists were based on faulty assumptions about the nature of inheritance. Most traits were simply not the product of single genes as had been supposed. Features such as intelligence, moral rectitude and personality were, if they had any genetic basis, the consequence of the action of many genes in concert with environmental influences. Consequently, it was extremely difficult to predict the outcome of any given union of parents. Even enthusiasts for negative eugenics realised that there were formidable problems. If a genetic abnormality caused an abnormality in the homozygous condition, heterozygous **carriers** could go undetected. It was not at all clear to the eugenicists what could be done about carriers.

By the late 1930s, scepticism over the viability of eugenic principles among Western biologists and social scientists turned to revulsion as it became clear to what depths the Nazis had sunk in their application of eugenic ideas. It is known that, while in prison, Hitler imbibed the ideas of eugenics from *The Principles of Heredity* and *Race Hygiene* by Eugene Fisher. In the hands of Hitler, the eugenic ideal of improving the national stock had become twisted to a concern with racial purity – the enhancement of the Aryan race and the outlawing of mixed race marriages between Aryans and the supposedly inferior Jews, Eastern Europeans and blacks. When the Nazis came to power in 1933, they set about the systematic forced sterilisation of schizophrenics, epileptics and the congenitally feeble-minded. Deformed or retarded children were sent to killing facilities, an estimated 5000 dying in this way. Seventy thousand mentally ill adults were also targeted and put to death (Steen, 1996). The horrific culmination of this reasoning was the Holocaust and the extermination of about 6 million Jews, homosexuals and others deemed unfit.

While support for biological accounts of human nature ebbed away in the 1930s and 40s in Britain and America, Konrad Lorenz in Vienna was developing his theories of instinct and was laying the foundations of ethology. The reception of Lorenz's ideas in the English-speaking world was heavily influenced by an essay review of his work written by the biological psychologist Daniel Lehrman, and it seems likely that Lehrman's appraisal was influenced by the apparent sympathy shown by Lorenz for Nazi ideology (Lehrman, 1953). Others were similarly concerned. When Lorenz won his Nobel prize in 1973, for example, Simon Wiesenthal, the Head of the Jewish Documentation Centre in Vienna, wrote to him to suggest that he should refuse it (Durant, 1981).

The precise linkage between Lorenz's science and his early attraction to the Nazi cause is not straightforward. There is no doubt that Lorenz initially showed sympathy for the Nazi ideal. He joined the Nazi party after the annexation of Austria by Germany (the *Anschluss*) and wrote articles for *Der Biologie*, a journal with explicit Nazi connections. In some aspects of his thinking, he displayed typical Nazi fears, such as the belief that urban man had unwisely acted to suspend the cleansing force of natural selection and consequently faced biological deterioration (Kalikow, 1983). The assertion that Nazi ideology moulded his scientific approach is, however, debatable. He might have developed his theory of instincts and the assertion that human aggression has an innate basis with or without the rise of the Nazis, and he certainly continued to develop these ideas after the war. More generally, it would also be wrong to overstate the connection between scientific Darwinism (as opposed to a more nebulous commitment to evolutionist thinking) and Nazi ideology. There is little evidence, for example, that the Nazis approved of the idea that the Aryan race evolved from ape-like ancestors roaming the African plains. The Nazis drew upon other intellectual traditions such as Hegel's philosophy of the state, the concept of superman found in Nietzsche and a romantic and racist attachment to the idea of the German *Volk*.

Marxists also distorted evolutionary theory for political purposes. In revolutionary Russia, Darwinian theories of natural selection never really took hold. The revolutionaries regarded natural selection as tainted with capitalist notions of competition. It was in this ideological climate that a quack geneticist called Trofim Lysenko (1898–1976) managed to steal the show. In place of natural selection, Lysenko asserted the mechanism of Lamarckism. Mendelian genetics was denounced as 'bourgeois', and its adherents were forced to recant or were exiled to Siberia to reconsider their position. Lysenko claimed that his philosophy and methods would bring about improvements in the Russian grain harvest. The effects were, however, disastrous. Only in the 1950s, after Stalin's death, did the science of genetics recover in Russia.

12.1.1 *Race, IQ and intelligence*

In the United States in the 1930s, support for eugenics ideas faded. In psychology also, the theory of instincts, which always seemed to have a slender empirical base, was gradually abandoned. A supposed link between race and intelligence was, however, more difficult to sever. One reason was that, by the 1920s, there had accumulated plenty of experimental data to suggest that there were biological differences in the innate mental capacities of different racial groups. Looking back, we can only groan at the crudity of the tests used and the mentality that lay behind their use. However, the sheer momentum built up by the process of intelligence testing, linked to immigration control, helps to explain the seeming paradox that as **behaviourism** was abandoning the concept of instincts, the idea of a link between race, biology and ability remained and was not really expunged from psychology until much later.

The greatest challenge to the validity of intelligence testing between racial groups came from Otto Klineberg, a psychologist at Columbia University. Klineberg became professionally acquainted with Boas and dedicated his book *Race Differences*, published in 1935, to him. Klineberg systematically and methodically set about to test suggestions of Nordic superiority and black inferiority. His results showed that, once background and upbringing were controlled for, the data were far more ambiguous. Any differences that did remain Klineberg attempted to explain through environmental influences. Despite Klineberg's well-intentioned and well-researched efforts, the connection asserted to exist between heredity and intelligence was weakened but never totally severed. Despite a preference for cultural and behaviourist explanations in the social sciences, anthropology and psychology in the 1940s, 50s and 60s, the issue of race and intelligence arose periodically to trouble the academic community (Harwood, 1976, 1977).

12.1.2 *A poisoned chalice*

Against this background, it is hardly surprising that biological theories of human behaviour are treated with great suspicion. It is a natural tendency of ideologies to seek support and corroboration from other fields of thought, and Darwinism has always provided an axe for virtually anyone to grind. Looking back, it now seems almost inevitable that when evolutionary explanations of human behaviour began to emerge in the 1970s, based on new evidence and theories in evolutionary biology, some hostility would result. Thus, in Wilson's *Sociobiology: The New Synthesis*, published in 1975, 95 per cent of the subject matter was concerned with animal behaviour and only 5 per cent with humans, yet the outcry was vociferous and at times hysterical. Referring to the discipline of sociobiology, the 'Science as Ideology Group of the British Society for Responsibility in Science' wrote that:

> Sociobiology arrives at a time when there are wide-ranging challenges to the existing social order are being made... It is of course racist and sexist and classist, imperialist and authoritarian too. (BSSRS, 1976, p. 348)

Attacks like this were common throughout the late 1970s and the early 1980s. The general thrust was that, by identifying universal human traits that were genetic rather than cultural, evolutionary theory applied to humans was guilty of propagating an oppressive biological determinism. Rose *et al.* (1985) saw the universities as having a special role in this process:

> Thus, universities serve as propagators and legitimators of the ideology of biological determinism. If biological determinism is a weapon in the struggle between classes, then the universities are weapons factories, and their teaching and research facilities are the engineers, designers, and production workers. (p. 30)

Such ideological reactions were common in the United Kingdom and America from those left of centre. In France, however, Wilson's book met with a far more favourable reception from the liberal intelligentsia. The eminent anthropologist Claude Levi-Strauss, for example, certainly no sociobiologist, deplored the fact that American liberals were denouncing sociobiology 'as neo-Fascist doctrine'. Looking to France, he observed wryly in 1983 that sociobiology had been embraced by the political left out of a 'neo-rousseauiste inspiration in an effort to integrate man in nature' (quoted in Degler, 1991, p. 319).

In the 1990s, the debate over the application of evolutionary ideas to human nature thankfully shifted somewhat from the zone of class conflict, and ideas began to be considered more for their scientific merits than their position in an ideological battleground. Some have even found support for a left of centre radicalism in the new evolutionary ideas. In an article on the need for a more Darwinian approach in anthropology, Knight and Maisels (1994) conclude:

> Regardless of precise political affiliations, Dawkins and his atheist allies are iconoclasts and anticlerical radicals, behind whose godless banner the left in all logic should rally. (p. 20)

The awkward and uncomfortable history of the political uptake of biology, the cranky and outrageous ideas that have resulted, and the hysterical reactions to even moderate attempts to probe the biological basis of human nature, should not blind us to the fact that there are serious and important issues at stake. Nor will it do simply to assert that scientists have no moral responsibility outside their subject area. At one level, it may seem only an academic exercise in the history of ideas, but ideas can ultimately affect lives. Evolutionary thinkers have a duty of care in this area, and we must carefully examine the tangled relationship between evolutionary thinking and moral, social and political thought as objectively as possible. In the next few sections, we will look at just a few of the specific issues that arise from the implications of Darwinism.

12.2 Social Darwinism and eugenics

12.2.1 Social Darwinism

Nowadays, the term 'social Darwinist' is one of abuse. Denouncing someone as a social Darwinist is often thought to be a sufficiently crushing argument in itself. But why exactly is social Darwinism an untenable exercise? Spencer's phrase 'survival of the fittest' has become a catch phrase for those who advocate the virtues of free competition. There may indeed be virtues, but Darwinism, to the disappointment of any contemporary would-be social Darwinists, must remain silent on the issue. At one level, it is not at all clear that nature runs strictly along red in tooth and claw lines anyway: animal groups show plenty of signs of co-operation, and even vampire bats share a meal with their needy brethren. If we look at some taxa, such as the ants, competition between individuals seems entirely suspended in favour of caring and sharing for the common good. If we wish to model human society on the natural world, it is difficult to know which group of organisms we should consider: the message from, for example, ants, bats and dandelions will be entirely different.

It could be retorted that genetic self-interest lies at the heart of all these manifestations of altruism and that we must allow this as a scientific statement. Perhaps we should. We could also allow the fact that nature (as far as we can tell) is not regulated by some external conscious agency and that indeed the purpose-less process of natural and sexual selection has led to such complex organisms as ourselves. But does it follow that society should also be left to the unregulated outcome of the effects of individuals all pursuing their selfish ends? The answer is no. To believe otherwise is to make a huge and invalid leap of logic. The Scottish philosopher David Hume is credited with first exposing this fallacy. In his *Treatise of Human Nature* (1964 [1739]), Hume pointed out that 'is' does not imply 'ought'. The way in which humans want their social world to operate is a matter of values; biology is no more reliable a guide to what values we should hold than, for example, chemistry or astronomy. The suggestion, contra Hume, that one can infer values from descriptive facts is now known as the naturalistic fallacy.

The invocation of Hume's law – the impossibility of deriving the 'ought' from the 'is' – is often thought to be sufficient to deal the death blow to social Darwinist reasoning, but we should be careful here that we do not shoot ourselves in the feet. As some stage, the Darwinian will want to give a naturalistic account of value and morality and this, in the absence of any transcendental notions of goodness, will presumably have to be based on a factual account of the natural world.

Midgley (1978), who is certainly no social Darwinist and is even sceptical about the full potential of the Darwinian paradigm, makes the point that values must at some level be related to facts. It is the factual nature of the human condition that enables us to express what human wants are and what are good things for humans. We value a society that allows couples to have children, for example, because this is allowing freedom of expression to our biological nature. We need to think carefully, therefore, about the reasoning underlying social Darwinism and the reasoning used to dismiss it.

We must in fact be aware of at least two layers to morality. One is the phenomenon of moral behaviour, of which a good Darwinian may be able to give a plausible account, in other words why people erect rules and choose to live by them and how such rules relate (or otherwise) to fitness gains in any given environment. Another layer is the question of whether such codes and rules are right. A large number of people, from T. H. Huxley onwards, have argued passionately that ethics transcends nature and have despaired at any attempt to draw ethical premises from evolutionary thought. A modern exponent of this view is the Harvard biologist Stephen Jay Gould. Gould, who has done much to expose the sexist and racist bias in some attempts to capture human nature, is of the opinion that 'evolution in general (and the theory of natural selection in particular) cannot legitimately buttress any particular moral or social philosophy' (Gould, 1998).

Returning to the logic of social Darwinism, we can show that the reasoning is fallacious, but we need to do better than simply to evoke Hume's law. What the social Darwinist does is to confuse the consequences with the value of natural processes. If fierce unbridled competition got us to our present state, there is no obvious reason why it should still serve our ends. Suppose, for example, that one could demonstrate that periodic famines on a global scale or massive doses of gamma rays from solar flares were instrumental in the course of the evolution that led us to *Homo sapiens*. I doubt if even the more ardent social Darwinist would suggest that famine and ionising radiation are to be welcomed as means by which we can improve the human stock.

The social Darwinist is also guilty of smuggling teleology in through the back door. Another reason why social Darwinism should be called social Spencerism is that it was Herbert Spencer rather than Darwin who kept ideas of progression in his system of thought. The abyss into which Darwin stared was always too much for Spencer, who clung to a belief in steady evolution towards perfection. The essential point here is that it is the very purposelessness of the natural world that makes it a doubly unreliable guide. Natural selection does not make organisms better in any absolute way: it merely rewards reproductive success. There is no progress measured on an absolute scale but merely change. The whole thing is not going anywhere.

12.2.2 *Eugenics*

Eugenics represents the other side of the coin from social Darwinism. Rather than let nature take its course, the eugenicist wanted to intervene to put it right. Eugenicists were concerned that the processes operating in urban societies were such that people producing the most offspring resided in lower socioeconomic groups and were therefore genetically inferior to those in the higher strata of society. Needless to say, those promulgating the idea regarded themselves as being genetically superior. The remedy for the eugenicist lay not in competition and *laissez-faire* – since civilisation for ethical reasons had already accepted the burden of helping the weak – but in active measures to encourage the spread of good genes (positive eugenics) and discourage the spread of weak ones (negative eugenics). The whole eugenics programme was so fraught with scientific, ethical and practical difficulties that no one today would seriously advocate the sort of measures proposed earlier this century. In fact, any hint of sympathy for eugenics ideas in the United Kingdom is regarded as a blight on the career of a politician. It is arguable that Margaret Thatcher became leader of the Conservative party in 1974 because of a speech made by her mentor, and the then favourite leadership candidate, Keith Joseph, in which he suggested that 'the balance of our human stock' was threatened by the rising proportion of children born into the lower social classes (Thatcher, 1995).

The fertility of different social groups is of course an empirical matter and could be settled by statistical means. It is also possible of course that certain genes are more frequently found in certain social groups than others. Problems start hereafter for who is to define what genes are desirable and worth increasing in frequency? At this point, we need to import an ethical system to help our judgements and, fortunately for our sanity, there is no general consensus on what qualities make up desirable human beings, and what standards there are shift with time and vary across cultures. Nor, I think, would most people wish to see any committee pronounce on this subject. As if this were not enough, there is then the practical problem of how the state could alter gene frequencies without an unacceptable infringement of other human values, or whether the state even has a responsibility to its gene pool that overrides its responsibility to the welfare of individuals and the preservation of individual freedoms.

In a profound sense, however, the eugenics programme is unrelated to the evolutionist's paradigm. From an evolutionary perspective, successful genes are simply successful genes. An ardent evolutionist desperate for meaning might be tempted to encourage the proliferation of fecund genes rather than any other qualities. Even this mad speculation is, however, cut short by the constant reminder (which we need in this territory) that evolution involves no sense of progress. Successful genes are no better on any absolute scale of values and certainly not on any human scale – we do not admire aphids because of their fecundity.

We must be alert to eugenics issues since gene technologies are increasingly delivering into our hands powerful tools to screen individuals for genetic defects. Prenatal screening enables doctors to assess whether the fetus is genetically defective for a wide range of conditions. If, for example, parents choose to terminate a pregnancy because of the condition of cystic fibrosis, this involves the judgement that a child with cystic fibrosis is too great a burden to justify its birth. Some have argued that this is a form of eugenics through the back door. The comparison with eugenics thinking is not, however, strictly accurate in these cases. Certainly, the motivation of the parents is not to eliminate the genes from the human population: their concerns are about the suffering of the child and the burden to the family. In fact, by such procedures it is extremely difficult to alter gene frequencies.

A child with cystic fibrosis is born when the relevant genes are homozygous in the recessive state; that is, it has two copies of the defective gene, one from each parent. If it has only one copy, it is said to be a carrier. People who are carriers live perfectly normal and healthy lives, never realising that they are carriers until they mate with another carrier. About 1 in 25 Caucasians are thought to be carriers of the recessive allele for cystic fibrosis. The chance of two carriers meeting is thus about $(\frac{1}{25})^2 = 1$ in 625, or 0.0016. The chance of a child from a union of these parents having both recessive alleles and hence displaying the condition is one quarter of $0.0016 = 0.0004$, or 1 in 2500. Hence about 1 in 2500 Caucasian children are born with cystic fibrosis. Simply removing those affected by cystic fibrosis, that is, those who are homozygous, will not remove the allele itself. In fact, the heterozygous condition only needs to carry a 2.3 per cent advantage compared with non-carriers for the recessive allele to persist indefinitely (see Strachan and Read, 1996). The only way for a true eugenics programme to work in this context would be to screen for carriers and discourage carriers from breeding with anyone. Such a programme would of course be impractical and ethically unacceptable.

12.3 Evolutionary biology and sexism

When sociobiology emerged in the 1970s, it was quickly denounced as sexist. 'Sexism' is in fact quite a complex term that needs to be unpacked carefully to examine this accusation. Evolutionary thinking could be sexist in the sense that concepts from socially constructed gender roles are transported into the natural world and have a distorting effect. To speak of the 'queen bee', for example, is really a metaphor that, if taken too literally, could give a very misleading effect of what actually happens in the hive, where, if anything, the queen seems to be controlled by her 'workers'. There are particular problems here in describing the sexual behaviour of animals, it being only too easy to transpose concepts from the social world to the biological and then back again. What may seem to be a dominant and resourceful male reigning over his harem could be a group of females with their own social bonds clubbing together to choose the best-looking male.

Another example concerns the way in which we give an account of how some species of ants make 'slaves' of others. In human slavery, members of the same species are violently coerced into labouring for others, yet the application of the word 'slave' to ants may be misleading. When ants make 'slaves', they capture immature members of a different species. The captured individuals then mature in the nest of their captors and perform housekeeping tasks, apparently without coercion. Perhaps a better metaphor here would be that of domestication.

The reception of scientific ideas can also be influenced by views on the social roles of males and females. We noted, for example, in Chapter 2 that in the patriarchal climate of Victorian Britain, where women were denied effective political power, Darwin's view that the female could, through her power of choice, exert an effect on the male was received with much scepticism. One might postulate then that ideas from evolutionary thinking are accepted by the scientific community if they conform to contemporary social expectations.

Both these points have epistemological and political dimensions and need to be considered carefully. There seems to be no doubt that, in constructing a knowledge of the world, scientists employ metaphors that betray a social origin, and that such metaphors may therefore condition a particular image of reality. Knowledge is rarely, if ever, value-neutral. It is produced by people with specific social, personal or professional interests. Even in the most abstruse field, someone decides that something is worth knowing about, and that invokes a value commitment. The important question here has to be whether or not our view of external reality is so distorted by this process that our image of the world is entirely a social construction. The answer, we think, has to be no. Unless you are a thorough-going relativist with respect to scientific knowledge (in which case you probably would not have read this far anyway), it has to be acknowledged that the world is not plastic enough to sustain any interpretation. Moreover, the checks and balances built into the methodology of modern science ensure that false images will eventually be exposed.

In a detailed paper, the Californian primatologist Craig Stanford has suggested that interpretations of the behaviours and social systems of chimpanzees (*Pan troglodytes*) and bonobos (*Pan paniscus*) have been heavily influenced by contemporary gender roles and images of men and women. Put briefly, chimps have been interpreted in masculine terms and bonobos in feminine ones. Stanford may be right, and the paper exists in the literature for all to judge (Stanford, 1998). The fact, however, that we have the intellectual tools to realise potential bias shows that we can transcend our prejudices. The very fact that we realise that analogies such as queen or harem or slave-making are simply that – imprecise metaphors – shows that we are not imprisoned by them. More generally, the very phrase 'natural selection' is a metaphorical extension of the way in which humans select, but no respectable biologist really believes that some conscious agency is at work doing the selecting.

One has to concede of course that particular lines of enquiry may be socially conditioned. At a trivial level, the funding arrangements of science will always ensure that social priorities enter into the direction of scientific research, for example. There are deeper ways too. The fact that, in his scientific speculations, Aristotle advocated the view that some men are fit for slavery, or that in sexual reproduction the male supplies the important organising form and the female supplies only the matter, no doubt reflects the sexist and slave-owning culture in which Aristotle lived. In recent years, there has been a great deal of work on female choice in the process of mate selection. One could speculate that this reflects the increase in social power of women in Western societies. This may be the case, and sociologists of science could be gainfully employed in establishing this. The crucial point, however, is that the results of the enquiry process are scrutinised by standard scientific procedures. In short, the outcome of the research is not logically predetermined by the motivating factors. To suppose that it is entails committing what Popper called the 'genetic fallacy' (nothing this time to do with genes): the belief that the origin of ideas impacts on their truth-value.

Another line of attack on evolutionary accounts of human nature is that it is sexist because it points to innate differences between the sexes. The concern here stems from the belief that to suggest differences in gender-specific behavioural dispositions means that (a) these dispositions are fixed and therefore incapable of moderation, and (b) the genetic basis of behaviour can be used to legitimate social roles. The first point to note here is that there are obvious physical differences between men and women that have a strong genetic basis. Men cannot bear babies or lactate. In relation to height and musculature, there are of course environmental influences, and girls who are well fed and nourished may grow to be larger than boys who are malnourished, but on average, when raised in similar conditions, men are slightly taller and more muscular than women. Girls, on the other hand, mature physically and emotionally faster than boys. These are not sexist statements in the sense of denigrating one sex or the other, nor are they sexist in that they entail distortions, deliberate or otherwise, of the world; they are descriptive statements about human development. If they are sexist, so too must large portions of the sciences of anatomy and physiology be. The facts could be used for sexist purposes, but that needs to be tackled on a different level and in no way challenges the data themselves.

I suspect that more concern is expressed over the supposed sexist implication of evolutionary accounts of behaviour than over physique because behaviour is what defines us as human. In terms of our bodies (apart from our extra-large brains), we are very similar to the great apes, but in behavioural terms we have a sophisticated culture that apes lack. The fear that a biology of mind destroys our humanity runs deep. For many, a belief in the autonomy of the mind and its susceptibility to beneficial moulding by culture represents the last raft of refuge from scientific attacks on the uniqueness of the human species. Copernicus and Darwin (and some would say Freud) effectively sank any claims that humans are the chosen species occupying a special place in creation. Those who have such worries should take heart: in a meaningful sense, evolutionary theory confirms that we are unique. It just adds the timely reminder that all species are unique.

It has been argued throughout this book that there are fundamental differences in the behavioural characteristics of human males and females. This should not concern us; instead, we should celebrate and take delight in the fact. Aristotle Onassis spoke up for about half of the human species when he said that 'If women didn't exist, all the money in the world would have no meaning.' It would be surprising in the extreme if genes conditioned our physique and the structure of our brains but stopped short at wiring us up for behaviour and handed it entirely over to culture. We are, however, not hard-wired: our genes long ago handed to us considerable autonomy. The sex drive, for example, is fundamental and strongly ingrained, but we can choose to override it or sublimate it. Chastity is a viable option for humans. Evolutionary reasoning may tell us that it could be a difficult option or that it is unlikely to have any strong genetic basis (we exclude here any accounts of the genetic basis of homosexuality), but evolutionary thinking ultimately makes no value judgement about chastity.

Evolutionary biology has of course been used to provide corroborating evidence in the legitimisation of the social roles of men and women. The arguments usually amount to the idea that certain contemporary roles are more suited to one sex or the other because of the ancestral division of labour that became encoded in our genes. It might be thought that this approach could provide valuable information in, for example, job selection, but in fact beyond surrogate motherhood or wet nursing, for which we could reasonably rule out men, any information we have on the evolutionary basis for sex differences is useless in this respect. In height and physique, for example, men and women are not strongly dimorphic. To use sex as a guide to these qualities would be useless given the overlap of the spread of values in male and female populations. Even strength is a quality increasingly less useful in a society in which muscle power is increasingly displaced by mental agility. The fact remains that virtually all modern social roles can be performed by both men and women, so sex alone is not a reliable criterion for assessing suitability for a particular role.

Other such extrapolations often fall prey to the naturalistic fallacy. To say, for example, that women or men *should* perform some tasks because they *did so* in the hunter-gatherer stage of evolution is to leap from facts to values. Western society has fortunately seen the sense of all this, and discrimination based on sex is largely outlawed.

Sexism is also most transparent when the science is bad. Consider the following section of a lecture by Darwin's follower George John Romanes (1887, p. 135).

> Woman is still regarded by public opinion all the world over as a psychological plant of tender growth which needs to be protected from the ruder blasts of social life in the conservatories of civilisation. ...Without ...recurring to the anatomical and physiological considerations which bar a priori any argument for the natural equality of the sexes... it is enough to repeat that women by tens of thousands, have enjoyed better educational as well as better social advantages than a Burns, a Keats, or a Faraday; and yet we have neither heard their voices not seen their work... we may predict with confidence that, even under the most favourable conditions as to culture and even supposing the mind of man to remain stationary... it must take many centuries for heredity to produce the missing five ounces of female brain.

Romanes here confuses anatomical and physiological differences with the notion of equality. He displays (not surprisingly for the time) a poor grasp of the social factors that mediate achievement in literature and science. Remarks such as those of Romanes are now virtually non-existent in academic circles; where they do arise, they are easily shot down. In fact, if we use the concept of encephalisation quotient (EQ), noted in Chapter 6 as a measure of intelligence, then although women have smaller brains than men, women's brains are larger than men's in terms of the size that would be predicted from the allometric line of brain size against body mass (Figure 6.10). Kappelman (1996), for example, gives an EQ for women of 4.64 and for men of 4.32. If we use EQ measures as an indication of intelligence, a modern-day Romanes has some explaining to do.

If anything, evolutionary accounts of human sexuality provide a strong antidote to sexism. There is no room in biology for the suggestion that one sex is in some way superior; the concept simply has no meaning. In sexual reproduction, each sex inherits half its genome from its mother and half from its father.

Whatever we think of the genes that meiotic shuffling has given us, we must give blame and thanks equally.

12.4 Evolutionary biology and racism

Racists have often turned to biology for support. Even before the evolutionary thinking of the 19th century, racists, particularly in the United States, used a mixture of biology and religion to justify their exploitation of African natives. It was both bad theology and bad biology. Following the advent of Darwinian thinking, the exponents of racism had to shift their ground but, unsurprisingly, came to similar conclusions as before: that some races were higher or more developed than others. This view crept into medicine. Down's syndrome, a problem caused by an error in the chromosomal inheritance of a child, was called 'mongolism' by its Victorian discoverer, John Langdon-Down. To him, it seemed an appropriate term; sufferers from this condition had slipped a few places in the evolutionary hierarchy to resemble a race lower than the Europeans – the Mongols.

In the 20th century, much emphasis has been placed on the vexed question of the heritability of IQ within and between racial groups. One may of course question the value of IQ and even question the ethical desirability of research in this area given the political context. Nevertheless, it has been done and probably will be done again in the future. The debate on this subject recently resurfaced with the publication of *The Bell Curve: Intelligence and Class Structure in American Life* by Herrnstein and Murray (1994). The authors were accused of scientific racism despite the fact that, in their overall view, the evidence that black/white differences in IQ had a genetic basis was ambiguous. These authors did, however, claim that lower social castes, which include a disproportionately large number of African-Americans, had a lower IQ and that IQ was to a large degree (40–80 per cent) inherited. The authors went on to argue that, since the lower socioeconomic groups tend to be more fertile, the 'cognitive capital' of the United States is facing decline.

This all of course sounds familiar. The scientific claims of Herrnstein and Murray with regard to the connection between IQ, social class and the heritability of IQ could, with some difficulty, be examined empirically if it is thought worth the effort. The most interesting feature of their work, however, is that the authors, like the eugenicists before them, draw political prescriptions from scientific findings. They suggest, for example, that more money should be put into programmes for gifted children and that the state should seriously question its commitment to equal opportunities. Similar arguments have been used in the past: if intelligence is largely inherited, why waste money trying to improve the skills of those with a lower IQ? Here, of course the logic, as is so often the case when we move from facts to policies, comes unstuck. It simply does not follow that the state should withdraw resources from those who are regarded as having a poorer genetic constitution. No one would reasonably argue, for example, that spending on health should be diverted away from those with genetic disabilities to those who are perfectly healthy.

The consensus view among Darwinians is that interracial biological differences are trivial. It is not, however, the only view among scientists. In the late 1980s,

J. Phillipe Rushton, a professor at the University of Western Ontario, achieved notoriety by publishing a series of articles purporting to show basic biological differences in IQ and personality between Africans, Caucasians and Orientals. Rushton's thinking derived from what are known as 'r/K' models in ecology. Put crudely, organisms can either be 'r' strategists or 'K' strategists. r strategists such as aphids go in for explosive breeding and produce masses of cheap, short-lived offspring. Most die, but enough survive to reproduce for the next round. Humans are a typical example of K strategists: we produce a few biologically expensive, long-lived offspring. Rushton claims that there are differentials among human K strategists such that a hierarchy exists. By looking at data on brain size, genital measurements, age of death and so on, Rushton suggests that Orientals are more 'K-like' than Caucasians, who are in turn more 'K-like' than Africans. Translated into behavioural terms, Rushton was implying that blacks invested more heavily in traits related to sexuality and procreation than whites, and consequently blacks were less intelligent, less sexuality restrained and less law-abiding.

The outcry was predictable and understandable. Rushton was vilified in the media, threatened with dismissal from his tenure and even investigated by a special force of the Ontario and Toronto police service on possible charges of spreading hate literature. Here is not the place to assess Rushton's work, but it serves as an illustration of the sensitivity of the issues and the need for reliable science and a scrutiny of the ethical value of research. The desirability of attempting to study interracial differences in a culture where racial harmony is an objective but not yet a reality is, to say the least, debatable. Numerous academics have questioned Rushton's methodology and assumptions as well as the reliability of his data. Fredric Weizmann in particular has been instrumental in exposing what many believe to be sloppy science (Weizmann *et al.*, 1990).

In the main, modern evolutionary thought and the science of genetics are destructive of racist ideas. It turns out that the concept of race is not a particularly useful one for the biologist. It was realised long ago that all races belong to the same species, *Homo sapiens*. (Given the fact that racism is a problem in our culture, one can only shudder at what the world would be like had another *Homo* species survived into the present epoch.) If we start with, for example, skin colour as a criterion for dividing people into groups, it transpires that only about 10 genes out of a total of at least 50 000 on the human genome are responsible for skin colour. We might then look for correlations between skin colour genes and others. When we do, patterns in the distribution of one set of genes are not matched by distributions in others. The human races are remarkably heterogeneous, possibly because of our relatively recent origin. Most of the genetic diversity between individuals occurs because they are individuals rather than because they are members of the same race. Put another way, most of the world's genetic diversity is found in any one race you choose. On the whole, the evolutionary approach to human behaviour is concerned with human universals – cross-cultural features that unite the different groups of the world and reveal our common evolutionary ancestry. The mental modules or Darwinian algorithms to which evolutionary psychologists refer were laid down before races differentiated.

It follows that the concerns of the eugenicists over the heritability of various traits is not of particular concern to the evolutionary theorist. The concept of heritability describes the percentage of variation between individuals that results from inheritance. Assuming for the moment that IQ has some validity, the heritability of this feature is a measure of the extent to which differences between individuals are attributable to genetics or environment. If we say that IQ has a heritability of 50 per cent, this means that half of the variation in IQ between, for example, two people is caused by genetic influences and half by environment. A heritability of 100 per cent would imply that all the difference between individuals is caused by genes, and 0 per cent would imply that any difference is entirely due to upbringing.

Now, in studying human nature from a Darwinian angle, we are dealing with low heritabilities. The premise is that all humans have mental hardware that predisposes them to behave in ways that are adaptively similar. This mental hardware is laid down by the genes, but the variance is small. As an analogy, consider the number of lungs (two) possessed by most people. The heritability of this is near zero: nearly all people are born with two lungs. If we examine people who have only one lung, it will usually be found to be a product of the environment – usually the surgeon's knife. The possession of two lungs is an inherited trait (very adaptive) but with low heritability. A feature such as eye colour will have nearly 100 per cent heritability, differences between people being almost entirely the result of genetic influences: the environment does not shape eye colour. This raises another point: features with low heritability tend to be more interesting. Heritability itself is not a good guide to establish whether something is under genetic control. We need say no more about the IQ heritability debate; it is not a part of the evolutionary paradigm applied to humans.

But racism exists and is in need of an explanation as well as a cure. We must consider the slightly frightening prospect that racism has some adaptive function. To offer an explanation is of course not to condone the behaviour. If we explain racism sociologically, for example, which is commonly done, this neither supports racism nor excuses it. Nor, crucially, does it undermine the sociological approach. If there is a biological basis to racism, it is something that we must face squarely.

Numerous attempts have been made to investigate whether the roots of racism lie in biology. One promising line of research stems from the inclusive fitness theory of Hamilton: inclusive fitness is increased if individuals are nice to those who bear copies of the same genes. Some have seen this as a basis for ethnicity: by distinguishing between those most likely to share copies of your genes and those less likely to do so, it is possible to distribute co-operative behaviour more effectively. Reviewing the evidence, however, Silverman (1987) concluded that most intergroup conflicts that have taken place are within an ethnic grouping rather than between such groups, and that such conflicts are explicable in terms of resource competition. Silverman also concluded that if racism were a fundamental part of the human psyche, any small gains to inclusive fitness that were gained by racial discriminations would be outweighed by a loss of the ability to form co-operative coalitions between groups as conditions changed. Thus, racism at this simple level would not be adaptive.

What may have been adaptive of course is the *post hoc* rationalisation that accompanies group conflict. As we have argued in Chapter 11, morality may be an adaptive device to ensure that we co-operate in conditions where it is favoured by fitness gains. If, however, circumstances favour the exploitation of a former ally, or favour cheating on obligations, racism could be a device to guard us from the full illogicality of our moral position. In this sense, racism may have acquired a function as a consequence rather than a cause of intergroup strife. Silverman's arguments are plausible and in a sense give reason for hope. Our self-deceptions are often fragile and open to elimination by education.

An exposé of the political abuse of evolutionary ideas may suggest that there is some sort of natural and therefore suspicious affinity between Darwinism and unpalatable political philosophies. Here, we should be wary. Scientific ideas can be taken up into pre-existing debates without inspiring them, and indeed with little logical connection to them. Sexism, racism, militarism and imperialism, for example, all existed before the Darwinian revolution and will probably persist for a long time thereafter. Looking again at the eugenics movement in America, it can be seen that it drew particular inspiration from the new science of Mendelian genetics, yet the early Mendelians rejected the notion that natural selection had been important in evolution. For the eugenicists, a belief in the necessity of artificial selection was not based on any acceptance of the power of natural selection. When it comes to notions of racial superiority, it was, as Bowler notes, Lamarckianism that was more easily incorporated into attempts to construct a racial hierarchy with Europeans at the top (Bowler, 1982).

In summary, this section should have demonstrated that Darwinism provides no ammunition for the eugenicist and little comfort for the racist. It does, however, provide an essential ingredient in our search for self-knowledge, and this theme is explored in the next section.

12.5 The limits of nature

12.5.1 Reductionism and determinism

In some circles, particularly where political correct thought control is in operation, the word '**reductionism**' has a strong pejorative tone. It carries the implication that one is doing damage to a complex topic by reducing it to its component parts. So is evolutionary reasoning reductionist, and is this a bad thing?

In a sense, all of the natural sciences are reductionist in that they subscribe, albeit in most cases indirectly, to ontological statements. In modern science, it is the accepted view that the building blocks of the universe are matter (in the form of particles) and energy, or quarks or strings or whatever the latest theory suggests. No serious biologists believe that, when they study life, they are studying some immaterial force, just as psychologists do not think that the human psyche consists of some non-physical entity that inhabits our brains. In this way, biology and evolutionary theory are reductionist. It happens, however, that thinking at the level of fundamental particles is not particularly useful for the evolutionist.

Evolutionary thought is no more or no less reductionist than, for example, geology or meteorology or economics, but its concepts and levels of explanation lie on a different plane from that of particle physics. It must be said that all of science is reductionist in that it reduces the bewildering variety of phenomena to the operation of fewer essential laws, principles or theories. This is the case for the evolutionary theorist as it is for the social scientist. Any subject which attempts to explain a set of facts by a larger set of principles that only have meaning and application in relation to those facts has missed the point of what to explain means.

The real danger and threat of reductionism arise from what might be called methodological imperialism – the view that only one type of explanation counts. Consider eating an apple. Physiology will tell you how it is digested, and environmental chemistry will tell you what happens to the carbon dioxide you breathe out as a result. Evolutionary theory will tell you why you find it sweet: you cannot manufacture your own vitamin C so you must obtain it from foodstuffs; consequently, there was a selective advantage long ago to finding the taste of fruits sweet. Try feeding an apple to domestic cats, animals that can manufacture their own vitamin C, to observe the contrast. Evolutionary theory also tells us why plants go to the trouble of packaging their seeds in a nutritious and expensive coat, but it does not have much to say on how the apple got into your hand, how it came to be in the shop or why the farmer bothered to grow it. Here, it must gracefully give ground to economics and perhaps human geography. Human knowledge is essentially pluralistic. If physicists ever reach the holy grail of a super-theory of everything, there will still be employment for literary critics, historians, economists, archaeologists and so on. Each domain of experience must call upon levels of explanation and notions of causality appropriate to that domain. Quantum gravity is of no use in explaining the fall of a man who jumps from a tower block.

Determinism is another concept that has had a bad press. The fear of determinism is related again to an anthropocentric view that humans tower over the rest of the animal world, aloof in their self-consciousness and confident in their power of free will and rational thought. Again, however, all science is based on the notion that events are caused. The whole of scientific progress is the story of how mysterious phenomena are brought within the fold of causal explanation. Social scientists, feminists and Marxists all assume that events and behaviours are determined: they do not just happen. The real fear is of course that **biological determinism** (now we're talking dirty) is somehow limiting, an affront to human dignity, but humans have always been deterministically limited. Science, biology in particular, clarifies and even helps to overcome those limits. Anatomy and physics tell us that we cannot fly unaided; biology tells us we cannot breathe underwater or on the moon. Men cannot have babies, nor can women father them.

Yet to explain, to reduce and to identify limits need not be limiting. With assistance, we can fly at twice the speed of sound 5 miles up, land on the moon and breath under water. And to those who point to the fate of Icarus, we should note that he knew too little science rather than too much.

12.5.2 *The perfectibility of man*

There is an age-old philosophical debate that goes back to the time of the Greeks concerning the origin of human vices and virtues. In the modern period, the debate was sharply defined by Hobbes and Rousseau. Hobbes, writing in England in the 1650s after the chaos of a civil war, argued that, in the natural state, the life of man was 'nasty, brutish and short'. Left to his own devices, man would live in a squalid state of perpetual struggle and conflict. The solution for Hobbes was for the state to impose order from above to curb the excesses of human nature. At the other end of the debating spectrum lies Jean-Jacques Rousseau. In his *Discourse on Inequality*, published in 1755, Rousseau argued that humans are by nature basically virtuous but are everywhere corrupted by civilisation. Rousseau gave Europeans the image of the noble savage living in a state of bliss before the arrival of civilisation. Rousseau's arguments were in part polemical and designed to expose the decadence in French culture, but his picture of the noble savage stuck and was profoundly influential. Ever since the time of Rousseau, weary Europeans have sought examples of the blissful and guiltless lives that Rousseau described.

The reality has, however, never really matched up to the expectations, but on one occasion it looked as though Rousseau's vision had been found. In 1925, Margaret Mead went to the Polynesian island of Samoa to study the life of the islanders. Mead spent just 5 months among the islanders before returning to New York. Her subsequent accounts in *Coming of Age in Samoa*, published in 1928, were seminal works. Mead claimed to have discovered a culture living in a state of grace, free from sexual jealousy or adolescent angst. Violence was extremely rare, and young people enjoyed a guilt-free, promiscuous lifestyle. Mead became a major celebrity; her books were best-sellers and became required reading for generations of undergraduates. She even had a crater on the planet Venus named after her.

Unfortunately, Mead was duped. At the onset of her career, she was strongly influenced by the anthropologist Franz Boas, who, appalled at the eugenic thinking that he encountered in his native Germany, propounded a culturalist view of human nature. Mead imbibed this, and her work was a product of her own expectations coupled with faulty data collection. Her errors were exposed by Derek Freeman, who, like Mead, spent time (five years) among the Samoans but who came to an entirely different conclusion. Mead had constructed her account of the care-free love lives of the Samoans from the reports of just two adolescent girls, Fa'apua'a and Fofoa. When Freeman interviewed the girls, by then old ladies, he heard how, in a state of embarrassment about Mead's questioning of their sex lives, they had made up fantastical stories of free love. So it was that a whole view of human nature in social anthropology was based on a prank by two young women (Freeman, 1996).

12.6 So human an animal

12.6.1 *Fine intentions*

It is easy to see why the left and the liberal intelligentsia should be so attracted to environmentalist conceptions of human nature. For a start, right of centre ideologies have often looked to a static human nature to support their claims. At a deeper level, however, there lies the often-unquestioned assumption that if human vices are the product of social circumstances, then by changing the circumstances, we can change human nature – for the better of course – and the perfectibility of man is at hand. Similarly, feminists have often argued that the unequal distribution of power between the sexes, the differences in historical cultural achievements between men and women, gender stereotypes and the 'glass ceiling' are products not of biological differences between the sexes but of socialisation in a patriarchal society. Change the society and we can change the roles.

Such thinking seems to have lain at the heart of the environmentalist programme of Boas. As a Jew, Boas found the anti-Semitism in Germany in the 1870s and 80s both discouraging and alarming. He foresaw a career path strewn with obstacles and disappointments merely as a result of his own racial identity. In contrast, Boas saw an America (before the proliferation of eugenic ideas and restrictive immigration policies) beckoning with an outlook that stressed equality of opportunity and intellectual freedom.

Boas almost single-handedly swung American anthropology away from explanations based on inherited mental traits towards cultural **relativism**. The transformation in anthropology was mirrored in the lives of individual social scientists. Carl Kelsey, who was a sociologist at the University of Pennsylvania, is a particularly interesting example. In his early career, Kelsey embraced Lamarckism and regarded the race problem in America as a product of inherent differences between blacks and whites brought about the exposure of thousands of generations to radically different environments. The downfall of Lamarckism that led some scientists to turn to eugenics as a method of effecting national improvement led others such as Kelsey to move in the opposite direction. If, as Boas had shown, nurture was instrumental in shaping character, Kelsey reasoned that social progress could be achieved by improving environmental conditions. Such a procedure had the additional merit of being faster than either selective breeding or waiting for the inheritance of acquired characteristics. Within this framework, Darwinism was an irrelevance. Some psychologists were quite open in their commitment to a science that was in keeping with liberal values. One such was Thomas Garth of the University of Texas, who in 1921 laid down a rule for students who were set upon examining racial differences:

> In no case may we interpret an action as the outcome of the exercise of an inferior psychical faculty if it can be interpreted as the outcome of the exercise of one which stands higher on the psychological scale, but is hindered by lack of training. (quoted in Degler, 1991, p. 190)

The rule is of course an amusing and ironic allusion to the canon laid down by Morgan 26 years earlier (see Chapter 1).

There is at present much controversy surrounding the proper application of Darwinism to human psychology. The journalist Andrew Brown draws a picture of two main warring camps, which he calls the Gouldians and the Dawkinsians, the former being sceptical and the latter optimistic about the whole project. The Gouldians include Stephen Jay Gould, Steven Rose and Richard Lewontin. Among the Dawkinsians, we find Richard Dawkins, John Maynard Smith, Daniel Dennett and Helena Cronin. If one wishes to examine the extra-scientific commitments that lie behind the rival philosophies of these groups, one may discern among the latter a distrust and rejection of the pretensions of organised religion. This is coupled with a belief that a science of human nature can guide social action and policy. As Brown notes, however, it is the background of the

Gouldians that is especially interesting (Brown, 1999). Of the three names noted, all are Jewish and all vaguely Marxist. Sociologically, this may be significant: Jewishness would give a strong motivation for being suspicious of any attempt to draw up a biological profile of human nature, and Marxism too, at least in the 20th century, became largely committed to a philosophy of biological egalitarianism and the belief that social existence determines human nature.

12.6.2 *Retrieving our humanity*

The history of ideas tells us that Darwinism is not the property of any single political **ideology**. It is a scientific view of nature that can be used to inform political discussions but one that does not translate easily into simple political remedies. It is simply misguided to imagine that the scientific enterprise of examining the evolutionary roots of human behaviour is somehow impugned by the errors of the past. In the coming years, skill will be needed to sift the legitimate from the spurious applications of Darwinism. We already factor a knowledge of human nature into our social systems in a myriad of ways. Consider the undeniable and biological propensity for humans to fall asleep. This is not something we learn – we are born with this tendency – but modern society relies upon the ability and willingness of some individuals to work through the night. A knowledge of biology tells us that there is a price to pay in terms of performance and fatigue, and elementary psychology tells us that we may need inducements to persuade people to work through 'unnatural' hours. But it can be done. Biology is not destiny, but it can provide a useful contour map.

This is the approach taken by the Australian philosopher Peter Singer, who argues that 'it is time to develop a Darwinian left' (Singer, 1998, p. 15). For Singer, Darwinism informs us of the price that we may have to pay to achieve desirable social goals. Uninformed state attempts to make socialist man have failed because they ignored human nature. For Singer, some aspects of human nature show little or no variation across culture and must consequently be taken account of in any social engineering. Singer's list includes concern for kin, an ability to enter into reciprocal relationships with non-kin, hierarchy and rank, and some traditional gender differences. To ignore these is, according to Singer, to risk disaster. The abolition of hierarchy in the name of equality, as attempted in the French and Russian revolutions for example, has all too often simply led to a new hierarchy. This, for Singer, is not an argument in favour of the *status quo*. The political reformer, like a good craftsman, should have a knowledge of the material with which he or she works. The trick is to work with the grain rather than against it.

A set of deeper problems arises with the view that the promise of a humane society lies only within an acceptance of the opinion that our humanity is culturally determined and defined. Supposing we could structure a society to shape people in the way in which we desire. Who draws up the blueprint for *Homo perfectus*? From where do we draw our notions on what constitutes ideal man? Reason by itself is not enough. Reason needs motives to act; it needs the will, beliefs, goals, ideals, something to serve, in other words human nature. Reason cannot lift itself up by its own bootlaces. Perfectly rational man would be a monster.

The blank slate approach to human nature that is still unquestioned in some branches of the social sciences would, if it were taken seriously, be a tyrant's licence to manipulate. Liberals would have to stand back powerless and impotent as a tyrannical state moulded its people into instruments of whatever crazy ideology was in fashion. There would be no basis for any objection since this would have been jettisoned when biology was thrown out: if human nature is anything that a society structures it to be, there is nothing to be abused.

Associated with the view that human nature is culturally determined is the philosophy of cultural relativism. If there is no fixed nature, there can be no single way of life conducive to its expression or fulfilment. Consequently, there is no judgemental moral high ground. Cultures in which the limbs of criminal offenders are severed, mixed-race marriages are forbidden and females undergo genital mutilation must be contemplated in silence. As the French philosopher Finkielkraut said, 'God is dead but the Volksgeist is strong' (Finkielkraut, 1988, p. 104).

We should remember that the Enlightenment project of progress through reason, science and the intellectual challenging of authority delivered human freedom precisely at the expense of culture. To resurrect culture as the new authority risks all that we have gained and threatens to tip us into a state of intellectual bankruptcy and moral free-fall. Fortunately for anyone so inflicted, Darwinism is the best antidote around to the fashionable fallacies of postmodernism.

In the coming decades of the 21st century, more pieces of the human jigsaw will be put in place with advances in the human genome project and neurobiology, and with Darwinism cutting ever deeper into the human psyche. One of the major challenges ahead will be knowing how to conduct research in this area wisely and ethically, and, just as importantly, how to integrate scientific findings into social life with intellectual and moral rigour.

Figure 12.1   A child chipping away at the Berlin Wall

Decades of Communism failed to mould human nature to eschew the freedoms and private property of the West. Political and ethical systems fail when they ignore the biological dimension to our common humanity

Looking back over the 20th century, historians will probably see a struggle for the ownership of human nature. They will note how many scientists and intellectuals, sometimes for the best of motives, allowed it to be snatched away by the social sciences and cultural relativists. It is time for an evolutionary understanding to reassert itself, and there are ample signs that this is just what is happening. For as Blaise Pascal noted 'If the earth moves, a decree from Rome cannot stop it'.

SUMMARY

▨ Human biology has often been a battleground where competing ideologies have struggled to secure scientific support for political actions. The whole procedure has usually involved errors of fact and logic.

▨ The evolutionary approach to human nature is largely divorced from eugenic concerns and the attempts of some psychologists to search for inherited racial differences.

▨ The plasticity of human nature has often been asserted from an ideological belief that the good society is more consistent with a science showing that human behaviour is a product of culture than one showing it to be understandable in evolutionary terms. A Darwinian approach to human nature is not only consistent with the search for the good society, but is also in fact a prerequisite for it.

KEY WORDS

Behaviourism ■ Biological determinism ■ Carrier ■ Dualism ■ Eugenics
Ideology ■ Reductionism ■ Relativism ■ Social Darwinism

FURTHER READING

Browne, K. (1998) *Women at Work*. London, Phoenix.
A book that will infuriate many. Browne gives an evolutionary analysis of why gender roles are as they are.

Gould, S. J. (1981) *The Mismeasure of Man*. London, Penguin.
A humane work that shows the folly of politically motivated science, in this case craniology and IQ measurements.

Moore, J. and Desmond, A. (1991) *Darwin*. London, Michael Joseph.
Lively, extremely well-written biography of Darwin. Contains little about the science of evolution but provides a penetrating analysis of the social context of Darwin's ideas.

Ruse, M. (1993) *The Darwinian Paradigm*. London, Routledge.
Ruse has written numerous works on the philosophical and political dimensions of Darwinism. This collection of essays is a good introduction to his thoughts.

Glossary

Adaptation A feature of an organism that has been shaped by natural selection such that it enhances the fitness of its possessor. Adaptation can also refer to the process by which the differential survival of genes moulds a particular trait so that it now appears to be designed for some particular survival-related purpose.

Adaptive significance The way in which the existence of a physical or behavioural feature can be related to the function it served and may continue to serve in helping an animal survive and reproduce.

Allele A particular form or variant of a single gene that exists at a given locus on the genome of an individual. There may be many forms of alleles within a population of one species. At each locus, a human possesses two alleles, one inherited from the father and one from the mother. An allele is, therefore, a sequence of nucleotides on the DNA molecule.

Allometry The relationship between the size of an organism as measured by, for example, length, volume or body mass, and the size of a single feature such as brain size. The relationship can often be expressed by mathematical allometric functions or by graphs showing allometric lines.

Altruism Self-sacrificing behaviour whereby one individual sacrifices some component of its reproductive value for another individual. Self-sacrifice for a relative is termed 'kin altruism'. What may appear to be help or self-sacrifice may be enlightened self-interest, as in mutualism or reciprocal altruism, where a favour is given in the expectation of some eventual return. Sacrifice at the level of phenotype can be interpreted in terms of the 'self-interest' of the genes involved.

Amino acid The molecular building blocks of proteins. There are 20 main amino acids in the proteins found in organisms. The particular sequence of amino acid in a protein determines its properties and is itself related to base sequences on DNA.

Anisogamy A situation in which the gametes from sexually reproducing species are of different sizes. Males produce small highly mobile gametes in large numbers; females produce fewer and larger eggs.

Asexual reproduction The production of offspring without sexual fertilisation of the eggs (*see also* Parthenogenesis).

Asymmetry A measure of the departure from symmetry of features that could in principle be symmetrical. Asymmetry is thought to be increased by poor condition and stress.

Australopithecines The earliest hominids that appeared on the African plains about 4 million years ago.

Autosome Any chromosome other than the sex chromosomes.

Bacteriophage A virus that infects bacteria. Also called a phage.

Base DNA consists of a phosphate–sugar backbone with attached nitrogenous bases. In DNA, a base can either be cytosine (C), guanine (G), adenine (A) or thymine (T). In RNA uracil (U) is substituted for thymine. The precise sequence of bases on the DNA serves as a set of instructions for building the cell.

Behaviourism The school of psychology largely founded by Watson that suggests that observable behaviour should be the subject matter of psychology.

Biological determinism A belief (often referred to pejoratively) that behaviour is caused by the biology of an individual, for example its physiology or genetics.

Carrier An individual who is heterozygous for an inherited trait (often a disorder) but who does not display the symptoms of the disorder.

Central dogma The idea that the information flow from DNA through RNA to the proteins of cells that make up an individual is one way and irreversible.

Cerebral cortex An area of the brain resembling a folded sheet of grey tissue that covers the rest of the brain. In humans, it is associated with such 'higher functions' as speech, language and reasoning.

Character A feature, trait or property of an organism.

Chromosome Structures in the nucleus of a cell that house DNA. Chromosome contains DNA and proteins bound to it. Chromosomes become visible to the optical microscope during meiosis and mitosis.

Cladistics A means of producing a phylogenetic classification. In cladistics, a group shares a more recent common ancestor than do members of a different group.

Clone A group of organisms that have exactly the same genome and are derived from the asexual reproduction of a parent.

Codon A set of three nucleotides along the DNA sequence that specify one amino acid.

Coefficient of relatedness (*r*) The probability that an allele chosen at random from one individual will also be present in another individual. Can also be thought of as the proportion of the total genome present in one individual that is present in another as the result of common ancestry.

Cognitive adaptation The view that the human mind and its modes of functioning have been shaped by natural selection. A cognitive adaptation is, therefore, an inherent disposition.

Commensalism A symbiotic relationship between two individuals of different species such that one member benefits without seriously affecting the fitness of the other.

Conspecifics Members of the same species.

Convergence A process whereby a similar character evolves independently in two or more species. The species may not necessarily share a recent ancestor but, in this case, natural selection has produced a similar solution to a problem. The term 'analogy' is also used in this sense, in contradistinction to homology, in which the character is similar because of a recent common ancestor. The wings of birds and bats are, for example, analogous but not homologous.

Crossing over The process during which, in meiosis, chromosomes in a diploid cell may exchange segments of DNA. This ensures a recombination of genetic material. The process results in highly variable gametes.

Culture An evolutionary account of culture is one of the most difficult problems facing Darwinism. As yet there is no consensus on the definition of the term. In common parlance, culture is usually taken to mean the knowledge, belief systems, art, morals and customs acquired by individuals as members of a society. More recently, evolutionists have attempted to remove behaviour from this definition, leaving behind the view that culture comprises socially transmitted information.

Cystic fibrosis A disease that occurs in people who possess two copies of a particular recessive allele, individuals with only one copy being carriers. The condition results in the secretion of abnormally thick mucus, increased sweat electrolytes and autonomic nervous system overactivity throughout the body.

Diploid A diploid cell is one that possesses two sets of chromosomes, one set obtained from the mother and one from the father. Humans are diploid organisms (*cf* Haploid).

Dishonest signal A manifestation designed to impress a potential mate or deter a potential rival, but one that falsely advertises the quality of the signaller.

DNA (deoxyribonucleic acid) The molecule that contains the information needed to build cells and control inheritance.

DNA hybridisation A technique used to estimate the degree of genetic similarity between individuals of the same or even different species. It relies upon the fact that single strands of DNA will bond with other single strands and that the strength of the bond is an indication of the complementarity of base sequences. It provides a useful measure of the evolutionary separation of two species.

Domain-specific mental modules It is one of the central postulates of many evolutionary psychologists that the human mind consists of discrete problem-solving areas or modules. It is suggested that these represent mental adaptations to specific problems that our ancestors faced.

Dominance hierarchy The ranking of individuals in a social group. The hierarchy is usually established by aggression and conflict, but once established it can be used to settle conflict issues without fighting.

Dominant allele An allele that is fully expressed in the phenotype. An allele A is said to be dominant if the phenotype in the heterozygotic condition, Aa, is the same as that in the homozygotic condition, AA. In this situation, allele a is said to be recessive. Alleles can be dominant, recessive or partly dominant.

Dualism The belief that humans have two aspects to their lives: physical and non-physical, matter and mind, or body and soul.

Environmentalism In the science of behaviour, the belief that social and cultural factors are paramount in determining (human) behaviour.

Environment of evolutionary adaptation (EEA) A concept highly favoured by evolutionary psychologists. That period in human evolution (over 30 000 years ago) during which the mind and body plans of humans were shaped and laid down by natural selection to solve survival problems operating then.

Eugenics A largely discredited set of beliefs that advocates selective breeding among humans to remove undesirable qualities and enhance the frequency of desirable genes.

Eusocial A term used to describe highly socialised societies, such as are found among ants and bees, in which some individuals forego reproduction to assist the reproductive efforts of other members of the social group.

Evolutionary stable strategy (ESS) A set of rules of behaviour that, once adopted by members of a group, is resistant to replacement by an alternative strategy.

Extrapair copulation Mating by a member of one sex with another outside what appears to be the stable pair bond in a supposed monogamous relationship.

Facial averageness By photographic and computer-assisted techniques the average appearance of a set of faces can be created. Average faces tend to be regarded as attractive, a finding that has spawned a number of sociobiological theories, but they may not be optimally attractive.

Fitness A term that, crucially important to evolutionary theory, continues to elude a precise and universally agreed definition. Fitness can be measured by the number of offspring that an individual leaves relative to other individuals of the same species. Direct fitness (sometimes called Darwinian fitness) can be thought of as being proportional to the number of genes contributed to the next generation by the production of direct offspring. Indirect fitness is proportional to the number of genes appearing in the next generation by an individual helping kin that also carry those genes. Inclusive fitness is the sum of direct and indirect fitness.

Fixed action pattern An innate or instinctive pattern of behaviour that is highly stereotyped and stimulated by some simple stimulus.

Fluctuating asymmetry (FA) A meaure of the symmetry of a bilateral character (for example, arm length) that fluctuates, it is thought, in response to internal and external stress factors such as parasitic infections. A high level of FA may indicate poor condition.

Founder effect The effect occurring when a new group of organisms is formed from a few in a larger population. The new group is likely to have less genetic variation and display an average genotype that may be shifted in some direction even though the shift was not the result of natural selection.

Function Sometimes used as shorthand for the adaptive value of a behavioural trait. The word has experienced an ambiguous usage in the human sciences. In the early years of the 20th century, a school of functionalism grew up in psychology and sociology that included William James (1842–1910) and John Dewey (1859–1952). The evolutionary adaptive significance of behaviours to individuals was eventually lost within this movement. In psychology, functionalism became concerned with how individuals became adapted or adjusted in their own lives rather than how traits that had been selected over time became manifest. In sociology and anthropology, functionalists looked at how current behaviours of social practices contributed to the stability of the current order. Bronislaw Malinowski (1884–1942), in his pioneering study of kinship among the Trobriand islanders, for example, tried to show how kinship served the social order as a whole rather than the genetic fitness of individuals. The functional approach also used analogies between organs of the body and parts of a social system. The situation is further confused by a modern school of functionalism, a merger of cognitive science and artificial intelligence, that uses the word in terms analogous to mathematical functions, the mind being interpreted in terms of computer-like inputs and outputs.

Gamete A sex cell. Gametes are said to be haploid in that they only contain one copy of any chromosome. A gamete can be an egg or a sperm.

Game theory A mathematical approach to establishing what behaviour is fitness-maximising by taking into account the pay-offs of particular strategies in the light of how other members of the group behave.

Gene A unit of hereditary information made up of specific nucleotide sequences of DNA.

Gene pool The entire set of alleles present in a population.

Genetic drift A change in the frequency of alleles in a population as a result of chance alone (as opposed to selection).

Genome The entire set of genes carried by an organism.

Genomic imprinting A mechanism by which cells express either the maternal or the paternal allele of a gene but not both.

Genotype The term can be used in two senses: (a) loosely as the genetic constitution of an individual, or (b) as the types of allele found at a locus on the genome.

Genus In classification, the genus is the taxonomic category (taxon) above the level of a species but below that of a family. Hominids (*Homo sapiens, Homo erectus* and so on), for example, belong to the genus *Homo*.

Good genes theory An approach to sexual selection suggesting that individuals choose mates according to the fitness potential of their genome.

Grooming Ostensibly the cleansing of skin, fur or feathers of an animal by itself or, more significantly, by another of the same species. The function of grooming may lie deeper than that of simple hygiene and could be an indication of the formation of alliances and the resolution of conflict.

Group selection Selection that operates between groups rather than individuals. The notion was attacked and shunned in the 1970s but may be making a comeback in studies on human evolution.

Haemophilia A sex-linked genetic disorder expressed in human males and characterised by excessive bleeding following injury.

Handicap principle An idea advanced by Zahavri in 1975 designed to account for what appear to be maladaptive features, or handicaps, of an organism, such as the long train of a peacock or the huge antlers of deer. Zahavri suggested that these features were honest advertisements of genetic quality since an animal, usually the male, must be strong in order to grow and bear such a burden.

Haploid A condition in which cells contain only one copy of any chromosome. Some organisms are haploid, but in humans only the gametes are haploid.

Herbivore An organism that eats only plants.

Heritability The extent to which a difference in a character between individuals in a population is caused by inherited differences in genotype. It is often expressed as a number between zero and 1 that refers to the proportion of variance in a character in a population that is ascribable to inherited genetic differences.

Heterozygote An individual that has two different alleles (for example, Aa) at a given genetic locus.

Homicide The deliberate killing of one human being by another.

Hominids Modern day humans and their ancestors in the genus *Homo*.

Homology A feature found in several species and present in their common ancestor.

Homozygote An individual having identical alleles for a given trait at a given locus (AA or aa).

Honest signal A signal that reliably communicates the quality of an individual in terms of its fitness.

Hypothesis A conjecture set forward as a provisional explanation for a phenomenon.

Ideology A set of beliefs, values and assumptions that structure understanding and inform policy decisions. An ideology usually supports the political interests of particular groups.

Immune system The complex variety of processes that enable the body to resist disease.

Inbreeding The production of offspring from two individuals who are genetically related to varying degrees.

Inclusive fitness Fitness that is measured by the number of copies of one's genes that appear in current or subsequent generations in offspring and non-offspring. Kin-directed altruism, for example, is said to increase inclusive fitness (*see also* Fitness).

Infanticide The deliberate killing of an infant shortly after its birth.

Innate releasing mechanism A hypothetical mechanism or model devised to help to explain how an innate response is triggered by a sign stimulus.

Intensionality A term used to express degrees of self-awareness and the awareness of the mental states of others. First-order intensionality is self-consciousness ('I know'); second-order, the awareness that others may have self-awareness ('I know that you know'); third-order, the knowledge that others may be aware of your thoughts ('I know that you know that I know') and so on.

Isogamy A condition where the gametes from each partner engaged in sexual reproduction are of equal size. Isogamy is common among protists and algae. High plants and all animals display anisogamy.

Kin discrimination The ability of an animal to react differently to other individuals depending on their degree of genetic relatedness.

Kin selection The suggestion that altruism can evolve because altruistic behaviour favours increases in the gene frequency of the genes responsible. A situation in which altruism towards relatives is favoured and spreads.

Lamarckian inheritance Shorthand for the inheritance of acquired characteristics. A mechanism (among others) proposed by the French evolutionist Lamarck whereby it is supposed that characters or modifications to characters acquired by the phenotype can be passed on to offspring through genetic inheritance. The mechanism is now rejected as being without foundation.

Lamarckism Doctrines associated with Jean Baptiste Lamarck. Usually taken to refer to his mechanism of inheritance.

Lek A display site where (usually) males display and females choose a mate.

Lineage A sequence showing how species are descended from one another.

Linkage disequilibrium A condition in which genes travel together in the process of inheritance. The result is that the frequency of a group of linked genes is different from what would be expected from random recombination.

Locus The particular site at which a particular gene is found on DNA.

Machiavellian intelligence The idea that one of the prime factors leading to the growth of intelligence in primates was the need for an individual to manipulate its social world through a mixture of cunning, deceit and political alliances.

Major histocompatibility complex (MHC) A set of genes coding for antigens responsible for the rejection of genetically different tissue. The antigens are known as histocompatibility (that is, tissue compatibility) antigens. There are at least 30 histocompatibility gene loci. The genes of the MHC are subject to simple Mendelian inheritance and are co-dominantly expressed; that is, alleles from both parents are equally expressed. Each cell, therefore, in any offspring has maternal and paternal MHC molecules on its surface. The human MHC is found on the short arm of chromosome 6.

Maternal–fetal conflict The theory that the fetus and the mother that carries it may have different genetic interests and so engage in a biological tug-of-war over the level of resources passing to the fetus as it develops.

Meiosis A type of cell division whereby diploid cells produce haploid gametes. The double set of chromosome in a diploid cell is thereby reduced to a single set in the resulting gametes. Crossing over and recombination occur during meiosis.

Meme An activity or unit of information that can be passed on by imitation.

Mendelian inheritance A mode of inheritance common to diploid species in which genes are passed to the offspring in the same form as they were inherited from the previous generation. At each locus, an individual has two genes (haploid), one inherited from its mother and one from its father.

Menopause The cessation of monthly ovulation experienced by women usually in their late forties. Women are infertile after the menopause. The adaptive significance of the menopause is probably related to the risks of childbirth and the need to care for existing children.

Mitosis Part of the process where one cell divides into another identical one with the same number of chromosomes as the original. The growth of an organism is through mitosis. Sexual reproduction requires meiosis.

Modularity The belief that the human brain consists of a number of discrete problem-solving units, or areas of specialised function, shaped by natural selection to solve specific problems.

Monism The belief that the universe is composed of one basic stuff. In the context of human behaviour (materialistic), monism asserts that behaviour has physical causes.

Monogamy The mating of a single male with a single female. In annual monogamy, the bond is dissolved each year and a fresh partner found. In perennial monogamy, the bond lasts for the reproductive life of the organisms.

Morgan's canon The assertion that explanations for the behaviour of animals should be kept as simple as possible. Higher mental activities should not be attributed if lower ones will suffice.

Mutation In modern genetics, a heritable change in the base sequences of the DNA of a genome. Most mutations are deleterious.

Mutualism A symbiotic relationship between individuals of two species such that both partners benefit.

Naturalism The belief that all phenomena can be explained by scientific laws and principles without recourse to the supernatural or entities outside the remit of science.

Nuptial gift An item of food transferred from an individual of one sex to one of the other prior to or during copulation.

Oestrus A period of heightened interest in copulation experienced by female animals.

Ontogeny Literally, coming into being. The development and growth of an organism from a fertilised cell, through the fetus to an adult.

Oogenesis The formation of female sex cells (ova) (*see also* Spermatogenesis).

Operant conditioning A type of learning whereby an action (operant) is carried out more frequently if it is rewarded.

Operational sex ratio The ration of sexually receptive males to females in a particular area or over a particular time.

Optimality The idea that the behaviour of animals will be that ideally suited to bring maximum gain for minimum cost, or more precisely where gain minus cost is maximised. The assumption that all behaviour must be optimal can be misleading.

Order A unit of classification above the family and below the class. Humans belong to the order of primates.

Ovarian cycle A cycle of events controlled by hormones in the ovary of mammals that leads to ovulation.

Ovulation The release of a female sex cell or ovum (plural ova) from an ovarian follicle.

Paradigm A cluster of ideas and theories that are consistent and form part of a distinct way of understanding the world. Evolutionary psychology can be regarded as a paradigm.

Parasite An organism in a symbiotic relationship with another (host) such that the parasite gains in fitness at the expense of the host.

Parental investment Actions that increase the survival chances of one set of offspring at the expense of the parent procuring more offspring.

Parent–offspring conflict Disputes between parents and their children over the allocation of resources. Such conflicts are the focus of attention by sociobiologists in that they may have a basis in the different reproductive interests of the parties concerned. Typically the best reproductive interests of young parents may be served by allocating resources to future children. A decision that may not be optimum from the point of view of an existing child.

Parsimony A principle used to construct phylogenetic trees. It involves an assumption that the branching pattern should involve the smallest number of evolutionary changes.

Parthenogenesis Asexual reproduction. The production of offspring by virgin birth.

Pathogen A disease-causing organism.

Phenotype The characteristics of an organism as they have been shaped by both the genotype and environmental influences.

Phenotypic plasticity A debated term. The average value for a phenotypic character between two individuals with identical genomes or two populations with similar gene frequencies may be different because of different environmental influences. Phenotypic plasticity is often used to refer to irreversible change, and phenotypic flexibility to refer to reversible change.

Phylogeny The branching history of a species showing its relationship to ancestral species. A phylogenetic tree can be used to infer relationships between existing species and their evolutionary history.

Phylum A taxonomic category above the class but below the kingdom. Humans belong to the order of primates, the class of mammals and the phylum Chordata.

Placenta The organ that provides nutrients and oxygen to an embryo.

Pleiotropy The ability of a single gene to have a number of different effects on the phenotype.

Polyandry A type of mating system such that a single female mates with more than one male in a given breeding season.

Polygamy A situation where one sex mates with more than one member of the other sex. Polygamy can be simultaneous where an individual has many partners at any one time or serial where the partners are spread over time.

Polygyny A mating system whereby a single male mates with more than one female in a given breeding season.

Polymorphism A condition in which a population may have more than one allele (at significant frequencies) at a particular locus. The different forms of the alleles may all be adaptive in their own right.

Population A group of individuals, usually geographically localised and interacting, that all belong to the same species.

Positivism The belief that science represents the positive and more advanced state of knowledge. Also the supposition that only objects that can be experienced directly form part of the proper process of scientific enquiry. Positivism seeks to repudiate metaphysics.

Prisoners' dilemma A model of the problem faced by individuals in knowing how to act to serve their own best interests in the context of uncertain knowledge of how others, who may be rivals, will also act. The analogy is drawn with two prisoners who, if they co-operate with each other, will receive a ligher sentence.

Prosimians A diverse group of small-bodied primates. Most species live in Africa, the best-known examples including bushbabies and lemurs. Thought to be more 'primitive' than the simians.

Proximate cause In behavioural terms, the immediate mechanism or stimulus that initiates or triggers a pattern of behaviour.

Recessive allele *See* Dominant allele.

Reciprocity The donation of assistance in the expectation that the favour will be returned at some future date.

Recombination An event whereby chromosomes cross over and exchange genetic material during meiosis. Recombination tends to break up genes that are linked together.

Reductionism The attempt to explain a wide range of phenomena by employing a smaller range of concepts and principles that is more basic.

Relativism The view that there is no privileged knowledge system that can claim supremacy over another. A relativist would argue that the validity of belief systems can not be decided by universal criteria.

RNA (Ribonucleic acid) A nucleic acid with a ribose sugar backbone. Plays a role in protein synthesis and serves as the genome of some viruses. Messenger RNA (mRNA) encodes information from the DNA and conveys it to other parts of the cell where proteins are built.

Selection The differential survival of organisms (or genes) in a population as a result of some selective force. 'Selectionist' thinking is the approach that looks for how features of organisms can be interpreted as the result of years of selection acting upon ancestral populations. Directional selection tends to favour an extreme measure of the natural variability in a population, and the average measure will gradually move in this direction. Disruptive selection tends to favour more than one phenotype. Stabilising selection tends to favour the means values currently found and ensures that variation is reduced (*see also* Sexual selection).

Sex chromosomes Chromosomes that determine whether an individual is male or female. In humans, a female possesses two X chromosomes and a male one X and one Y chromosome.

Sex ratio The ratio of males to females at any one time. At birth, the ratio for humans is about 1.06.

Sex role reversal A situation in which the 'normal' roles of males and females are reversed. Usually applied to cases where females compete with other females for access to males.

Sexual dimorphism Differences in morphology, physiology or behaviour between the sexes in a single species.

Sexual selection Selection that takes place as a result of mating behaviour. **Intrasexual** selection occurs as a result of competition between members of the same sex, **intersexual** selection as a result of choices made by one sex for the features of another.

Siblicide The killing of a sibling (brother or sister) by a brother or sister.

Simians The New World monkeys, Old World monkeys and apes.

Social Darwinism A rather loose collection of ideas and philosophies that took root in the United States and Europe during and after the last quarter of the 19th century. Social Spencerism would be a more accurate description since, like the British philosopher Herbert Spencer, social Darwinists advocated minimum state control in the economy and free competition.

Speciation The creation of a new species through the splitting of an existing species into two or more new species.

Species Using the biological species concept, a species is a set of organisms that possess similar inherited characteristics and, crucially, have the potential to interbreed to produce fertile offspring.

Spermatogenesis The formation of male sex cells (sperm).

Sperm competition Competition between sperm from two or more males that are present in the reproductive tract of the female.

Strategy A pattern of behaviour or rules guiding behaviour shaped by natural selection to increase the fitness of an animal. A strategy is not taken to be a set of conscious decisions in non-human animals. A strategy may also be flexible in that different biotic and abiotic factors may trigger different forms of optimising behaviour.

Symbiosis A relationship between organisms of different species that live in close association and interact (*see also* Mutualism, Parasite and Commensalism.)

Symmetry A state where the physical features of an organism on one side of its body are matched in size and shape by those on the other. Symmetry may be an indication of physiological health since stress increases asymmetry. Animals may, therefore, use symmetry as an honest signal of fitness.

Taxon (plural taxa) A named group in classification. This may be a species, a genus, a family, an order or some other category.

Taxonomy The theory and practice of classifying organisms.

Teleology The belief that nature has purposes, that events are shaped by intended outcomes.

Testis (plural testes) The male sex organ that produces sperm and their associated hormones.

Theory of mind The suggestion that an important component of the mental and emotional life of humans and some other primates is the ability to be self-aware and to appreciate that others also have awareness. The theory of mind also implies that an individual is capable of distinguishing between the real intentional or emotional states of others and those that may be feigned. Having a theory of mind is an essential component of Machiavellian intelligence.

Tit for tat A strategy that can be played by individuals stuck in a prisoners' dilemma. The strategy is never to be the first to defect but to follow defection or cheating by the other player with a retaliatory move. The tit for tat strategy, on average, works well against a variety of other strategies.

Trivers–Willard hypothesis The suggestion that, under conditions where the variance in reproductive success is greater for one of the sexes, parents will bias their investment towards offspring of one sex. Typically, among polygynous mammals, where the variance in reproductive success is greater for males than females, males in good condition have a good chance of leaving more offspring than females in good condition. Parents who are in above average condition would then be expected to produce more males than females, thus biasing the sex ratio. Trivers and Willard suggested that this idea could be applied to human societies, taking wealth and socioeconomic status as an indication of condition.

Ultimate explanation The explanation for the behaviour of an organism that reveals its adaptive value. Hence the ultimate cause of a trait can be related to its adaptive function.

Waist to hip ratio (WHR) The circumference of the waist divided by that of the hips.

Zygote A fertilised cell formed by the fusion of two gametes. Monozygotic refers to (identical) twins that result from the separation of a single fertilised egg or zygote and are, therefore, identical in terms of their DNA. Dizygotic refers to twins that result from two independent fertilisations of two eggs by two sperm and are, therefore, non-identical.

FURTHER READING

Many of the terms above are uncontentious but some have disputed meanings. The reader is referred to Keller, E. F. and Lloyd, E. A. (eds) 1992, *Keywords in Evolutionary Biology*, Cambridge, MA, Harvard University Press. This book considers 37 key terms in some depth.

References

Alexander, R. D. (1979) *Darwinism and Human Affairs*. Seattle, University of Washington Press.

Alexander, R. (1987) *The Biology of Moral Systems*. New York, Aldine de Gruyter.

Alexander, R. and Noonan, K. (1979) 'Concealment of ovulation, parental care and human social evolution' in Chagon, N. I. A. and Irons, W. (eds) *Evolutionary Biology and Human Social Behaviour: An Anthropological Perspective*. North Scituate, MA, Duxbury.

Alexander, R. D., Hoogland, J. H., Howard, R. D. *et al.* (1979) 'Sexual dimorphisms and breeding systems in pinnipeds, ungulates, primates and humans' in Chagon, N. I. A. and Irons, W. (eds) *Evolutionary Biology and Human Social Behaviour: An Anthropological Perspective*. North Scituate, MA, Duxbury.

Andersson, M. (1982) 'Female choice selects for extreme tail length in a widowbird'. *Nature* 299: 818–20.

Andersson, M. (1994) *Sexual Selection*. Princeton, NJ, Princeton University Press.

Archer, J. (1992) *Ethology and Human Development*. Hemel Hempstead, Harvester Wheatsheaf.

Ashworth, T. (1980) *Trench Warfare, 1914–1918: The Live and Let Live System*. New York, Holmes & Meier.

Askenmo, C. E. H. (1984) 'Polygyny and nest site selection in the pied flycatcher'. *Animal Behaviour* 32: 972–80.

Austad, S. N. and Sunquist, M. E. (1986) 'Sex ratio manipulation in the common opossum'. *Nature* 324: 58–60.

Axelrod, R. (1984) *The Evolution of Co-operation*. New York, Basic Books.

Axelrod, R. and Hamilton, W. D. (1981) 'The evolution of co–operation'. *Science* 211: 1390–6.

Badcock, C. (1991) *Evolution and Individual Behaviour: An Introduction to Human Sociobiology*. Oxford, Blackwell.

Bailey, J. M. (1998) 'Can behaviour genetics contribute to evolutionary behavioural science?' in Crawford, C. and Krebs, D. L. (eds) *Handbook of Evolutionary Psychology*. Mahwah, NJ, Lawrence Erlbaum.

Baker, R. R. (1996) *Sperm Wars*. London, Fourth Estate.

Baker, R. R. and Bellis, M. A. (1989) 'Number of sperm in human ejaculates varies in accordance with sperm competition theory'. *Animal Behaviour* 37: 867–9.

Baker, R. R. and Bellis, M. A. (1995) *Human Sperm Competition*. London, Chapman & Hall.

Bakker, T. C. M. (1993) 'Positive genetic correlation between female preference and preferred male ornament in sticklebacks'. *Nature* 363: 255–7.

Barash, D. (1982) *Sociobiology and Behaviour*. New York, Elsevier.

Barber, N. (1995) 'The evolutionary psychology of physical attractiveness: sexual selection and human morphology'. *Ethology and Sociobiology* 16: 395–424.

Barkow, J. H. (1992) 'Beneath new culture is old psychology: gossip and social stratification' in Barkow, J. H. Cosmides, L. and Tooby, J. (eds) *The Adapted Mind*. Oxford, Oxford University Press.

Barkow, J. H. Cosmides, L. and Tooby, J. (1992) (eds) *The Adapted Mind*. Oxford, Oxford University Press.

Bateman, A. J. (1948) 'Intra-sexual selection in Drosophila'. *Heredity* 2: 349–68.

Bateson, P. (1982) 'Preferences for cousins in Japanese quail'. *Nature* 295: 236–7.

Beach, F. A. (1950) 'The Snark was a Boojum'. *American Psychologist* 5: 115–24.

Bell, G. (1982) *The Masterpiece of Nature*. London, Croom Helm.

Benshoof, L. and Thornhill, R. (1979) 'The evolution of monogamy and concealed ovulation in humans'. *Journal of Social and Biological Structures* 2: 95–106.

Bernstein, H., Hopf, F. A. and Michod, R. E. (1989) 'The evolution of sex: DNA repair hypothesis' in Vogel, C. Rasa, A. and Voland, E. (eds) *The Sociobiology of Sexual and Reproductive Strategies*. London, Chapman & Hall.

Berreman, G. D. (1962) 'Pahari polyandry: a comparison'. *American Anthropologist* 64: 60–75.

Betzig, L. (1982) 'Despotism and differential reproduction'. *Ethology and Sociobiology* 3: 209–21.

Betzig, L. (1986) *Despotism and Differential Reproduction: A Darwinian View of History*. Hawthorne, NY, Aldine de Gruyter.

Betzig, L. (1992) 'Roman polygyny'. *Ethology and Sociobiology* 13: 309–49.

Betzig, L. (1998) 'Not whether to count babies, but which' in Crawford, C. and Krebs, D. L. (eds) *Handbook of Evolutionary Psychology*. Mahwah, NJ, Lawrence Erlbaum.

Birkhead, T. R. and Moller, A. P. (1992) *Sperm Competition in Birds: Evolutionary Causes and Consequences*. London, Academic Press.

Birkhead, T. R. and Parker, G. A. (1997) 'Sperm competition and mating systems' in Krebs, J. R. and Davies, N. B. (eds) *Behavioural Ecology*. Oxford, Blackwell Science.

Birkhead, T. R., Moore, H. D. M. and Bedford, J. M. (1997) 'Sex, science and sensationalism'. *Trends in Ecology and Evolution* 12(3): 121–2.

Blackmore, S. (1999a) 'The forget meme not theory'. *The Times Higher Educational Supplement*, 26 February, pp. 20–1.

Blackmore, S. (1999b) *The Meme Machine*. Oxford, Oxford University Press.

Boaz, N. T. and Almquist, A. J. (1997) *Biological Anthropology*. Englewood Cliffs, NJ, Prentice Hall.

Borgerhoff-Mulder, M. (1988) 'Bridewealth variability among the Kipsigis' in Betzig, L. (ed.) *Human Reproductive Behaviour: A Darwinian Perspective*. Oxford, Oxford University Press.

Borgerhoff-Mulder, M. (1990) 'Kipsigis women prefer wealthy men: evidence for female choice in mammals?' *Behavioural Ecology and Sociobiology* 27: 255–64.

Boring, E. G. (1950) *A History of Experimental Psychology*. Englewood Cliffs, NJ, Prentice Hall.

Bourke, A. F. G. (1997) 'Sociality and kin selection in insects' in Krebs, J. R. and Davies, N. B. (eds) *Behavioural Ecology*, 3rd edn. Oxford, Blackwell Science.

Bourke, A. F. and Franks, N. R. (1995) *Social Evolution in Ants*. Princeton, NJ, Princeton University Press.

Bowler, P. (1982) *Evolution: The History of an Idea*. Berkeley, University of California Press.

Bowler, J. (1995) *Is God a Virus?* London, Society for the Promotion of Christian Knowledge.

British Society for Social Responsibility in Science (BSSRS) (1976) 'The new synthesis is an old story'. *New Scientist* (May 13): 346–8.

Brown, A. (1999) *Darwin Wars. How Stupid Genes became Selfish Gods.* London, Simon & Schuster.

Buckle, L., Gallup, G. G. and Rodd, Z. (1996) 'Marriage as a reproductive contract: patterns of marriage, divorce, amd remarriage'. *Ethology and Sociobiology* **17**: 363–77.

Burkhardt, R. W. (1983) 'The development of an evolutionary ethology' in Bendall, D. S. (ed.) *Evolution from Molecules to Men.* Cambridge, Cambridge University Press.

Burley, N. (1979) 'The evolution of concealed ovulation'. *American Naturalist* **114**: 835–58.

Buss, D. M. (1989) 'Sex differences in human mate preferences: evolutionary hypotheses tested in 37 cultures'. *Behavioural and Brain Sciences* **12**: 1–49.

Buss, D. M. (1994) *The Evolution of Desire.* New York, HarperCollins.

Buss, D. M. (1999) *Evolutionary Psychology.* Needham Heights, MA, Allyn & Bacon.

Buss, D. and Barnes, M. (1986) 'Preferences in human mate selection'. *Journal of Personality and Social Psychology* **50**: 559–70.

Buss, D. M., Larsen, R. J., Westen, D. and Semmelroth, J. (1992) 'Sex differences in jealousy: evolution, physiology and psychology'. *Psychological Science* **3**(4): 251–5.

Byrne, R. (1995) *The Thinking Ape.* Oxford, Oxford University Press.

Byrne, R. W. and Whiten, A. (1988) *Machiavellian Intelligence: Social Expertise and the Evolution of Intellect in Monkeys, Apes and Humans.* Oxford, Clarendon Press.

Cashden, E. (1989) 'Hunters and gatherers: economic behaviour in bands' in Plattner, S. (ed.) *Economic Anthropology.* Stanford, CT, Stanford University Press.

Christenfeld, N. J. S. and Hill, E. A. (1995) 'Whose baby are you?' *Nature* **378**: 669.

Clutton-Brock, T. H. (1994) 'The costs of sex' in Short, R. V. and Balaban, E. (eds) *The Differences Between the Sexes.* Cambridge. Cambridge University Press.

Clutton-Brock, T. and Godfray, C. (1991) 'Parental investment' in Krebs, J. R. and Davies, N. B. (eds) *Behavioural Ecology.* Oxford, Blackwell Scientific.

Clutton-Brock, T. H. and Harvey, P. H. (1976) 'Evolutionary rules and primate societies' in Bateson, P. P. G. and Hinde, R. A. (eds) *Growing Points in Ethology.* Cambridge, Cambridge University Press.

Clutton-Brock, T. H. and Vincent, A. C. J. (1991) 'Sexual selection and the potential reproductive rates of males and females'. *Nature* **351**: 58–60.

Clutton-Brock, T. H., Albon, S. D. and Guinness, F. E. (1986) 'Great expectations: maternal dominance, sex ratios and offspring reproductive success in red deer'. *Animal Behaviour* **34**: 460–71.

Conover, M. R. (1990) 'Evolution of a balanced sex ratio by frequency-dependent selection in fishes'. *Science* **250**: 1556–8.

Cooper, R. M. and Zubek, J. P. (1958) 'Effects of enriched and restricted early environments on the learning ability of bright and dull rats'. *Canadian Journal of Psychology* **12**: 159–64.

Crawford, C. B. (1993) 'The future of sociobiology: counting babies or studying proximity mechanisms'. *Trends in Evolution and Ecology* **8**: 184–7.

Crawford, C. (1998) 'Environments and adaptations: then and now' in Crawford, C. and Krebs, D. L (eds) *Handbook of Evolutionary Psychology.* Mahwah, NJ, Lawrence Erlbaum.

Creel, S. and Creel, N. M. (1995) 'Communal hunting and pack size in African wild dogs *Lycaon pictus*'. *Animal Behaviour* **50**: 1325–39.

Cronin, H. (1991) *The Ant and the Peacock.* Cambridge, Cambridge University Press.

Crook, J. H. and Crook, S. J. (1988) 'Tibetan polyandry: problems of adaption and fitness' in Betzig, L., Borgehoff-Mulder, M. and Turke, P. (eds) *Human Reproductive Behaviour*. Cambridge, Cambridge University Press.

Crow, J. F. (1997) 'The high spontaneous mutation rate. Is it a health risk ?' *Proceedings of the National Academy of Sciences of the USA*, **94**: 8380–6.

Cziko, G. (1995) *Without Miracles*. Cambridge MA, MIT Press.

Daly, M. (1997) 'Introduction' in Bock, G. R. and Cardew, G. (eds) *Characterising Human Psychological Adaptations*. Ciba Foundation Symposium 208, Chichester, Wiley.

Daly, M. and Wilson, M. (1988a) *Homicide*. Hawthorne, NY, Aldine de Gruyter.

Daly, M. and Wilson, M. (1988b) 'Evolutionary social psychology and family homicide'. *Science* **242**: 519–24.

Daly, M. and Wilson, M. I. (1999) 'Human evolutionary psychology and animal behaviour'. *Animal Behaviour* **57**: 509–19.

Darwin, C. (1858) 'Letter to Charles Lyell 18th June 1858' in Burkhardt, F. and Smith, S. (eds) *The Correspondence of Charles Darwin*. Cambridge, Cambridge University Press, 1991.

Darwin, C. (1859a) 'Letter to Alfred Russel Wallace, 13th November, 1859' in Burkhardt, F. and Smith, S. (eds) *The Correspondence of Charles Darwin*. Cambridge, Cambridge University Press, 1991.

Darwin, C. (1859b) *On the Origin of Species by Means of Natural Selection*. London, John Murray.

Darwin, C. (1860) 'Letter to Charles Lyell, 25th February, 1860' in Burkhardt, F. and Smith, S. (eds) *The Correspondence of Charles Darwin*. Cambridge, Cambridge University Press, 1991.

Darwin, C. (1871) *The Descent of Man and Selection in Relation to Sex*. London, John Murray.

Darwin, C. (1881) 'Letter to W. Graham' in Darwin, F. *(ed.) Autobiography of Charles Darwin*. London, Watts, 1929.

Darwin, F. (ed.) (1887) *The Life and Letters of Charles Darwin*. London, John Murray.

Davies, N. B. (1992) *Dunnock Behaviour and Social Evolution*. Oxford, Oxford University Press.

Davies, N. B. and Houston, A. (1986) 'Reproductive success of dunnocks (*Prunella modularis*) in a variable mating system'. *Journal of Animal Ecology* **55**: 139–54.

Davies, P., Fetzer, H. and Foster, T. (1995) 'Logical reasoning and domain specificity: a critique of the social exchange theory of reasoning'. *Biology and Philosophy* **10**(1): 1–37.

Dawkins, M. S. (1986) *Unravelling Animal Behaviour*. Harlow, Longman.

Dawkins, R. (1976) *The Selfish Gene*. Oxford, Oxford University Press.

Dawkins, R. (1982) *The Extended Phenotype*. Oxford, W.H. Freeman.

Dawkins, R. (1986) *The Blind Watchmaker*. London, Longman.

Dawkins, R. (1989) *The Selfish Gene*, 2nd edn. Oxford, Oxford University Press.

Dawkins, R. (1995) *River Out of Eden*. London, Weidenfeld & Nicholson.

Dawkins, R. and Carlisle, T. R. (1976) 'Parental investment, mate desertion and a fallacy'. *Nature* **262**: 131–2.

Dawkins, R. and Treisman, M. (1976) 'The "costs of meiosis": is there any ?' *Journal of Theoretical Biology* **63**: 479–84.

Deacon, T. W. (1992) 'The human brain' in Jones, S., Martin, R. and Pilbeam, D. (eds) *The Cambridge Encyclopedia of Human Evolution*. Cambridge, Cambridge University Press.

Deacon, T. (1997) *The Symbolic Species*. London, Penguin.

Degler, C. N. (1991) *In Search of Human Nature. The Decline and Revival of Darwinism in American Social Thought*. Oxford, Oxford University Press.

Dennett, D. C. (1995) *Darwin's Dangerous Idea*. New York, Simon & Schuster.

Desmond, A. and Moore. J. (1991) *Darwin*. London, Michael Joseph.

Dewsbury, D. A. (1984) *Comparative Psychology in the Twentieth Century*. Stroudsburg, PA, Hutchinson Ross.

Dewsbury, D. A. (1988) 'A test of the role of copulatory plugs in sperm competition in deer mice (*Peromyscus maniculatus*)'. *Journal of Mammals* **69**: 854–7.

Dewsbury, D. A. (1990) *Contemporary Issues in Comparative Psychology*. Sunderland, MA, Sinauer Associates.

Diamond, J. (1991) *The Rise and Fall of the Third Chimpanzee*. London, Vintage.

Doolittle, W. and Sapienza, C. (1980) 'Selfish genes, the phenotype paradigm and genome evolution'. *Nature* **284**: 601–3.

Downhower, J. F. and Armitage, K. B. (1971) 'The yellow-bellied marmot and the evolution of polygyny'. *American Naturalist* **105**: 355–70.

Dugatin, L. A. (1996) 'Interface between culturally based preferences and genetic preferenecs: female choice in *Poecilla reticulata*'. *Proceedings of the National Academy of Sciences of the USA* **93**(7): 2770–3.

Dunbar, R. (1980) 'Determinants and evolutionary consequences of dominance among female Gelada baboons'. *Behavioural Ecology and Sociobiology* **7**: 253–65.

Dunbar, R. I. M. (1993) 'Coevolution of neocortical size, group size and language in humans'. *Behavioural and Brain Sciences* **16**: 681–735.

Dunbar, R. (1995) 'Are you lonesome tonight?' *New Scientist* **145**(1964): 12–16.

Dunbar, R. (1996a) *Grooming, Gossip and the Evolution of Language*. London, Faber and Faber.

Dunbar, R. I. M. (1996b) 'Determinants of group size in primates: a general model' in Runciman, W. G., Maynard Smith, J. and Dunbar, R. I. M. (eds) *Evolution of Social Behaviour in Primates and Man*. Oxford, Oxford University Press.

Dunbar, R. I. M. and Aiello, L. C. (1993) 'Neocortex size, group size, and the evolution of language'. *Current Anthropology* **34**(2): 184–93.

Dunbar, R. I. M., Duncan, N. D. C. and Nettle, D. (1994) 'Size and structure of freely forming conversational groups'. *Human Nature* **6**(1): 67–78.

Dunham, C., Myers, F., Bernden, N. *et al.* (1991) *Mamatoto: A Celebration of Birth*. London, Virago.

Durant, J. (1981) 'Innate character in animals and man: a perspective on the origins of ethology' in Webster, C. *Biology, Medicine and Society 1840–1940*. Cambridge, Cambridge University Press.

Durant, J. R. (1986) 'The making of ethology: the association for the study of animal behaviour, 1936–1986'. *Animal Behaviour* **34**: 1601–16.

Durham, W. (1991) *Coevolution: Genes, Culture and Human Diversity*. Stanford, Stanford University Press.

Eibl-Eibesfeldt, I. (1989) *Human Ethology*. Hawthorne, NY, Aldine de Gruyter.

Einon, D. (1998) 'How many children can one man have?' *Evolution and Human Behaviour* **19**: 413–26.

Emlen, S. T. (1995) 'An evolutionary theory of the family'. *Proceedings of the National Academy of Sciences of the USA* **92**: 8092–9.

Endler, J. H. and Houde, A. E. (1995) 'Geographic variation in female preferences for male traits in *Poecilia reticulata*'. *Evolution* **49**: 456–8.

Erickson, C. J. and Zenone, P. G. (1976) 'Courtship differences in male ring doves: avoidance of cuckoldry?' *Science* **192**: 1353–4.

Eyre-Walker, A. and Keightley, P. D. (1999) 'High genomic deleterious mutation rates in hominids'. *Nature* **397**: 344–7.

Falk, D. (1983) 'Cerebral cortices of East African early hominids'. *Science* **221**: 1072–4.

Feldman, M. W. and Laland, K. N. (1996) 'Gene-culture coevolutionary theory'. *Trends in Evolution and Ecology* **11**: 453–7.

Finkielkraut, A. (1988) *The Undoing of Thought*. London, Claridge Press.

Fisher, A. E. (1962) 'Effects of stimulus variation on sexual satiation in the male rat'. *Journal of Comparative Physiological Psychology* **55**: 614–20.

Fisher, R. A. (1930) *The Genetical Theory of Natural Selection*. Oxford, Clarendon Press.

Flinn, M. (1988) 'Step-parent/step-offspring interactions in a Caribbean village'. *Ethology and Sociobiology* **9**: 335–69.

Foley, R. (1987) *Another Unique Species*. Harlow, Longman.

Foley, R. A. (1989) 'The evolution of hominid social behaviour' in Standen, V. and Foley, R. A. (eds) *Comparative Socioecology*. Oxford, Blackwell Scientific.

Foley, R. (1992) 'Studying human evolution by analogy' in Jones, S., Martin, R. and Pilbeam, D. (eds) *The Cambridge Encyclopedia of Human Evolution*. Cambridge, Cambridge University Press.

Foley, R. A. (1996) 'An evolutionary and chronological framework for human social behaviour' in Runciman, W. G., Maynard Smith, J. and Dunbar, R. (eds) *Evolution of Social Behaviour Patterns in Primates and Man*. Oxford, Oxford University Press.

Freeman, D. (1996) *Margaret Mead and the Heretic*. London, Penguin.

Friday, A. E. (1992) 'Human evolution: the evidence from DNA sequencing' in Jones, S., Martin, R. and Pilbeam, D. (eds) *The Cambridge Encyclopedia of Human Evolution*. Cambridge, Cambridge University Press.

Gallup, G. G. (1970) 'Chimpanzees: self-recognition'. *Science* **167**: 86–7.

Gangestad, S. W. and Buss, D. M. (1993) 'Pathogen prevalence and human mate preference'. *Ethology and Sociobiology* **14**: 89–96.

Gangestad, S. W. and Thornhill, R. (1994) 'Facial attractiveness, developmental stability and fluctuating asymmetry'. *Ethology and Sociobiology* **15**: 73–85.

Garber, P. A. (1989) 'Role of spatial memory in primate foraging patterns: *Saguinus mystax* and *Saguinus fuscicollis*'. *American Journal of Primatology* **19**: 203–16.

Gaulin, S. J. C. and Boster. J. S. (1990) 'Dowry as female competition'. *American Anthropologist* **92**: 994–1005.

Geary, D. C. (1998) *Male, Female. The Evolution of Human Sex Differences*. Washington DC, American Psychological Association.

Ghiselin, M. T. (1974) *The Economy of Nature and the Evolution of Sex*. Berkeley, University of California Press.

Gigerenzer, G. and Hug, K. (1992) 'Domain-specific reasoning: social contracts, cheating and perspective change'. *Cognition* **43**: 127–71.

Goodenough, J., McGuire, B. and Wallace, R. (1993) *Perspectives on Animal Behaviour*. New York, John Wiley.

Gopnik, M., Dalalakis, J., Fukuda, S. E. *et al.* (1996) 'Genetic language impairment: unruly grammars'. *Proceedings of the British Academy* **88**: 223–49.

Gottlieb, G. (1971) *Development of Species Identification in Birds*. Chicago, University of Chicago Press.

Gould, S. J. (1981) *The Mismeasure of Man*. London, Penguin.

Gould, S. J. (1998) 'Let's leave Darwin out of it'. *New York Times*, 29 May.

Gould, S. J. and Lewontin, R. C. (1979) 'The spandrels of San Marco and the Panglossian paradigm: a critique of the adaptionist programme'. *Proceedings of the Royal Society of London* **205**: 581–98.

Grammer, K. (1992) 'Variations on a theme: age dependent mate selection in humans'. *Behaviour and Brain Sciences* **15**: 100–2.

Greenberg, L. (1979) 'Genetic component of bee odor in kin recognition'. *Science* **206**: 1095–7.

Greenless, I. A. and McGrew, W. C. (1994) 'Sex and age differences in preferences and tactics of mate attraction: analysis of published advertisements'. *Ethology and Sociobiology* **15**: 59–72.

Greenwood, P. J. (1980) 'Mating systems, phiopatry and dispersal in birds and mammals'. *Animal Behaviour* **28**: 1140–62.

Grosberg, R. K. and Quinn, J. F. (1986) 'The genetic control and consequences of kin recognition by the larvae of a colonial marine invertebrate'. *Nature* **322**: 456–9.

Gruber, H. E. (1974) *Darwin on Man: A Psychological Study of Scientific Creativity*, (together with Darwin's early and unpublished notebooks transcribed and annotated by Paul H. Barrett). London, Wildwood House.

Gruber, H. E. (1981) *Darwin on Man: A Psychological Study of Scientific Creativity*, 2nd edn. Chicago, University of Chicago Press.

Guttentag, M. and Secord, P. (1983) *Too Many Women?* Beverly Hills, CA, Sage.

Gwynne, D. T. (1988) 'Courtship feeding and the fitness of female katydids (*Orthoptera: Tettigoniidae*)'. *Evolution* **42**: 545–55.

Haig, D. (1993) 'Genetic conflicts in human pregnancy'. *Quarterly Review of Biology* **68**(4): 495–532.

Haig, D. (1997) 'The social gene' in Krebs, J. R. and Davies, N. B. (eds) *Behavioural Ecology*. Oxford, Blackwell Scientific.

Haldane, J. B. S. (1932) *The Causes of Evolution*. London, Longmans.

Hall, M. and Halliday, T. (1992) *Behaviour and Evolution*, Book 1, *Biology: Brain and Behaviour*. Milton Keynes, Open University Press.

Hamilton, W. D. (1964) 'The genetical evolution of social behaviour, I and II'. *Journal of Theoretical Biology* **7**: 1–52.

Hamilton, W. D. and Zuk, M. (1982) 'Heritable true fitness and bright birds: a role for parasites?' *Science* **218**: 384–7.

Harcourt, A. H., Harvey, P. H., Larson, S. G. *et al.* (1981) 'Testis weight, body weight and breeding system in primates'. *Nature* **293**: 55–7.

Hardin, G. (1968) 'The tragedy of the commons'. *Science* **162**: 1243–8.

Hartung, J. (1976) 'On natural selection and the inheritance of wealth'. *Current Anthropology* **17**: 607–22.

Hartung, J. (1982) 'Polygyny and the inheritance of wealth'. *Current Anthropology* **23**: 1–12.

Harvey, P. H. and Bradbury, J. W. (1991) 'Sexual selection' in Krebs, J. R. and Davis, W. B. (eds) *Behavioural Ecology*. Oxford, Blackwell Scientific.

Harvey, P. H. and May, R. M. (1989) 'Out for the sperm count'. *Nature* **337**: 508–9.

Harvey, P. H. and Reynolds, J. D. (1994) 'Sexual selection and the evolution of sex differences' in Short, R. V. and Balaban, E. (eds) *The Differences Between the Sexes* Cambridge, Cambridge Univesity Press.

Harwood, J. (1977) 'The race–intelligence controversy: a sociological approach, II – external factors'. *Social Studies of Science* **7**: 1–30.

Herrnstein, R. and Murray, C. (1994) *The Bell Curve: Intelligence and Class Structure in American Life*. New York, Simon & Schuster.

Hill, K. (1982) 'Hunting and human evolution'. *Journal of Human Evolution* **11**: 521–44.

Hill, K. and Hurtado, M. (1996) *Demographic/Life History of Ache Foragers*. Hawthorne, NY, Aldine de Gruyter.

Hill, K. and Kaplan, H. (1988) 'Tradeoffs in male and female reproductive strategies among the Ache.' in Betzig, L., Borgehoff-Mulder, M. and Turke, P. (eds) *Human Reproductive Behaviour*. Cambridge, Cambridge University Press.

Hinde, R. A. (1982) *Ethology*. Oxford, Oxford University Press.

Holloway, R. (1983) 'Human paleontological evidence relevant to language behaviour'. *Human Neurobiology* 2: 105–14.

Holmes, W. G. and Sherman, P. W. (1982) 'The ontogeny of kin recognition in two species of ground squirrels'. *American Zoologist* 22: 491–517.

Hoogland, J. L. (1983) 'Nepotism and alarm calls in the black-tailed prairie dog (*Cynomys ludovicianus*). *Animal Behaviour* 31: 472–9.

Hosken, F. P. (1979) *The Hosken Report: Genital and Sexual Mutilation of Females*. Lexicon, MA, Women's International Network News.

Howell, N. (1979) *The Demography of the Dobe !Kung*. New York, Academic Press.

Hrdy, S. B. (1979) 'Infanticide among animals: a review, classification and examination of the implications for the reproductive strategies of females'. *Ethology and Socio-biology* 1: 13–40.

Hull, D. L. (1981) 'Units of evolution: a metaphysical essay' in Jensen, V. J. and Harre, R. (eds) *The Philosophy of Evolution*. Brighton, Harvester.

Hume, D. (1964[1739]) *A Treatise of Human Nature*, (ed.) Selby-Bigge, L. A. Oxford, Clarendon Press.

Hurst, L. D. (1990) 'Parasite diversity and the evolution of diploidy, multicellularity and anisogamy'. *Journal of Theoretical Biology* 144: 429–43.

Irons, W. (1979) 'Natural selection, adaptation and human social behaviour' in Chagnon, N. A. and Irons, W. (eds) *Evolutionary Biology and Human Social Behaviour: An Anthropological Perspective*. North Scituate, MA, Duxbury.

Jackson, L. A. (1992) *Physical Appearance and Gender: Sociobiological and Sociocultural Perspectives*. Albany, State University of New York Press.

Jarman, P. J. (1974) 'The social organization of antelope in relation to their ecology'. *Behaviour* 48: 215–67.

Jerison, H. J. (1973) *Evolution of the Brain and Intelligence*. New York, Academic Press.

Kalikow, T. (1983) 'Konrad Lorenz's ethological theory: explanation and ideology 1938–1943'. *Journal of the History of Biology* 16: 39–73.

Kappelman, J. (1996) 'The evolution of body mass and relative brain size in fossil hominids'. *Journal of Human Evolution* 30: 243–76.

Keller, L. and Ross, K. G. (1998) 'Selfish genes: a green beard effect'. *Nature* **394**: 573–5.

Kenrick, D. T., Sadalla, E., Groth, G. *et al.* (1996) 'Evolution, traits, and the stages of human courtship: qualifying the parental investment model'. *Journal of Personality* **58**: 97–116.

Kipling, R. (1967) *Just So Stories*. London, Macmillan.

Kirkwood, T. B. L. (1977) 'Evolution of aging'. *Nature* 270: 301–4.

Kitchen, D. (1974) 'Social behaviour and ecology of the pronghorn'. *Wildlife Monographs* 38: 1–96.

Klineberg, D. (1935) *Race Differences*. New York, Harper & Bros.

Knight, C. and Maisels, C. (1994) 'Fertility rights'. *Times Higher Educational Supplement* (Sep) **23**: 20.

Knight, C., Power, C. and Watts, I. (1995) 'The human symbolic revolution: a Darwinian account'. *Cambridge Archaeological Journal* 5: 75–114.

Krebs, J. R. and Davies, N. B. (1987) *An Introduction to Behavioural Ecology*. Oxford, Blackwell Scientific.

Krebs, J. R. and Davies, N. B. (eds) (1991) *Behavioural Ecology*. Oxford, Blackwell Scientific.

Kvarnemo, C. and Ahnesjo, I. (1996) 'The dynamics of operational sex ratios and competition for mates'. *Trends in Evolution and Ecology* **11**: 4–7.

Lack, D. (1943) *The Life of the Robin*. London, Penguin.

Laitman, J. T. (1984) 'The anatomy of human speech'. *Natural History* (Aug): 20–7.

Lande, R. (1981) 'Models of speciation by sexual selection on polygenic traits'. *Proceedings of the National Academy of Sciences of the USA* **78**: 3721–5.

Langlois, J. H. (1987) 'Infant preferences for attractive faces: rudiments of a stereotype'. *Developmental Psychology* **23**: 363–9.

Langlois, J. H. and Roggman, L. A. (1990) 'Attractive faces are only average'. *Psychological Science* **1**: 115–21.

Langlois, J. H., Roggman, L. A. and Reiser-Danner, L. A. (1990) 'Infants' differential social responses to attractive and unattractive faces'. *Developmental Psychology* **26**: 153–9.

Leakey, R. (1994) *The Origin of Humankind*. London, Weidenfeld & Nicolson.

Lehrman, D. S. (1953) 'A critique of Konrad Lorenz's theory of instinctive behaviour'. *Quarterly Review of Biology* **28**: 337–63.

Lincoln, G. A. (1972) 'The role of antlers in the behaviour of red deer'. *Journal of Experimental Zoology* **182**: 233–50.

Lorenz, K. (1953) *King Solomon's Ring*. London, Reprint Society.

Lotem, A., Nakamura, H. and Zahavi, A. (1995) 'Constraints on egg discrimination in cuckoo–host evolution'. *Animal Behaviour* **49**: 1185–209.

Low, B. (1989) 'Cross-cultural patterns in the training of children: an evolutionary perspective'. *Journal of Comparative Psychology* **103**(4): 311–19.

Low, B., Alexander, R. D. and Noonan, K. M. (1987) 'Human hips, breasts and buttocks: is fat deceptive?' *Ethology and Sociobiology* **8**: 249–57.

Lown, B. A. (1975) 'Comparative psychology twenty five years after'. *American Psychologist* **30**: 858–9.

Lumsden, C. J. and Wilson, E. O. (1981) *Genes, Mind and Culture*. Cambridge, MA, Harvard University Press.

Lyell, C. (1830–33) *Principles of Geology*. London, Murray.

McClintock, M. K. (1971) 'Menstrual synchrony and suppression'. *Nature* **229**: 229–45.

McGill, T. E. (1965) *Readings in Animal Behaviour*. New York, Rinehart & Winston.

MacKenzie, D. (1976) 'Eugenics in Britain'. *Social Studies of Science* **6**: 499–532.

MacLean, P. D. (1972) 'Cerebral evolution and emotional processes: new findings on the striatal complex'. *Annals of the New York Academy of Sciences* **193**: 137–49.

Mace, R. and Cowlishaw, G. (1996) 'Cross-cultural patterns of marriage and inheritance: a phylogenetic approach'. *Ethology and Sociobiology* **17**: 87–97.

Manning, J. T. (1995) 'Fluctuating asymmetry and body weight in men and women: implications for sexual selection'. *Ethology and Sociobiology* **16**: 145–53.

Manning, J. T. and Chamberlain, A. T. (1994) 'Fluctuating asymmetry in gorilla canines: a sensitive indicator of environmnetal stress'. *Proceedings of the Royal Society of London B.* **255**: 189–193.

Manning, J. T., Scutt, D., Whitehouse, G. H. *et al.* (1996) 'Asymmetry and the menstrual cycle in women'. *Ethology and Sociobiology* **17**: 129–43.

Manning, J. T., Koukourakis, K. and Brodie, D. A. (1997) 'Fluctuating asymmetry, metabolic rate and sexual selection in human males'. *Evolution and Human Behaviour* **18**: 15–21.

Maynard Smith, J. (1974) 'The theory of games and the evolution of animal conflicts'. *Journal of Theoretical Biology* 47: 209–21.

Maynard Smith, J. (1977) 'Parental investment: a prospective analysis'. *Animal Behaviour* 25: 1–9.

Maynard Smith, J. (ed.) (1982) *Evolution Now: A Century after Darwin*. London, Macmillan.

Maynard Smith, J. (1989) *Evolutionary Genetics*. Oxford, Oxford University Press.

Maynard Smith, J. and Brown, R. L. W. (1986) 'Competition and body size'. *Theoretical Population Biology* 30: 166–79.

Mead, M. (1928) *Coming of Age in Samoa: A Psychological Study of Primitive Youth for Western Civilization*. New York, William Morrow.

Mead, M. (1935) *Sex and Temperament in Three Primitive Societies*. London, Routledge.

Mealey, L., Daood, C. and Krage, M. (1996) 'Enhanced memory for faces of cheaters'. *Ethology and Sociobiology* 17: 119–28.

Metcalf, R. A. and Whitt, G. S. (1977) 'Relative inclusive fitness in the social wasp *Polistes metricus*'. *Behavioural Ecology and Sociobiology* 2: 353–60.

Midgely, M. (1978) *Beast and Man: The Biological Roots of Human Nature*. London, Methuen.

Miller, G. (1996) 'Sexual selection in human evolution: review and prospects' in Crawford, C. and Krebs, D. (eds) *Evolution and Human Behaviour: Ideas, Issues, Applications*. New York, Lawrence Erlbaum.

Miller, G. F. (1998) 'How mate choice shaped human nature: a review of sexual selection and human evolution' in Crawford, C. and Krebs, D. L. (eds) *Handbook of Evolutionary Psychology*. Mahwah, NJ, Lawrence Erlbaum.

Miller, G. F. (1999a) 'Sexual selection for cultural displays' in Dunbar, R., Knight, C. and Power, C. (eds) *The Evolution of Culture*. Edinburgh, Edinburgh University Press.

Miller, G. F. (1999b) 'Evolution and consumerism'. *Prospect* (Feb): 18–23.

Mock, D. W. (1984) 'Siblicidal aggression and resource monopolization in birds'. *Science* 225: 731–3.

Mock, D. W. and Parker, G. A. (1997) *The Evolution of Sibling Rivalry*. Oxford, Oxford University Press.

Moller, A. P. (1987) 'Behavioural aspects of sperm competition in swallows (*Hirundo rustica*)'. *Behaviour* 100: 92–104.

Morgan, C. Lloyd (1894) *An Introduction to Comparative Psychology*. London, Walter Scott.

Mueller, U. (1993) 'Social status and sex'. *Nature* 363: 490.

Nowak, M. and Sigmund, K. (1998) 'Evolution of indirect reciprocity by image scoring'. *Nature* 393: 573–6.

Oldroyd, D. R. (1980) *Darwinian Impacts: An Introduction to the Darwinian Revolution*. Buckingham, Open University Press.

Olsson, M., Shine, R., Madsen, T. *et al.* (1996) 'Sperm selection by females'. *Nature* 383: 585.

Orians, G. H. (1969) 'On the evolution of mating systems in birds and mammals'. *American Naturalist* 103: 589–603.

Oring, L. and Lank, D. B. (1986) 'Polyandry in spotted sandpipers: the impact of environment and experience' in Rubenstein, D. I. and Wrangham, R. W. (eds) *Ecological Aspects of Social Evolution*. Princeton, Princeton University Press.

Pagel, M. (1998) 'Mother and father in surprise genetic agreement'. *Nature* 397: 19–20.

Paley, W. (1802) *Natural Theology; or, Evidences of the Existence and Attributes of the Deity, Collected from the Appearances of Nature*, 5th edn. London, printed for R. Faulder, 1803.

Parker, G. A. (1970) 'Sperm competition and its evolutionary consequences in the insects'. *Biology Review* **45**: 525–67.

Parker, G. (1983) 'Arms race in evolution: an ESS to the opponent-independent costs game'. *Journal of Theoretical Biology* **101**: 619–48.

Parker, G. A. (1982) 'Why are there so many tiny sperm? Sperm competition and the maintenance of two sexes'. *Journal of Theoretical Biology* **96**: 281–94.

Parker, G. A., Baker, R. R. and Smith, V. G. F. (1972) 'The origin and evolution of gamete dimorphism and the male–female phenomenon'. *Journal of Theoretical Biology* **36**: 529–53.

Parker, S. (1976) 'The precultural basis of the incest taboo: towards a biosocial theory'. *American Anthropologist* **78**: 285–305.

Passingham, R. (1988) *The Human Primate*. Oxford, W.H. Freeman.

Pearce, D. (1987) *Blueprint for a Green Economy*. London, Penguin.

Perrett, D. I., May, K. A. and Yoshikawa, S. (1994) 'Facial shape and judgements of female attractiveness'. *Nature* **368**: 239–42.

Perrett, D. I., Lee, K. J. and Penton-Voak, I. *et al.* (1998) 'Effects of sexual dimorphism on facial attractiveness'. *Nature* **394**: 884–7.

Petrie, M. (1983) 'Female moorhens compete for small fat males'. *Science* **220**: 413–15.

Petrie, M. (1994) 'Improved growth and survival of offspring of peacocks with more elaborate trains'. *Nature* **371**: 598–9.

Petrie, M., Halliday, T. and Saunders, C. (1991) 'Peahens prefer peacocks with elaborate trains'. *Animal Behaviour* **41**: 323–31.

Pinker, S. (1994) *The Language Instinct*. London, Penguin.

Pinker, S. and Bloom, P. (1990) 'Natural language and natural selection'. *Behavioural and Brain Sciences* **13**: 707–84.

Plavcan, J. M. and van Schaik, C. P. (1997) 'Interpreting hominid behaviour on the basis of sexual dimorphism'. *Journal of Human Evolution* **32**: 345–74.

Pleszczynska, W. K. (1978) 'Microgeographic prediction of polygyny in the lark bunting'. *Science* **164**: 1170–2.

Plomin, R. (1990) *Behavioural Genetics. A Primer*. New York, Freeman.

Popper, K. (1959) *The Logic of Scientific Discovery*. London, Hutchinson.

Porter, R. H., Tepper, V. J. and White, D. M. (1981) 'Experimental influences on the development of huddling preferences and sibling recognition in spiny mice'. *Developmental Psychobiology* **14**: 375–82.

Potts, B. (1995) 'Queen Victoria's gene'. *Biological Sciences Review* (November): 18–22.

Profet, M. (1993) 'Menstruation as a defense against pathogens transported by sperm'. *Quarterly Review of Biology* **68**: 335–86.

Pusey, A. E. and Packer, C. (1994) 'Non-offspring nursing in social carnivores: minimizing the cost'. *Behavioural Ecology* **5**: 362–74.

Ralls, K. (1976) 'Mammals in which females are larger than males'. *Quarterly Review of Biology* **51**: 245–76.

Reynolds, J. D. and Harvey, P. H. (1994) 'Sexual selection and the evolution of sex differences' in Short, R. V. and Balban, E. (eds) *The Differences Between the Sexes*. Cambridge, Cambridge University Press.

Richards, J. R. (1987) *Darwin and the Emergence of Evolutionary Theories of Mind and Behaviour*. Chicago, Chicago University Press.

Ridley, M. (1993) *The Red Queen*. London, Viking.

Ridley, M. (1996) *The Origins of Virtue*. London, Viking.

Roberts, J. M. and Lowe, C. R. (1975) 'Where have all the conceptions gone?' *Lancet* 1: 498–9.

Romanes, G. J. (1887) 'Mental differences between men and women' in Morgan, C. L. (ed.) *Essays by George John Romanes (1897)*. London, Longmans Green.

Romanes, G. J. (1888) *Mental Evolution in Man*. New York, Appleton.

Rose, S., Kamin, L. J. and Lewontin, R. C. (1985) *Not in Our Genes*. London, Penguin.

Rothenbuhler, W. (1964) 'Behaviour genetics of nest cleaning in honey bees'. *American Zoologist* 4: 111–23.

Saino, N., Bolzern, A. M. and Moller, A. P. (1997) 'Immunocompetence, ornamentation, and viability of male barn swallows (*Hirundo rustica*)'. *Proceedings of the National Academy of Sciences of the USA* 94: 549–52.

Samuels, R. (1998) 'Evolutionary psychology and the massive modularity hypothesis'. *British Journal for the Philosophy of Science* 49: 575–602.

Schroder, I. (1993) 'Concealed ovulation and clandestine copulation: a female contribution to human evolution'. *Ethology and Sociobiology* 14: 381–9.

Scott, J. P. (1973) 'The organisation of comparative psychology'. *Annals of the New York Academy* 223: 7–40.

Shackelford, T. K. and Larsen, R. J. (1999) 'Facial attractiveness and physical health'. *Evolution and Human Behaviour* 20: 71–6.

Shapiro, L. and Epstein, W. (1998) 'Evolutionary theory meets cognitive psychology: a more selective perspective'. *Mind and Language* 13(2): 171–94.

Sherman, P. W. and Reeve, H. K. (1997) 'Forward and backward: alternative approaches to studying human behaviour' in Betzig. L. (ed.) *Human Nature*. Oxford, Oxford University Press.

Short, R. V. (1994) 'Why sex?' in Short, R. V. and Balaban, E. (eds) *The Differences Between the Sexes*. Cambridge, Cambridge University Press.

Short, R. V. and Balaban, E. (eds) (1994) *The Differences Between the Sexes*. Cambridge, Cambridge University Press.

Sieff, D. F. (1990) 'Explaining based sex ratios in human populations'. *Current Anthropology* 31(1): 25–48.

Sigmund, K. (1993) *Games of Life. Explorations in Ecology, Evolution and Behaviour*. Oxford, Oxford University Press.

Silberglied, R. E., Shepherd, J. G. and Dickinson, J. L. (1984) 'Eunuchs: the role of apyrene sperm in Lepidoptera'. *American Naturalist* 123: 255–65.

Sillen-Tullberg, B. and Moller, A. (1993) 'The relationship between concealed ovulation and mating systems in anthropoid primates: a phylogenetic analysis'. *American Naturalist* 141: 1–25.

Silverman, I. (1987) 'Race, race differences, and race relations: perspectives from psychology and sociobiology' in Crawford, C., Smith, M. and Krebs, D. (eds) *Sociobiology and Psychology: Ideas, Issues and Applications*. Mahwah, NJ, Lawrence Erlbaum.

Singer, P. (1998) Evolutionary workers' party. *Times Higher Educational Supplement* (15 May): 15.

Singh, D. (1993) 'Adative significance of female attractiveness'. *Journal of Personality and Social Psychology* 65: 293–307.

Singh, D. (1995) 'Female judgement of male attractiveness and desirability for relationships: role of waist-to-hip ratios and financial status'. *Journal of Personality and Social Psychology* 69(6): 1089–101.

Singh, D. and Bronstad, P. M. (1997) 'Sex differences in the anatomical locations of human body scarification and tattooing as a function of pathogen prevalence'. *Evolution and Human Behaviour* 18: 403–16.

Singh, D. and Luis, S. (1995) 'Ethnic and gender consensus for the effect of waist-to-hip ratio on judgement of women's attractiveness'. *Human Nature* 6(1): 51–65.

Skinner, B. F. (1957) *Verbal Behaviour*. New York, Appleton-Century-Crofts.

Slater, P. J. B. (1994) 'Kinship and altruism' in Slater, P. J. B. and Halliday, T. R. (eds) *Behaviour and Evolution*. Cambridge, Cambridge University Press.

Smith, M., Kish, B. J. and Crawford, C. B. (1987) 'Inheritance of wealth as human kin investment'. *Ethology and Sociobiology* 8: 171–82.

Smith, R. L. (1984) 'Human sperm competition' in Smith, R. L. (ed.) *Sperm Competition and the Evolution of Animal Mating Systems*. London, Academic Press.

Smith, R. (1997) *The Fontana History of the Human Sciences*. London, Fontana.

Spencer, H. (1855) *Principles of Psychology*. London, Longman, Brown, Green and Longmans.

Stanford, C. B. (1998) 'The social behaviour of chimpanzees and bonobos'. *Current Anthropology* 39(4): 399–419.

Steen, R. G. (1996) *DNA and Destiny. Nature and Nurture in Human Behaviour*. London, Plenum Press.

Stern, K. and McClintock, M. K. (1998) 'Regulation of ovulation by human pheremones'. *Nature* 392: 177–9.

Sternglanz, S. H., Gray, J. L. and Murakami, M. (1977) 'Adult preferences for infantile facial features: an ethological approach'. *Animal Behaviour* 25: 108–15.

Strachan, T. and Read, A. P. (1996) *Human Molecular Genetics*. Oxford, Bios Scientific.

Strassman, B. (1992) 'The function of menstrual taboos among the Dogon: defense against cuckoldry'. *Human Nature* 3: 89–131.

Strassman, B. (1996) 'Menstrual huts visits by Dogon women: a hormonal test distinguishes deceipt from honest signalling'. *Behavioural Ecology* 7(3): 304–15.

Struhsaker, T. (1969) 'Correlates of ecology and social organization among African cercopithecines'. *Folia Primatologica* 11: 80–118.

Sudbury, P. (1998) *Human Molecular Genetics*. London, Addison-Wesley.

Symington, M. M. (1987) 'Sex ratio and maternal rank in wild spider monkeys: when daughters disperse'. *Behavioural Ecology and Sociobiology* 20: 421–5.

Symons, D. (1979) *The Evolution of Human Sexuality*. Oxford, Oxford University Press.

Symons, D. (1992) 'On the use and misuse of Darwinism' in Barkow, J. H., Cosmides, L. and Tooby, J. (eds) *The Adapted Mind*. Oxford, Oxford University Press.

Symons, D. and Ellis, B. (1989) 'Human male–female differences in sexual desire' in Rasa, A. E., Vogel, C. and Voland, E. (eds) *The Sociobiology of Sexual and Reproductive Strategies*. London, Chapman & Hall.

Thatcher, M. (1995) 'Ted. How I beat him and became leader'. *Sunday Times*. 11 June.

Thornhill, R. and Gangestad, S. W. (1993) 'Human facial beauty: averageness, symmetry and parasite resistance'. *Human Nature* 4: 237–69.

Thornhill, R. and Gangestad, S. W. (1994) 'Human fluctuating asymmetry and sexual behaviour'. *Psychological Science* 5: 297–302.

Thornhill, R., Gangestad, S. W. and Comer, R. (1996) 'Human female orgasm and male fluctuating asymmetry'. *Animal Behaviour* 50: 1601–15.

Thorpe, W. H. (1961) *Bird Song*. London, Cambridge University Press.

Tinbergen, N. (1952) 'The curious behaviour of the stickleback'. *Scientific American* 187: 22–6.

Tinbergen, N. (1963) 'On the aims and methods of ethology'. *Zeitschrift für Tierpsychologie* 20: 410–33.

Tomasello, M. and Call, J. (1997) *Primate Cognition*. Oxford, Oxford University Press.

Tooby, J. and Cosmides, L. (1990) 'The past explains the present: adaptations and the structure of ancestral environments'. *Ethology and Sociobiology* **11**: 375–424.

Tooby, J. and Cosmides, L. (1992) 'Cognitive adaptations for social exchange' in Barkow, J. H., Cosmides, L. and Tooby, J. (eds) *The Adapted Mind*. Oxford, Oxford University Press.

Tooby, J. and Cosmides, L. (1992) 'The psychological foundations of culture' in Barkow, J. H., Cosmides, L. and Tooby, J. (eds) *The Adapted Mind*. Oxford, Oxford University Press.

Trivers, R. L. (1971) 'The evolution of reciprocal altruism'. *Quarterly Review of Biology* **46**: 35–57.

Trivers, R. L. (1972) 'Parental investment and sexual selection' in Campbell, B. (ed.) *Sexual Selection and the Descent of Man*. Chicago, Aldine.

Trivers, R. L. (1974) 'Parent–offspring conflict'. *American Zoologist* **14**: 249–64.

Trivers, R. (1981) 'Sociobiology and politics' in White, E. (ed.) *Sociobiology and Human Politics*. Lexington, MA, Lexington Books.

Trivers, R. (1985) *Social Evolution*. Menlow Park, CA, Benjamin/Cummings.

Trivers, R. L. and Willard, D. E. (1973) 'Natural selection of parental ability to vary the sex ratio of offspring'. *Science* **179**: 90–2.

Tryon, R. C. (1940) 'Genetic differences in maze-learning ability in rats'. *Yearbook of the National Society for the Study of Education* **39**: 111–19.

Turner, P. E. and Chao, L. (1999) 'Prisoners' dilemma in an RNA virus'. *Nature* **398**: 441–3.

Tutin, C. (1979) 'Responses of chimpanzees to copulation, with special reference to interference by immature individuals'. *Animal Behaviour* **27**: 845–54.

Van Valen, L. (1973) 'A new evolutionary law'. *Evolutionary Theory* **I**: 1–30.

Voland, E. and Engel, C. (1989) 'Women's reproduction and longevity in a premodern population' in Rasa, E., Vogel, C. and Voland, E. (eds) *The Sociobiology of Sexual and Reproductive Strategies*. London, Chapman & Hall.

Voss, R. (1979) 'Male accessory glands and the evolution of copulatory plugs in rodents'. *Museum of Zoology University of Michegan. Occasional Papers* **968**: 1–27.

Waal, F. B. M. de (1997) 'The chimpanzees service economy: food for grooming'. *Evolution and Human Behaviour* **18**: 375–86.

Wallace, A. R. (1905) *My Life: A Record of Events and Opinions*. London, Chapman & Hall.

Warner, H., Martin, D. E. and Keeling, M. E. (1974) 'Electroejaculation of the great apes'. *Annals of Biomedical Engineering* **2**: 419–32.

Wason, P. (1966) 'Reasoning' in Foss, B. M. (ed.) *New Horizons in Psychology*. Harmondsworth, Penguin.

Watson, J. B. (1930) *Behaviourism*. New York, Norton.

Watson, J. P. and Crick, F. H. C. (1953) 'A structure for deoxyribonucleic acid'. *Nature* **171**: 737–8.

Watson, P. J. (1990) 'Female-enhanced male competition determines the first mate and principal sire in the spider *Linyphia litigiosa*'. *Behavioural Ecology and Sociobiology* **26**: 77–90.

Wedekind, C., Seebeck, T., Bettens, F. *et al*. (1995) 'MHC-dependent mate preferences in humans'. *Proceedings of the Royal Society of London* B **260**: 245–9.

Weizmann, F., Wiener, N. I., Wiesenthal, D. L. *et al*. (1990) 'Differential K theory and racial hierachies'. *Canadian Psychology* **31**: 1–12.

West-Eberhard, M. J. (1975) 'The evolution of social behaviour by kin selection'. *Quarterly Review of Biology* **50**: 1–33.

Westendorp, R. G. J. and Kirkwood, T. B. L. (1998) 'Human longevity at the cost of reproductive success'. *Nature* **396**: 743–6.

Westermarck, E. A. (1891) *The History of Human Marriage*. New York, Macmillan.

White, R. (1985) 'Thoughts on social relationships and language in hominid evolution'. *Journal of Social and Personal Relationships* **2**: 95–115.

Wilkinson, G. (1984) 'Reciprocal food sharing in vampire bats'. *Nature* **308**: 181–4.

Wilkinson, G. S. (1990) 'Food sharing in vampire bats'. *Scientific American* **262**: 76–82.

Williams, G. C. (1966) *Adaptation and Natural Selection*. Princeton, Princeton University Press.

Williams, G. C. (1975) *Sex and Evolution* in *Monographs in Population Biology*. Princeton, Princeton University Press.

Wilson, D. S. (1992a) 'Group selection' in Keller, E. F. and Lloyd, E. A. (eds) *Keywords in Evolutionary Biology*. Cambridge, MA, Harvard University Press.

Wilson, D. S. (1994) 'Adaptive genetic variation and human evolutionary psychology'. *Ethology and Sociobiology* **15**: 219–35.

Wilson, D. S. and Sober, E. (1994) 'Re-introducing group slection to the human behavioural sciences'. *Behavioural and Brain Sciences* **17**: 585–654.

Wilson, E. O. (1975) *Sociobiology: The New Synthesis*. Cambridge, MA, Harvard University Press.

Wilson, E. O. (1998) *Consilience: The Unity of Knowledge*. London, Little, Brown.

Wilson, E. O. and Ruse, M. (1985) 'The evolution of ethics'. *New Scientist* (Oct 17): 50–2.

Wilson, H. C. (1992b) 'A critical review of menstrual synchrony research'. *Psychoneuroendocrinology* **16**: 353–9.

Wilson, M. and Daly, M. (1992) 'The man who mistook his wife for a chattel' in Tooby, J. and Cosmides, L. (eds) *The Adapted Mind*. Oxford, Oxford University Press.

Wirtz, P. (1997) 'Sperm selection by females'. *Trends in Ecology and Evolution* **12**(5): 172–3.

Wispe, L. G. and Thompson, J. W. (1976) 'The war between worlds: biological versus social evolution and some related issues'. *American Psychologist* **31**: 346.

Wolf, A. P. (1970) 'Childhood association and sexual attraction: a further test of the Westermarck hypothesis'. *American Anthropologist* **72**: 503–15.

Wright, R. (1994) *The Moral Animal: Evolutionary Psychology and Everyday Life*. London, Little, Brown.

Wynne-Edwards (1962) *Animal Dispersion in Relation to Social Behaviour*. Edinburgh, Oliver & Boyd.

Young, J. Z. (1981) *The Life of Vertebrates*. Oxford, Oxford University Press.

Yu, D. W. and Shepard, G. H. (1998) 'Is beauty in the eye of the beholder?' *Nature* **396**: 321–2.

Zahavi, A. (1975) 'Mate selection – a selection for handicap'. *Journal of Theoretical Biology* **53**: 205–14.

Index

Scientific names for species and titles of books are given in *italics*. Page numbers of Glossary definitions are given in **bold**